SCHAUM'S OUTLINE OF

THEORY AND PROBLEMS

OF

TENSOR CALCULUS

•

DAVID C. KAY, Ph.D.
Professor of Mathematics
University of North Carolina at Asheville

McGRAW-HILL

New York San Francisco Washington, D.C. Auckland Bogotá Caracas London Madrid Mexico City Milan Montreal New Dehli San Juan Singapore Sydney Tokyo Toronto

DAVID C. KAY is currently Professor and Chairman of Mathematics at the University of North Carolina at Asheville; formerly he taught in the graduate program at the University of Oklahoma for 17 years. He received his Ph.D. in geometry at Michigan State University in 1963. He is the author of more than 30 articles in the areas of distance geometry, convexity theory, and related functional analysis.

Schaum's Outline of Theory and Problems of
TENSOR CALCULUS

19 20 CUS CUS 0

ISBN 0-07-033484-6

Sponsoring Editor, David Beckwith
Production Supervisor, Denise Puryear
Editing Supervisor, Marthe Grice

Library of Congress Cataloging-in-Publication Data

Kay, David C.
Schaum's outline of theory and problems of tensor calculus.

(Schaum's Outline series)
1. Calculus of tensors—Problems, exercises, etc.
I. Title. II. Title: Theory and problems of tensor calculus.
QA433.K39 1988 515'.63 87-32515
ISBN 0-07-033484-6

McGraw-Hill
A Division of The McGraw-Hill Companies

Preface

This Outline is designed for use by both undergraduates and graduates who find they need to master the basic methods and concepts of tensors. The material is written from both an elementary and applied point of view, in order to provide a lucid introduction to the subject. The material is of fundamental importance to theoretical physics (e.g., field and electromagnetic theory) and to certain areas of engineering (e.g., aerodynamics and fluid mechanics). Whenever a change of coordinates emerges as a satisfactory way to solve a problem, the subject of tensors is an immediate requisite. Indeed, many techniques in partial differential equations are tensor transformations in disguise. While physicists readily recognize the importance and utility of tensors, many mathematicians do not. It is hoped that the solved problems of this book will allow all readers to find out what tensors have to offer them.

Since there are two avenues to tensors and since there is general disagreement over which is the better approach for beginners, any author has a major decision to make. After many hours in the classroom it is the author's opinion that the tensor component approach (replete with subscripts and superscripts) is the correct one to use for beginners, even though it may require some painful initial adjustments. Although the more sophisticated, noncomponent approach is necessary for modern applications of the subject, it is believed that a student will appreciate and have an immensely deeper understanding of this sophisticated approach to tensors after a mastery of the component approach. In fact, noncomponent advocates frequently surrender to the introduction of components after all; some proofs and important tensor results just do not lend themselves to a completely component-free treatment. The Outline follows, then, the traditional component approach, except in the closing Chapter 13, which sketches the more modern treatment.

The author has been strongly influenced over the years by the following major sources of material on tensors and relativity:

J. Gerretsen, *Lectures on Tensor Calculus and Differential Geometry*, P. Noordhoff: Goningen, 1962.

I. S. Sokolnikoff, *Tensor Analysis and Its Applications*, McGraw-Hill: New York, 1950.

Synge and Schild, *Tensor Calculus*, Toronto Press: Toronto, 1949.

W. Pauli, Jr., *Theory of Relativity*, Pergamon: New York, 1958.

R. D. Sard, *Relativistic Mechanics*, W. A. Benjamin: New York, 1970.

Bishop and Goldberg, *Tensor Analysis on Manifolds*, Macmillan: New York, 1968.

Of course, the definitive work from the geometrical point of view is L. P. Eisenhart, *Riemannian Geometry*, Princeton University Press: Princeton, N.J., 1949.

The author would like to acknowledge significant help in ferreting out typographical errors and other imperfections by the readers: Ronald D. Sandstrom, Professor of Mathematics at Fort Hays State University, and John K. Beem, Professor of Mathematics at the University of Missouri. Appreciation is also extended to the editor, David Beckwith, for many helpful suggestions.

DAVID C. KAY

Contents

Chapter 1

The Einstein Summation Convention

1.1 INTRODUCTION

A study of tensor calculus requires a certain amount of background material that may seem unimportant in itself, but without which one could not proceed very far. Included in that prerequisite material is the topic of the present chapter, the summation convention. As the reader proceeds to later chapters he or she will see that it is this convention which makes the results of tensor analysis surveyable.

1.2 REPEATED INDICES IN SUMS

A certain notation introduced by Einstein in his development of the Theory of Relativity streamlines many common algebraic expressions. Instead of using the traditional sigma for sums, the strategy is to allow the repeated subscript to become itself the designation for the summation. Thus,

$$a_1x_1 + a_2x_2 + a_3x_3 + \cdots + a_nx_n \equiv \sum_{i=1}^{n} a_ix_i$$

becomes just a_ix_i, where $1 \leqq i \leqq n$ is adopted as the universal range for summation.

EXAMPLE 1.1 The expression $a_{ij}x_k$ does not indicate summation, but both $a_{ii}x_k$ and $a_{ij}x_j$ do so over the respective ranges $1 \leqq i \leqq n$ and $1 \leqq j \leqq n$. If $n = 4$, then

$$a_{ii}x_k \equiv a_{11}x_k + a_{22}x_k + a_{33}x_k + a_{44}x_k$$
$$a_{ij}x_j \equiv a_{i1}x_1 + a_{i2}x_2 + a_{i3}x_3 + a_{i4}x_4$$

Free and Dummy Indices

In Example 1.1, the expression $a_{ij}x_j$ involves two sorts of indices. The index of summation, j, which ranges over the integers $1, 2, 3, \ldots, n$, cannot be preempted. But at the same time, it is clear that the use of the particular character j is inessential; e.g., the expressions $a_{ir}x_r$ and $a_{iv}x_v$ represent exactly the same sum as $a_{ij}x_j$ does. For this reason, j is called a *dummy* index. The index i, which may take on any particular value $1, 2, 3, \ldots, n$ independently, is called a *free* index. Note that, although we call the index i "free" in the expression $a_{ij}x_j$, that "freedom" is limited in the sense that generally, unless $i = k$,

$$a_{ij}x_j \neq a_{kj}x_j$$

EXAMPLE 1.2 If $n = 3$, write down explicitly the equations represented by the expression $y_i = a_{ir}x_r$.

Holding i fixed and summing over $r = 1, 2, 3$ yields

$$y_i = a_{i1}x_1 + a_{i2}x_2 + a_{i3}x_3$$

Next, setting the free index $i = 1, 2, 3$ leads to three separate equations:

$$y_1 = a_{11}x_1 + a_{12}x_2 + a_{13}x_3$$
$$y_2 = a_{21}x_1 + a_{22}x_2 + a_{23}x_3$$
$$y_3 = a_{31}x_1 + a_{32}x_2 + a_{33}x_3$$

Einstein Summation Convention

Any expression involving a twice-repeated index (occurring twice as a subscript, twice as a superscript, or once as a subscript and once as a superscript) shall automatically stand for its sum over the values $1, 2, 3, \ldots, n$ *of the repeated index*. Unless explicitly stated otherwise, the single exception to this rule is the character n, which represents the range of all summations.

Remark 1: Any free index in an expression shall have the same range as summation indices, unless stated otherwise.

Remark 2: No index may occur *more than twice* in any given expression.

EXAMPLE 1.3 (*a*) According to Remark 2, an expression like $a_{ii}x_i$ is without meaning. (*b*) The meaningless expression $a^i_i x_i x_i$ might be presumed to represent $a^i_i (x_i)^2$, which is meaningful. (*c*) An expression of the form $a_i(x_i + y_i)$ is considered well-defined, for it is obtained by composition of the meaningful expressions $a_i z_i$ and $x_i + y_i = z_i$. In other words, the index i is regarded as occuring *once* in the term $(x_i + y_i)$.

1.3 DOUBLE SUMS

An expression can involve more than one summation index. For example, $a_{ij}x_i y_j$ indicates a summation taking place on both i and j simultaneously. If an expression has two summation (dummy) indices, there will be a total of n^2 terms in the sum; if there are three indices, there will be n^3 terms; and so on. The expansion of $a_{ij}x_i y_j$ can be arrived at logically by first summing over i, then over j:

$$\begin{aligned} a_{ij}x_i y_j &= a_{1j}x_1 y_j + a_{2j}x_2 y_j + a_{3j}x_3 y_j + \cdots + a_{nj}x_n y_j && \text{[summed over } i\text{]} \\ &= (a_{11}x_1 y_1 + a_{12}x_1 y_2 + \cdots + a_{1n}x_1 y_n) && \text{[summed over } j\text{]} \\ &\quad + (a_{21}x_2 y_1 + a_{22}x_2 y_2 + \cdots + a_{2n}x_2 y_n) \\ &\quad + (a_{31}x_3 y_1 + a_{32}x_3 y_2 + \cdots + a_{3n}x_3 y_n) \\ &\quad \cdots\cdots\cdots\cdots\cdots\cdots\cdots\cdots \\ &\quad + (a_{n1}x_n y_1 + a_{n2}x_n y_2 + \cdots + a_{nn}x_n y_n) \end{aligned}$$

The result is the same if one sums over j first, and then over i.

EXAMPLE 1.4 If $n = 2$, the expression $y_i = c^r_i a_{rs} x_s$ stands for the two equations:

$$\begin{aligned} y_1 &= c^1_1 a_{11}x_1 + c^2_1 a_{21}x_1 + c^1_1 a_{12}x_2 + c^2_1 a_{22}x_2 \\ y_2 &= c^1_2 a_{11}x_1 + c^2_2 a_{21}x_1 + c^1_2 a_{12}x_2 + c^2_2 a_{22}x_2 \end{aligned}$$

1.4 SUBSTITUTIONS

Suppose it is required to substitute $y_i = a_{ij}x_j$ in the equation $Q = b_{ij}y_i x_j$. Disregard of Remark 2 above would lead to an absurd expression like $Q = b_{ij}a_{ij}x_j x_j$. The correct procedure is first to identify any dummy indices in the expression to be substituted that coincide with indices occurring in the main expression. Changing these dummy indices to characters not found in the main expression, one may then carry out the substitution in the usual fashion.

STEP 1 $Q = b_{ij}y_i x_j, \quad y_i = a_{ij}x_j$ [dummy index j is duplicated]

STEP 2 $y_i = a_{ir}x_r$ [change dummy index from j to r]

STEP 3 $Q = b_{ij}(a_{ir}x_r)x_j = a_{ir}b_{ij}x_r x_j$ [substitute and rearrange]

EXAMPLE 1.5 If $y_i = a_{ij}x_j$, express the quadratic form $Q = g_{ij}y_i y_j$ in terms of the x-variables.

First write: $y_i = a_{ir}x_r$, $y_j = a_{js}x_s$. Then, by substitution,

$$Q = g_{ij}(a_{ir}x_r)(a_{js}x_s) = g_{ij}a_{ir}a_{js}x_r x_s$$

or $Q = h_{rs}x_r x_s$, where $h_{rs} \equiv g_{ij}a_{ir}a_{js}$.

1.5 KRONECKER DELTA AND ALGEBRAIC MANIPULATIONS

A much used symbol in tensor calculus has the effect of annihilating the "off-diagonal" terms in a double summation.

Kronecker Delta

$$\delta_{ij} \equiv \delta^i_j \equiv \delta^{ij} \equiv \begin{cases} 1 & i = j \\ 0 & i \neq j \end{cases} \tag{1.1}$$

Clearly, $\delta_{ij} = \delta_{ji}$ for all i, j.

EXAMPLE 1.6 If $n = 3$,

$$\begin{aligned} \delta_{ij}x_ix_j &= 1x_1x_1 + 0x_1x_2 + 0x_1x_3 + 0x_2x_1 + 1x_2x_2 + 0x_2x_3 + 0x_3x_1 + 0x_3x_2 + 1x_3x_3 \\ &= (x_1)^2 + (x_2)^2 + (x_3)^2 = x_ix_i \end{aligned}$$

In general, $\delta_{ij}x_ix_j = x_ix_i$ and $\delta^r_j a_{ir}x_i = a_{ij}x_i$.

EXAMPLE 1.7 Suppose that $T^i = g^i_r a_{rs} y_s$ and $y_i = b_{ir}x_r$. If further, $a_{ir}b_{rj} = \delta_{ij}$, find T^i in terms of the x_r.

First write $y_s = b_{st}x_t$. Then, by substitution,

$$T^i = g^i_r a_{rs} b_{st} x_t = g^i_r \delta_{rt} x_t = g^i_r x_r$$

Algebra and the Summation Convention

Certain routine manipulations in tensor calculus can be easily justified by properties of ordinary sums. However, some care is warranted. For instance, the identity (*1.2*) below not only involves the distributive law for real numbers, $a(x + y) \equiv ax + ay$, but also requires a rearrangement of terms utilizing the associative and commutative laws. At least a mental verification of such operations must be made, if false results are to be avoided.

EXAMPLE 1.8 The following *nonidentities* should be carefully noted:

$$a_{ij}(x_i + y_j) \not\equiv a_{ij}x_i + a_{ij}y_j$$
$$a_{ij}x_iy_j \not\equiv a_{ij}y_ix_j$$
$$(a_{ij} + a_{ji})x_iy_j \not\equiv 2a_{ij}x_iy_j$$

Listed below are several valid identities; they, and others like them, will be used repeatedly from now on.

$$a_{ij}(x_j + y_j) \equiv a_{ij}x_j + a_{ij}y_j \tag{1.2}$$
$$a_{ij}x_iy_j \equiv a_{ij}y_jx_i \tag{1.3}$$
$$a_{ij}x_ix_j \equiv a_{ji}x_ix_j \tag{1.4}$$
$$(a_{ij} + a_{ji})x_ix_j \equiv 2a_{ij}x_ix_j \tag{1.5}$$
$$(a_{ij} - a_{ji})x_ix_j \equiv 0 \tag{1.6}$$

Solved Problems

REPEATED INDICES

1.1 Use the summation convention to write the following, and assign the value of n in each case:

$$(a)\quad a_{11}b_{11}+a_{21}b_{12}+a_{31}b_{13}+a_{41}b_{14}$$
$$(b)\quad a_{11}b_{11}+a_{12}b_{12}+a_{13}b_{13}+a_{14}b_{14}+a_{15}b_{15}+a_{16}b_{16}$$
$$(c)\quad c^i_{11}+c^i_{22}+c^i_{33}+c^i_{44}+c^i_{55}+c^i_{66}+c^i_{77}+c^i_{88}\qquad (1\leqq i\leqq 8)$$

(*a*) $a_{i1}b_{1i}$ $(n=4)$; (*b*) $a_{1i}b_{1i}$ $(n=6)$; (*c*) c^i_{jj} $(n=8)$.

1.2 Use the summation convention to write each of the following systems, state which indices are free and which are dummy indices, and fix the value of n:

$$(a)\quad \begin{aligned} c_{11}x_1+c_{12}x_2+c_{13}x_3&=2\\ c_{21}x_1+c_{22}x_2+c_{23}x_3&=-3\\ c_{31}x_1+c_{32}x_2+c_{33}x_3&=5\end{aligned}\qquad (b)\quad a^1_jx_1+a^2_jx_2+a^3_jx_3+a^4_jx_4=b_j\quad (j=1,2)$$

(*a*) Set $d_1=2$, $d_2=-3$, and $d_3=5$. Then one can write the system as $c_{ij}x_j=d_i$ $(n=3)$. The free index is i and the dummy index is j.

(*b*) Here, the range of the free index does not match that of the dummy index $(n=4)$, and this fact must be indicated:

$$a^i_jx_i=b_j\qquad (j=1,2)$$

The free index is j and the dummy index is i.

1.3 Write out explicitly the summations

$$c_i(x_i+y_i)\qquad c_jx_j+c_ky_k$$

where $n=4$ for both, and compare the results.

$$\begin{aligned} c_i(x_i+y_i)&=c_1(x_1+y_1)+c_2(x_2+y_2)+c_3(x_3+y_3)+c_4(x_4+y_4)\\ &=c_1x_1+c_1y_1+c_2x_2+c_2y_2+c_3x_3+c_3y_3+c_4x_4+c_4y_4\\ c_jx_j+c_ky_k&=c_1x_1+c_2x_2+c_3x_3+c_4x_4+c_1y_1+c_2y_2+c_3y_3+c_4y_4\end{aligned}$$

The two summations are identical except for the order in which the terms occur, constituting a special case of (*1.2*).

DOUBLE SUMS

1.4 If $n=3$, expand $Q=a^{ij}x_ix_j$.

$$\begin{aligned} Q&=a^{1j}x_1x_j+a^{2j}x_2x_j+a^{3j}x_3x_j\\ &=a^{11}x_1x_1+a^{12}x_1x_2+a^{13}x_1x_3+a^{21}x_2x_1+a^{22}x_2x_2+a^{23}x_2x_3+a^{31}x_3x_1+a^{32}x_3x_2+a^{33}x_3x_3\end{aligned}$$

1.5 Use the summation convention to write the following, and state the value of n necessary in each case:

$$(a)\quad a_{11}b_{11}+a_{21}b_{12}+a_{31}b_{13}+a_{12}b_{21}+a_{22}b_{22}+a_{32}b_{23}+a_{13}b_{31}+a_{23}b_{32}+a_{33}b_{33}$$
$$(b)\quad g^1_{11}+g^1_{12}+g^1_{21}+g^1_{22}+g^2_{11}+g^2_{12}+g^2_{21}+g^2_{22}$$

(*a*) $a_{i1}b_{1i} + a_{i2}b_{2i} + a_{i3}b_{3i} \equiv a_{ij}b_{ji} \quad (n = 3)$.

(*b*) Set $c_i = 1$ for each $i \quad (n = 2)$. Then the expression may be written

$$g^i_{11}c_i + g^i_{12}c_i + g^i_{21}c_i + g^i_{22}c_i = (g^i_{11} + g^i_{12} + g^i_{21} + g^i_{22})c_i$$
$$= (g^i_{jk}c_jc_k)c_i = g^i_{jk}c_ic_jc_k$$

1.6 If $n = 2$, write out explicitly the triple summation $c_{rst}x^ry^sz^t$.

Any expansion technique that yields all $2^3 = 8$ terms will do. In this case we shall interpret the triplet *rst* as a three-digit integer, and list the terms in increasing order of that integer:

$$c_{rst}x^ry^sz^t = c_{111}x^1y^1z^1 + c_{112}x^1y^1z^2 + c_{121}x^1y^2z^1 + c_{122}x^1y^2z^2$$
$$+ c_{211}x^2y^1z^1 + c_{212}x^2y^1z^2 + c_{221}x^2y^2z^1 + c_{222}x^2y^2z^2$$

1.7 Show that $a_{ij}x_ix_j = 0$ if $a_{ij} \equiv i - j$.

Because, for all i and j, $a_{ij} = -a_{ji}$ and $x_ix_j = x_jx_i$, the "off-diagonal" terms $a_{ij}x_ix_j \quad (i < j$; no sum) and $a_{ji}x_jx_i \quad (j > i$; no sum) cancel in pairs, while the "diagonal" terms $a_{ii}(x_i)^2$ are zero to begin with. Thus the sum is zero.

The result also follows at once from (*1.5*).

1.8 If the a_{ij} are constants, calculate the partial derivative

$$\frac{\partial}{\partial x_k}(a_{ij}x_ix_j)$$

Reverting to Σ-notation, we have:

$$\sum_{i,j} a_{ij}x_ix_j = \sum_{\substack{i\neq k\\ j\neq k}} a_{ij}x_ix_j + \sum_{\substack{i=k\\ j\neq k}} a_{ij}x_ix_j + \sum_{\substack{i\neq k\\ j=k}} a_{ij}x_ix_j + \sum_{\substack{i=k\\ j=k}} a_{ij}x_ix_j$$
$$= C + \Big(\sum_{j\neq k} a_{kj}x_j\Big)x_k + \Big(\sum_{i\neq k} a_{ik}x_i\Big)x_k + a_{kk}(x_k)^2$$

where C is independent of x_k. Differentiating with respect to x_k,

$$\frac{\partial}{\partial x_k}\Big(\sum_{i,j} a_{ij}x_ix_j\Big) = 0 + \sum_{j\neq k} a_{kj}x_j + \sum_{i\neq k} a_{ik}x_i + 2a_{kk}x_k$$
$$= \sum_j a_{kj}x_j + \sum_i a_{ik}x_i$$

or, going back to the Einstein summation convention,

$$\frac{\partial}{\partial x_k}(a_{ij}x_ix_j) = a_{ki}x_i + a_{ik}x_i = (a_{ik} + a_{ki})x_i$$

SUBSTITUTIONS, KRONECKER DELTA

1.9 Express $b^{ij}y_iy_j$ in terms of x-variables, if $y_i = c_{ij}x_j$ and $b^{ij}c_{ik} = \delta^j_k$.

$$b^{ij}y_iy_j = b^{ij}(c_{ir}x_r)(c_{js}x_s) = (b^{ij}c_{ir})x_rc_{js}x_s = \delta^j_rx_rc_{js}x_s = x_jc_{js}x_s = c_{ij}x_ix_j$$

1.10 Rework Problem 1.8 by use of the product rule for differentiation and the fact that

$$\frac{\partial x_p}{\partial x_q} = \delta_{pq}$$

$$\frac{\partial}{\partial x_k}(a_{ij}x_ix_j) = a_{ij}\frac{\partial}{\partial x_k}(x_ix_j) = a_{ij}\left(x_j\frac{\partial x_i}{\partial x_k} + x_i\frac{\partial x_j}{\partial x_k}\right)$$
$$= a_{ij}(x_j\delta_{ik} + x_i\delta_{jk}) = a_{kj}x_j + a_{ik}x_i$$
$$= (a_{ik} + a_{ki})x_i$$

1.11 If $a_{ij} = a_{ji}$ are constants, calculate

$$\frac{\partial^2}{\partial x_k \partial x_l}(a_{ij}x_ix_j)$$

Using Problem 1.8, we have

$$\frac{\partial^2}{\partial x_k \partial x_l}(a_{ij}x_ix_j) = \frac{\partial}{\partial x_k}\left[\frac{\partial}{\partial x_l}(a_{ij}x_ix_j)\right] = \frac{\partial}{\partial x_k}[(a_{lj} + a_{jl})x_j]$$
$$= \frac{\partial}{\partial x_k}(2a_{il}x_i) = 2a_{il}\frac{\partial}{\partial x_k}(x_i) = 2a_{il}\delta_{ki} = 2a_{kl}$$

1.12 Consider a system of linear equations of the form $y^i = a^{ij}x_j$ and suppose that (b_{ij}) is a matrix of numbers such that for all i and j, $b_{ir}a^{rj} = \delta_i^j$ [that is, the matrix (b_{ij}) is the inverse of the matrix (a^{ij})]. Solve the system for x_i in terms of the y^j.

Multiply both sides of the ith equation by b_{ki} and sum over i:

$$b_{ki}y^i = b_{ki}a^{ij}x_j = \delta_k^j x_j = x_k$$

or $x_i = b_{ij}y^j$.

1.13 Show that, generally, $a_{ijk}(x_i + y_j)z_k \neq a_{ijk}x_iz_k + a_{ijk}y_jz_k$.

Simply observe that on the left side there are no free indices, but on the right, j is free for the first term and i is free for the second.

1.14 Show that $c_{ij}(x_i + y_i)z_j \equiv c_{ij}x_iz_j + c_{ij}y_iz_j$.

Let us prove (*1.2*); the desired identity will then follow upon setting $a_{ij} \equiv c_{ji}$.

$$a_{ij}x_j + a_{ij}y_j \equiv \sum_j a_{ij}x_j + \sum_j a_{ij}y_j = \sum_j (a_{ij}x_j + a_{ij}y_j)$$
$$= \sum_j a_{ij}(x_j + y_j) \equiv a_{ij}(x_j + y_j)$$

Supplementary Problems

1.15 Write out the expression a_ib_i $(n = 6)$ in full.

1.16 Write out the expression R^i_{jki} $(n = 4)$ in full. Which are free and which are dummy indices? How many summations are there?

1.17 Evaluate $\delta^i_j x_i$ (n arbitrary).

1.18 For n arbitrary, evaluate (*a*) δ_{ii}, (*b*) $\delta_{ij}\delta_{ij}$, (*c*) $\delta_{ij}\delta^j_k c_{ik}$.

1.19 Use the summation convention to indicate $a_{13}b_{13} + a_{23}b_{23} + a_{33}b_{33}$, and state the value of n.

1.20 Use the summation convention to indicate

$$a_{11}(x_1)^2 + a_{12}x_1x_2 + a_{13}x_1x_3 + a_{21}x_2x_1 + a_{22}(x_2)^2 + a_{23}x_2x_3 + a_{31}x_3x_1 + a_{32}x_3x_2 + a_{33}(x_3)^2$$

1.21 Use the summation convention and free subscripts to indicate the following linear system, stating the value of n:

$$y_1 = c_{11}x_1 + c_{12}x_2$$
$$y_2 = c_{21}x_1 + c_{22}x_2$$

1.22 Find the following partial derivative if the a_{ij} are constants:

$$\frac{\partial}{\partial x_k}(a_{11}x_1 + a_{12}x_2 + a_{13}x_3) \quad (k = 1, 2, 3)$$

1.23 Use the Kronecker delta to calculate the partial derivative if the a_{ij} are constants:

$$\frac{\partial}{\partial x_k}(a_{ij}x_j)$$

1.24 Calculate

$$\frac{\partial}{\partial x_k}[a_{ij}x_i(x_j)^2]$$

where the a_{ij} are constants such that $a_{ij} = a_{ji}$.

1.25 Calculate

$$\frac{\partial}{\partial x_l}(a_{ijk}x_ix_jx_k)$$

where the a_{ijk} are constants.

1.26 Solve Problem 1.11 without the symmetry condition on a_{ij}.

1.27 Evaluate: (*a*) $b^i_jy_i$ if $y_i = T^{ij}_i$, (*b*) $a_{ij}y_j$ if $y_i = b_{ij}x_j$, (*c*) $a_{ijk}y_iy_jy_k$ if $y_i = b_{ij}x_j$.

1.28 If $\varepsilon_i = 1$ for all i, prove that

(*a*) $(a_1 + a_2 + \cdots + a_n)^2 \equiv \varepsilon_i\varepsilon_ja_ia_j$

(*b*) $a_i(1 + x_i) \equiv a_i\varepsilon_i + a_ix_i$

(*c*) $a_{ij}(x_i + x_j) \equiv 2a_{ij}\varepsilon_ix_j$ if $a_{ij} = a_{ji}$

Chapter 2

Basic Linear Algebra for Tensors

2.1 INTRODUCTION

Familiarity with the topics in this chapter will result in a much greater appreciation for the geometric aspects of tensor calculus. The main purpose is to reformulate the expressions of linear algebra and matrix theory using the summation convention.

2.2 TENSOR NOTATION FOR MATRICES, VECTORS, AND DETERMINANTS

In the ordinary matrix notation (a_{ij}), the first subscript, i, tells what row the number a_{ij} lies in, and the second, j, designates the column. A fuller notation is $[a_{ij}]_{mn}$, which exhibits the number of rows, m, and the number of columns, n. This notation may be extended as follows.

Upper-Index Matrix Notation

$$[a^i_j]_{mn} \equiv \begin{bmatrix} a^1_1 & a^1_2 & a^1_3 \ldots a^1_n \\ a^2_1 & a^2_2 & a^2_3 \ldots a^2_n \\ \cdots & \cdots & \cdots \\ a^m_1 & a^m_2 & a^m_3 \ldots a^m_n \end{bmatrix} \qquad [a^{ij}]_{mn} \equiv \begin{bmatrix} a^{11} & a^{12} & a^{13} \ldots a^{1n} \\ a^{21} & a^{22} & a^{23} \ldots a^{2n} \\ \cdots & \cdots & \cdots \\ a^{m1} & a^{m2} & a^{m3} \ldots a^{mn} \end{bmatrix}$$

Note that, for *mixed indices* (one upper, one lower), it is the upper index that designates the row, and the lower index, the column. In the case of pure superscripts, the scheme is identical to the familiar one for subscripts.

EXAMPLE 2.1

$$[c^i_j]_{23} \equiv \begin{bmatrix} c^1_1 & c^1_2 & c^1_3 \\ c^2_1 & c^2_2 & c^2_3 \end{bmatrix} \qquad [d^j_i]_{23} \equiv \begin{bmatrix} d^1_1 & d^1_2 & d^1_3 \\ d^2_1 & d^2_2 & d^2_3 \end{bmatrix} \equiv [d^i_j]_{23}$$

$$[x^r_s]_{14} \equiv [x^1_1 \quad x^1_2 \quad x^1_3 \quad x^1_4] \qquad [y^{pq}]_{42} \equiv \begin{bmatrix} y^{11} & y^{12} \\ y^{21} & y^{22} \\ y^{31} & y^{32} \\ y^{41} & y^{42} \end{bmatrix}$$

Vectors

A real n-dimensional *vector* is any column matrix $\mathbf{v} = [x_{ij}]_{n1}$ with real components $x_i \equiv x_{i1}$; one usually writes simply $\mathbf{v} = (x_i)$. The collection of all real n-dimensional vectors is the n-dimensional real vector space denoted $\mathbf{R}^n$.

Vector sums are determined by coordinatewise addition, as are matrix sums: if $A \equiv [a_{ij}]_{mn}$ and $B \equiv [b_{ij}]_{mn}$, then

$$A + B \equiv [a_{ij} + b_{ij}]_{mn}$$

Scalar multiplication of a vector or matrix is defined by

$$\lambda[a_{ij}]_{mn} \equiv [\lambda a_{ij}]_{mn}$$

Basic Formulas

The essential formulas involving matrices, vectors, and determinants are now given in terms of the summation convention.

Matrix multiplication. If $A \equiv [a_{ij}]_{mn}$ and $B \equiv [b_{ij}]_{nk}$, then

$$AB = [a_{ir}b_{rj}]_{mk} \tag{2.1a}$$

Analogously, for mixed or upper indices,

$$AB \equiv [a^i_j]_{mn}[b^i_j]_{nk} = [a^i_r b^r_j]_{mk} \qquad AB \equiv [a^{ij}]_{mn}[b^{ij}]_{nk} = [a^{ir}b^{rj}]_{mk} \tag{2.1b}$$

wherein i and j are not summed on.

Identity matrix. In terms of the Kronecker deltas, the identity matrix of order n is

$$I = [\delta_{ij}]_{nn} \equiv [\delta^i_j]_{nn} \equiv [\delta^{ij}]_{nn}$$

which has the property $IA = AI = A$ for any square matrix A of order n.

Inverse of a square matrix. A square matrix $A \equiv [a_{ij}]_{nn}$ is *invertible* if there exists a (unique) matrix $B \equiv [b_{ij}]_{nn}$, called the *inverse* of A, such that $AB = BA = I$. In terms of components, the criterion reads:

$$a_{ir}b_{rj} = b_{ir}a_{rj} = \delta_{ij} \tag{2.2a}$$

or, for mixed or upper indices,

$$a^i_r b^r_j = b^i_r a^r_j = \delta^i_j \qquad a^{ir}b^{rj} = b^{ir}a^{rj} = \delta^{ij} \tag{2.2b}$$

Transpose of a matrix. Transposition of an arbitrary matrix is defined by $A^T \equiv [a_{ij}]^T_{mn} = [a'_{ij}]_{nm}$, where $a'_{ij} = a_{ji}$ for all i, j. If $A^T = A$ (that is, $a_{ij} = a_{ji}$ for all i, j), then A is called *symmetric*; if $A^T = -A$ (that is, $a_{ij} = -a_{ji}$ for all i, j), then A is called *antisymmetric* or *skew-symmetric*.

Orthogonal matrix. A matrix A is *orthogonal* if $A^T = A^{-1}$ (or if $A^TA = AA^T = I$).

Permutation symbol. The symbol $e_{ijk\ldots w}$ (with n subscripts) has the value zero if any pair of subscripts are identical, and equals $(-1)^p$ otherwise, where p is the number of subscript transpositions (interchanges of consecutive subscripts) required to bring $(i\,j\,k\ldots w)$ to the natural order $(1\,2\,3\ldots n)$.

Determinant of a square matrix. If $A \equiv [a_{ij}]_{nn}$ is any square matrix, define the scalar

$$\det A \equiv e_{i_1 i_2 i_3 \ldots i_n} a_{1i_1} a_{2i_2} a_{3i_3} \cdots a_{ni_n} \tag{2.3}$$

Other notations are $|A|$, $|a_{ij}|$, and $\det(a_{ij})$. The chief properties of determinants are

$$|AB| = |A|\,|B| \qquad |A^T| = |A| \tag{2.4}$$

Laplace expansion of a determinant. For each i and j, let M_{ij} be the determinant of the square matrix of order $n-1$ obtained from $A \equiv [a_{ij}]_{nn}$ by deleting the ith row and jth column; M_{ij} is called the *minor* of a_{ij} in $|A|$. Define the *cofactor* of a_{ij} to be the scalar

$$A_{ij} = (-1)^k M_{ij} \qquad \text{where} \qquad k = i + j \tag{2.5}$$

Then the Laplace expansions of $|A|$ are given by

$$\begin{aligned} |A| &= a_{1j}A_{1j} = a_{2j}A_{2j} = \cdots = a_{nj}A_{nj} && \text{[row expansions]} \\ |A| &= a_{i1}A_{i1} = a_{i2}A_{i2} = \cdots = a_{in}A_{in} && \text{[column expansions]} \end{aligned} \tag{2.6}$$

Scalar product of vectors. If $\mathbf{u} = (x_i)$ and $\mathbf{v} = (y_i)$, then

$$\mathbf{uv} \equiv \mathbf{u} \cdot \mathbf{v} \equiv \mathbf{u}^T\mathbf{v} = x_i y_i \tag{2.7}$$

If $\mathbf{u} = \mathbf{v}$, the notation $\mathbf{uu} \equiv \mathbf{u}^2 \equiv \mathbf{v}^2$ will often be used. Vectors $\mathbf{u}$ and $\mathbf{v}$ are *orthogonal* if $\mathbf{uv} = 0$.

Norm (length) of a vector. If $\mathbf{u} = (x_i)$, then

$$\|\mathbf{u}\| \equiv \sqrt{\mathbf{u}^2} = \sqrt{x_i x_i} \tag{2.8}$$

Angle between two vectors. The angle θ between two nonzero vectors, $\mathbf{u} = (x_i)$ and $\mathbf{v} = (y_i)$, is defined by

$$\cos\theta \equiv \frac{\mathbf{uv}}{\|\mathbf{u}\|\,\|\mathbf{v}\|} = \frac{x_i y_i}{\sqrt{x_j x_j}\,\sqrt{y_k y_k}} \qquad (0 \leqq \theta \leqq \pi) \tag{2.9}$$

It follows that $\theta = \pi/2$ if $\mathbf{u}$ and $\mathbf{v}$ are nonzero orthogonal vectors.

Vector product in $\mathbf{R}^3$. If $\mathbf{u} = (x_i)$ and $\mathbf{v} = (y_i)$, and if the standard basis vectors are designated

$$\mathbf{i} \equiv (\delta_{i1}) \qquad \mathbf{j} \equiv (\delta_{i2}) \qquad \mathbf{k} \equiv (\delta_{i3})$$

then

$$\mathbf{u} \times \mathbf{v} \equiv \begin{vmatrix} \mathbf{i} & \mathbf{j} & \mathbf{k} \\ x_1 & x_2 & x_3 \\ y_1 & y_2 & y_3 \end{vmatrix} = \begin{vmatrix} x_2 & x_3 \\ y_2 & y_3 \end{vmatrix}\mathbf{i} - \begin{vmatrix} x_1 & x_3 \\ y_1 & y_3 \end{vmatrix}\mathbf{j} + \begin{vmatrix} x_1 & x_2 \\ y_1 & y_2 \end{vmatrix}\mathbf{k} \tag{2.10a}$$

Expressing the second-order determinants by means of (*2.3*), one can rewrite (*2.10a*) in terms of components only:

$$\mathbf{u} \times \mathbf{v} = (e_{ijk} x_j y_k) \tag{2.10b}$$

2.3 INVERTING A MATRIX

There are a number of algorithms for computing the inverse of $A \equiv [a_{ij}]_{nn}$, where $|A| \neq 0$ (a necessary and sufficient condition for A to be invertible). When n is large, the method of elementary row operations is efficient. For small n, it is practical to apply the explicit formula

$$A^{-1} = \frac{1}{|A|}\,[A_{ij}]_{nn}^T \tag{2.11a}$$

which follows from the generalized Laplace expansion theorem (Problem 2.10) and (*2.2a*). Thus, for $n = 2$,

$$\begin{bmatrix} a_{11} & a_{12} \\ a_{21} & a_{22} \end{bmatrix}^{-1} = \frac{1}{|A|}\begin{bmatrix} a_{22} & -a_{12} \\ -a_{21} & a_{11} \end{bmatrix} \tag{2.11b}$$

in which $|A| = a_{11}a_{22} - a_{12}a_{21}$; and, for $n = 3$,

$$\begin{bmatrix} a_{11} & a_{12} & a_{13} \\ a_{21} & a_{22} & a_{23} \\ a_{31} & a_{32} & a_{33} \end{bmatrix}^{-1} = \frac{1}{|A|}\begin{bmatrix} A_{11} & A_{21} & A_{31} \\ A_{12} & A_{22} & A_{32} \\ A_{13} & A_{23} & A_{33} \end{bmatrix} \tag{2.11c}$$

in which

$$A_{11} = a_{22}a_{33} - a_{23}a_{32} \qquad A_{21} = -(a_{12}a_{33} - a_{13}a_{32}) \qquad \ldots$$

2.4 MATRIX EXPRESSIONS FOR LINEAR SYSTEMS AND QUADRATIC FORMS

Because of the product rule for matrices and the rule for matrix equality, one can write a system of equations such as

$$\begin{aligned} 3x - 4y &= 2 \\ -5x + 8y &= 7 \end{aligned}$$

in the matrix form

$$\begin{bmatrix} 3 & -4 \\ -5 & 8 \end{bmatrix}\begin{bmatrix} x \\ y \end{bmatrix} = \begin{bmatrix} 2 \\ 7 \end{bmatrix}$$

In general, any $m \times n$ system of equations

$$a_{ij}x_j = b_i \qquad (1 \leqq i \leqq m) \tag{2.12a}$$

can be written in the matrix form

$$A\mathbf{x} = \mathbf{b} \tag{2.12b}$$

where $A \equiv [a_{ij}]_{mn}$, $\mathbf{x} \equiv (x_i)$, and $\mathbf{b} \equiv (b_i)$. One advantage in doing this is that, if $m = n$ and A is invertible, the solution of the system can proceed entirely by matrices: $\mathbf{x} = A^{-1}\mathbf{b}$.

Another useful fact for work with tensors is that a quadratic form Q (a homogeneous second-degree polynomial) in the n variables $x_1, x_2, \ldots, x_n$ also has a strictly matrix representation:

$$Q = a_{ij}x_ix_j = \mathbf{x}^TA\mathbf{x} \tag{2.13}$$

where the row matrix $\mathbf{x}^T$ is the transpose of the column matrix $\mathbf{x} = (x_i)$ and where $A \equiv [a_{ij}]_{nn}$.

EXAMPLE 2.2

$$[x_1 \quad x_2 \quad x_3]\begin{bmatrix} a_{11} & a_{12} & a_{13} \\ a_{21} & a_{22} & a_{23} \\ a_{31} & a_{32} & a_{33} \end{bmatrix}\begin{bmatrix} x_1 \\ x_2 \\ x_3 \end{bmatrix} = [x_1 \quad x_2 \quad x_3]\begin{bmatrix} a_{1j}x_j \\ a_{2j}x_j \\ a_{3j}x_j \end{bmatrix} = [x_i(a_{ij}x_j)] = a_{ij}x_ix_j$$

The matrix A that produces a given quadratic form is not unique. In fact, the matrix $B = \frac{1}{2}(A + A^T)$ may always be substituted for A in (2.13); i.e., the matrix of a quadratic form may always be assumed *symmetric*.

EXAMPLE 2.3 Write the quadratic equation

$$3x^2 + y^2 - 2z^2 - 5xy - 6yz = 10$$

using a symmetric matrix.

The quadratic form (2.13) is given in terms of the nonsymmetric matrix

$$A = \begin{bmatrix} 3 & -5 & 0 \\ 0 & 1 & -6 \\ 0 & 0 & -2 \end{bmatrix}$$

The symmetric equivalent is obtained by replacing each off-diagonal element by one-half the sum of that element and its mirror image in the main diagonal. Hence, the desired representation is

$$[x \quad y \quad z]\begin{bmatrix} 3 & -5/2 & 0 \\ -5/2 & 1 & -3 \\ 0 & -3 & -2 \end{bmatrix}\begin{bmatrix} x \\ y \\ z \end{bmatrix} = 10$$

2.5 LINEAR TRANSFORMATIONS

Of utmost importance for the study of tensor calculus is a basic knowledge of transformation theory and changes in coordinate systems. A set of linear equations like

$$\begin{aligned} \bar{x} &= 5x - 2y \\ \bar{y} &= 3x + 2y \end{aligned} \tag{I}$$

defines a *linear transformation* (or *linear mapping*) from each point (x, y) to its corresponding image $(\bar{x}, \bar{y})$. In matrix form, a linear transformation may be written $\bar{\mathbf{x}} = A\mathbf{x}$; if, as in (I), the mapping is one-one, then $|A| \neq 0$.

There is always an *alias-alibi* aspect of such transformations: When $(\bar{x}, \bar{y})$ is regarded as defining *new coordinates* (a new name) for (x, y), one is dealing with the alias aspect; when $(\bar{x}, \bar{y})$ is regarded as a *new position* (place) for (x, y), the alibi aspect emerges. In tensor calculus, one is generally more interested in the alias aspect: the two coordinate systems related by $\bar{\mathbf{x}} = A\mathbf{x}$ are referred to as the *unbarred* and the *barred* systems.

EXAMPLE 2.4 In order to find the image of the point (0, −1) under (*I*), merely set $x = 0$ and $y = -1$; the result is

$$\bar{x} = 5(0) - 2(-1) = 2 \qquad \bar{y} = 3(0) + 2(-1) = -2$$

Hence, $\overline{(0,-1)} = (2,-2)$. Similarly, we find that $\overline{(2,1)} = (8,8)$.

If we regard $(\bar{x}, \bar{y})$ merely as a different coordinate system, we would say that two fixed points, P and Q, have the respective coordinates (0, −1) and (2, 1) in the unbarred system, and (2, −2) and (8, 8) in the barred system.

Distance in a Barred Coordinate System

What is the expression for the (invariant) distance between two points in terms of differing aliases? Let $\bar{\mathbf{x}} = A\mathbf{x}$ ($|A| \neq 0$) define an invertible linear transformation between unbarred and barred coordinates. It is shown in Problem 2.20 that the desired distance formula is

$$d(\bar{\mathbf{x}}, \bar{\mathbf{y}}) = \sqrt{(\bar{\mathbf{x}} - \bar{\mathbf{y}})^T G (\bar{\mathbf{x}} - \bar{\mathbf{y}})} = \sqrt{g_{ij}\, \Delta\bar{x}_i\, \Delta\bar{x}_j} \tag{2.14}$$

where $[g_{ij}]_{nn} \equiv G = (AA^T)^{-1}$ and $\bar{\mathbf{x}} - \bar{\mathbf{y}} = (\Delta\bar{x}_i)$. If A is orthogonal (a rotation of the axes), then $g_{ij} = \delta_{ij}$, and (*2.14*) reduces to the ordinary form

$$d(\bar{\mathbf{x}}, \bar{\mathbf{y}}) = \|\bar{\mathbf{x}} - \bar{\mathbf{y}}\| = \sqrt{\Delta\bar{x}_i\, \Delta\bar{x}_i}$$

[cf. (*2.8*)].

EXAMPLE 2.5 Calculate the distance between points P and Q of Example 2.4 in terms of their barred coordinates. Verify that the same distance is found in the unbarred coordinate system.

First calculate the matrix $G = (AA^T)^{-1} = (A^{-1})^T A^{-1}$ (see Problem 2.13):

$$A = \begin{bmatrix} 5 & -2 \\ 3 & 2 \end{bmatrix}, \quad |A| = 10 - (-6) = 16 \quad \Rightarrow \quad A^{-1} = \frac{1}{16}\begin{bmatrix} 2 & 2 \\ -3 & 5 \end{bmatrix}$$

and
$$G = \frac{1}{16}\begin{bmatrix} 2 & 2 \\ -3 & 5 \end{bmatrix}^T \cdot \frac{1}{16}\begin{bmatrix} 2 & 2 \\ -3 & 5 \end{bmatrix} = \frac{1}{256}\begin{bmatrix} 2 & -3 \\ 2 & 5 \end{bmatrix}\begin{bmatrix} 2 & 2 \\ -3 & 5 \end{bmatrix} = \frac{1}{256}\begin{bmatrix} 13 & -11 \\ -11 & 29 \end{bmatrix}$$

Hence $g_{11} = 13/256$, $g_{12} = g_{21} = -11/256$, and $g_{22} = 29/256$. Now, with $\bar{\mathbf{x}} - \bar{\mathbf{y}} = [2-8 \quad -2-8]^T = [-6 \quad -10]^T$, (*2.14*) gives:

$$\begin{aligned} d^2 &= g_{ij}\, \Delta\bar{x}_i\, \Delta\bar{x}_j \\ &= \frac{13}{256}(-6)^2 + 2 \cdot \frac{-11}{256}(-6)(-10) + \frac{29}{256}(-10)^2 \\ &= \frac{13(36) - 22(60) + 29(100)}{256} = 8 \end{aligned}$$

In the unbarred system, the distance between $P(0, -1)$ and $Q(2, 1)$ is given, in agreement, by the Pythagorean theorem:

$$d^2 = (0-2)^2 + (-1-1)^2 = 8$$

2.6 GENERAL COORDINATE TRANSFORMATIONS

A general mapping or transformation T of $\mathbf{R}^n$ may be indicated in functional (vector) or in component form:

$$\bar{\mathbf{x}} = T(\mathbf{x}) \qquad \text{or} \qquad \bar{x}_i = T_i(x_1, x_2, \ldots, x_n)$$

In the alibi description, any point $\mathbf{x}$ in the *domain* of T (possibly the whole of $\mathbf{R}^n$) has as its image the point $T(\mathbf{x})$ in the *range* of T. Considered as a coordinate transformation (the alias description), T sets up, for each point P in its domain, a correspondence between (x_i) and $(\bar{x}_i)$, the coordinates of P in two different systems. As explained below, T may be interpreted as a coordinate transformation only if a certain condition is fulfilled.

Bijections, Curvilinear Coordinates

A map T is called a *bijection* or a *one-one mapping* if it maps each pair of distinct points $\mathbf{x} \neq \mathbf{y}$ in its domain into distinct points $T(\mathbf{x}) \neq T(\mathbf{y})$ in its range. Whenever T is bijective, we call the image $\bar{\mathbf{x}} = T(\mathbf{x})$ a set of *admissible coordinates* for $\mathbf{x}$, and the aggregate of all such coordinates (alibi: the range of T), a *coordinate system*.

Certain coordinate systems are named after the characteristics of the mapping T. For example, if T is linear, the $(\bar{x}_i)$-system is called *affine*; and if T is a rigid motion, $(\bar{x}_i)$ is called *rectangular* or *cartesian*. [It is presumed in making this statement that the original coordinate system (x_i) is the familiar cartesian coordinate system of analytic geometry, or its natural extension to vectors in $\mathbf{R}^n$.] Nonaffine coordinate systems are generally called *curvilinear coordinates*; these include polar coordinates in two dimensions, and cylindrical and spherical coordinates in three dimensions.

2.7 THE CHAIN RULE FOR PARTIAL DERIVATIVES

In working with curvilinear coordinates, one needs the Jacobian matrix (Chapter 3) and, therefore, the chain rule of multivariate calculus. The summation convention makes possible a compact statement of this rule: If $w = f(x_1, x_2, x_3, \ldots, x_n)$ and $x_i = x_i(u_1, u_2, \ldots, u_m)$ $(i = 1, 2, \ldots, n)$, where all functions involved have continuous partial derivatives, then

$$\frac{\partial w}{\partial u_j} = \frac{\partial f}{\partial x_i}\frac{\partial x_i}{\partial u_j} \qquad (1 \leq j \leq m) \tag{2.15}$$

Solved Problems

TENSOR NOTATION

2.1 Display explicitly the matrices (*a*) $[b_i^j]_{42}$, (*b*) $[b_j^i]_{24}$, (*c*) $[\delta^{ij}]_{33}$.

$$(a)\quad [b_i^j]_{42} = \begin{bmatrix} b_1^1 & b_2^1 \\ b_1^2 & b_2^2 \\ b_1^3 & b_2^3 \\ b_1^4 & b_2^4 \end{bmatrix} \qquad (b)\quad [b_j^i]_{24} = \begin{bmatrix} b_1^1 & b_2^1 & b_3^1 & b_4^1 \\ b_1^2 & b_2^2 & b_3^2 & b_4^2 \end{bmatrix}$$

$$(c)\quad [\delta^{ij}]_{33} = \begin{bmatrix} \delta^{11} & \delta^{12} & \delta^{13} \\ \delta^{21} & \delta^{22} & \delta^{23} \\ \delta^{31} & \delta^{32} & \delta^{33} \end{bmatrix} = \begin{bmatrix} 1 & 0 & 0 \\ 0 & 1 & 0 \\ 0 & 0 & 1 \end{bmatrix}$$

From (*a*) and (*b*) it is evident that merely interchanging the indices i and j in a matrix $A \equiv [a_{ij}]_{mn}$ does not necessarily yield the transpose, A^T.

2.2 Given

$$A = \begin{bmatrix} a & -a & -a \\ 2b & b & -b \\ 4c & 2c & -2c \end{bmatrix} \qquad B = \begin{bmatrix} 2 & 4 & -6 \\ -1 & -2 & 3 \\ 3 & 6 & -9 \end{bmatrix}$$

verify that $AB \neq BA$.

$$AB = \begin{bmatrix} 2a + a - 3a & 4a + 2a - 6a & -6a - 3a + 9a \\ 4b - b - 3b & 8b - 2b - 6b & -12b + 3b + 9b \\ 8c - 2c - 6c & 16c - 4c - 12c & -24c + 6c + 18c \end{bmatrix} = \begin{bmatrix} 0 & 0 & 0 \\ 0 & 0 & 0 \\ 0 & 0 & 0 \end{bmatrix} \equiv \mathrm{O}$$

but
$$BA = \begin{bmatrix} 2a+8b-24c & -2a+4b-12c & -2a-4b+12c \\ -a-4b+12c & a-2b+6c & a+2b-6c \\ 3a+12b-36c & -3a+6b-18c & -3a-6b+18c \end{bmatrix} \neq O$$

Thus, the commutative law ($AB = BA$) fails for matrices. Further, $AB = O$ does not imply that $A = O$ or $B = O$.

2.3 Prove by use of tensor notation and the product rule for matrices that $(AB)^T = B^TA^T$, for any two conformable matrices A and B.

Let $A \equiv [a_{ij}]_{mn}$, $B \equiv [b_{ij}]_{nk}$, $AB \equiv [c_{ij}]_{mk}$, and, for all i and j,

$$a'_{ij} = a_{ji} \qquad b'_{ij} = b_{ji} \qquad c'_{ij} = c_{ji}$$

Hence, $A^T = [a'_{ij}]_{nm}$, $B^T = [b'_{ij}]_{kn}$, and $(AB)^T = [c'_{ij}]_{km}$. We must show that $B^TA^T = [c'_{ij}]_{km}$. By definition of matrix product, $B^TA^T = [b'_{ir}a'_{rj}]_{km}$, and since

$$b'_{ir}a'_{rj} = b_{ri}a_{jr} = a_{jr}b_{ri} = c_{ji} = c'_{ij}$$

the desired result follows.

2.4 Show that any matrix of the form $A = B^TB$ is symmetric.

By Problem 2.3 and the involutory nature of the transpose operation,

$$A^T = (B^TB)^T = B^T(B^T)^T = B^TB = A$$

2.5 From the definition, (*2.3*), of a determinant of order 3, derive the Laplace expansion by cofactors of the first row.

In the case $n = 3$, (*2.3*) becomes

$$\begin{vmatrix} a_{11} & a_{12} & a_{13} \\ a_{21} & a_{22} & a_{23} \\ a_{31} & a_{32} & a_{33} \end{vmatrix} \equiv |a_{ij}| = e_{ijk}a_{1i}a_{2j}a_{3k}$$

Since $e_{ijk} = 0$ if any two subscripts coincide, we write only terms for which (ijk) is a permutation of (123):

$$\begin{aligned} |a_{ij}| &= e_{123}a_{11}a_{22}a_{33} + e_{132}a_{11}a_{23}a_{32} + e_{213}a_{12}a_{21}a_{33} \\ &\qquad + e_{231}a_{12}a_{23}a_{31} + e_{312}a_{13}a_{21}a_{32} + e_{321}a_{13}a_{22}a_{31} \\ &= a_{11}a_{22}a_{33} - a_{11}a_{23}a_{32} - a_{12}a_{21}a_{33} + a_{12}a_{23}a_{31} + a_{13}a_{21}a_{32} - a_{13}a_{22}a_{31} \\ &= a_{11}(a_{22}a_{33} - a_{23}a_{32}) - a_{12}(a_{21}a_{33} - a_{23}a_{31}) + a_{13}(a_{21}a_{32} - a_{22}a_{31}) \end{aligned}$$

But, for $n = 2$, (*2.3*) gives

$$\begin{vmatrix} a_{22} & a_{23} \\ a_{32} & a_{33} \end{vmatrix} \equiv +A_{11} = e_{12}a_{22}a_{33} + e_{21}a_{23}a_{32} = a_{22}a_{33} - a_{23}a_{32}$$

and the analogous expansions of $-A_{12}$ and $+A_{13}$. Hence,

$$|a_{ij}| = a_{11}A_{11} + a_{12}A_{12} + a_{13}A_{13} = a_{1j}A_{1j}$$

as in (*2.6*).

2.6 Evaluate:

$$(a)\ \begin{vmatrix} b & -2a \\ -2c & b \end{vmatrix} \qquad (b)\ \begin{vmatrix} 5 & -2 & 15 \\ -10 & 0 & 10 \\ 15 & 0 & 30 \end{vmatrix}$$

(*a*)
$$\begin{vmatrix} b & -2a \\ -2c & b \end{vmatrix} = b \cdot b - (-2a)(-2c) = b^2 - 4ac$$

(*b*) Because of the zeros in the second column, it is simplest to expand by that column:

$$\begin{vmatrix} 5 & -2 & 15 \\ -10 & 0 & 10 \\ 15 & 0 & 30 \end{vmatrix} = -(-2)\begin{vmatrix} -10 & 10 \\ 15 & 30 \end{vmatrix} + 0\begin{vmatrix} 5 & 15 \\ 15 & 30 \end{vmatrix} - 0\begin{vmatrix} 5 & 15 \\ -10 & 10 \end{vmatrix}$$

$$= 2\begin{vmatrix} -10 & 10 \\ 15 & 30 \end{vmatrix} = 2(10)(15)\begin{vmatrix} -1 & 1 \\ 1 & 2 \end{vmatrix} = 300(-2-1) = -900$$

2.7 Calculate the angle between the following two vectors in $\mathbf{R}^5$:

$$\mathbf{x} = (1, 0, -2, -1, 0) \qquad \text{and} \qquad \mathbf{y} = (0, 0, 2, 2, 0)$$

We have:

$$\mathbf{xy} = (1)(0) + (0)(0) + (-2)(2) + (-1)(2) + (0)(0) = -6$$
$$\mathbf{x}^2 = 1^2 + 0^2 + (-2)^2 + (-1)^2 + 0^2 = 6$$
$$\mathbf{y}^2 = 0^2 + 0^2 + 2^2 + 2^2 + 0^2 = 8$$

and (*2.9*) gives

$$\cos\theta = \frac{-6}{\sqrt{6}\cdot\sqrt{8}} = -\frac{\sqrt{3}}{2} \qquad \text{or} \qquad \theta = \frac{5\pi}{6}$$

2.8 Find three linearly independent vectors in $\mathbf{R}^4$ which are orthogonal to the vector $(3, 4, 1, -2)$.

It is useful to choose vectors having as many zero components as possible. The components $(0, 1, 0, 2)$ clearly work, and $(1, 0, -3, 0)$ also. Finally, $(0, 0, 2, 1)$ is orthogonal to the given vector, and seems not to be dependent on the first two chosen. To check independence, suppose scalars x, y, and z exist such that

$$x\begin{bmatrix} 0 \\ 1 \\ 0 \\ 2 \end{bmatrix} + y\begin{bmatrix} 1 \\ 0 \\ -3 \\ 0 \end{bmatrix} + z\begin{bmatrix} 0 \\ 0 \\ 2 \\ 1 \end{bmatrix} = \begin{bmatrix} 0 \\ 0 \\ 0 \\ 0 \end{bmatrix} \qquad \text{or} \qquad \begin{aligned} x(0) + y(1) \quad + z(0) &= 0 \\ x(1) + y(0) \quad + z(0) &= 0 \\ x(0) + y(-3) + z(2) &= 0 \\ x(2) + y(0) \quad + z(1) &= 0 \end{aligned}$$

This system has the sole solution $x = y = z = 0$, and the vectors are independent.

2.9 Prove that the vector product in $\mathbf{R}^3$ is anticommutative: $\mathbf{x} \times \mathbf{y} = -\mathbf{y} \times \mathbf{x}$.

By (*2.10b*),

$$\mathbf{x} \times \mathbf{y} = (e_{ijk}x_j y_k) \qquad \text{and} \qquad \mathbf{y} \times \mathbf{x} = (e_{ijk}y_j x_k)$$

But $e_{ikj} = -e_{ijk}$, so that

$$e_{ijk}y_j x_k = e_{ikj}y_k x_j = -e_{ijk}y_k x_j = -e_{ijk}x_j y_k$$

INVERTING A MATRIX

2.10 Establish the generalized Laplace expansion theorem: $a_{rj}A_{sj} = |A|\,\delta_{rs}$.

Consider the matrix

$$A^* = \begin{bmatrix} a_{11} & a_{12} & \cdots & a_{1n} \\ \cdots & \cdots & \cdots & \cdots \\ a_{r1} & a_{r2} & \cdots & a_{rn} \\ \cdots & \cdots & \cdots & \cdots \\ a_{r1} & a_{r2} & \cdots & a_{rn} \\ \cdots & \cdots & \cdots & \cdots \\ a_{n1} & a_{n2} & \cdots & a_{nn} \end{bmatrix} \begin{matrix} \\ \\ \textit{row } r \\ \\ \textit{row } s \\ \\ \\ \end{matrix}$$

which is obtained from matrix A by replacing its sth row by its rth row ($r \neq s$). By (2.6), applied to row r of A^*,

$$\det A^* = a_{rj}A^*_{rj} \qquad \text{(not summed on } r\text{)}$$

Now, because rows r and s are identical, we have for all j,

$$A^*_{rj} = (-1)^p A^*_{sj} = (-1)^p A_{sj} \qquad (p \equiv r - s)$$

Therefore, $\det A^* = (-1)^p a_{rj}A_{sj}$. But it is easy to see (Problem 2.31) that, with two rows the same, $\det A^* = 0$. We have thus proved that

$$a_{rj}A_{sj} = 0 \qquad (r \neq s)$$

and this, together with (2.6) for the case $r = s$, yields the theorem.

2.11 Given a matrix $A \equiv [a_{ij}]_{nn}$, with $|A| \neq 0$, use Problem 2.10 to show that

$$AB = I \qquad \text{where} \qquad B = \frac{1}{|A|}[A_{ij}]^T_{nn}$$

Since the (i, j)-element of B is $A_{ji}/|A|$,

$$AB = [a_{ik}(A_{jk}/|A|)]_{nn} = \frac{1}{|A|}[|A|\delta_{ij}]_{nn} = \frac{|A|}{|A|}[\delta_{ij}]_{nn} = I$$

[It follows from basic facts of linear algebra that also $BA = I$; therefore, $A^{-1} = B$, which establishes (2.11a).]

2.12 Invert the matrix

$$A = \begin{bmatrix} -2 & 0 & 1 \\ 3 & 1 & 0 \\ 2 & -2 & 3 \end{bmatrix}$$

Use (2.11c). To evaluate $|A|$, add twice the third column to the first and then expand by the first row:

$$|A| = \begin{vmatrix} 0 & 0 & 1 \\ 3 & 1 & 0 \\ 8 & -2 & 3 \end{vmatrix} = 1 \cdot \begin{vmatrix} 3 & 1 \\ 8 & -2 \end{vmatrix} = -6 - 8 = -14$$

Then, computing cofactors as we go,

$$A^{-1} = \frac{1}{-14}\begin{bmatrix} 3 & -2 & -1 \\ -9 & -8 & 3 \\ -8 & -4 & -2 \end{bmatrix} = \begin{bmatrix} -3/14 & 1/7 & 1/14 \\ 9/14 & 4/7 & -3/14 \\ 4/7 & 2/7 & 1/7 \end{bmatrix}$$

2.13 Let A and B be invertible matrices of the same order. Prove that (a) $(A^T)^{-1} = (A^{-1})^T$ (i.e., the operations of transposition and inversion commute); (b) $(AB)^{-1} = B^{-1}A^{-1}$.

(a) Transpose the equations $AA^{-1} = A^{-1}A = I$, recalling Problem 2.3, to obtain

$$(A^{-1})^T A^T = A^T(A^{-1})^T = I^T = I$$

which show that A^T is invertible, with inverse $(A^T)^{-1} = (A^{-1})^T$.

(b) By the associative law for matrix multiplication,

$$(AB)(B^{-1}A^{-1}) = A(BB^{-1})A^{-1} = AIA^{-1} = AA^{-1} = I$$

and, similarly,

$$(B^{-1}A^{-1})(AB) = I$$

Hence, $(AB)^{-1} = B^{-1}A^{-1}$.

LINEAR SYSTEMS; QUADRATIC FORMS

2.14 Write the following system of equations in matrix form, then solve by using the inverse matrix:

$$3x - 4y = -18$$
$$-5x + 8y = 34$$

The matrix form of the system is

$$\begin{bmatrix} 3 & -4 \\ -5 & 8 \end{bmatrix}\begin{bmatrix} x \\ y \end{bmatrix} = \begin{bmatrix} -18 \\ 34 \end{bmatrix} \qquad (1)$$

The inverse of the 2×2 coefficient matrix is:

$$\begin{bmatrix} 3 & -4 \\ -5 & 8 \end{bmatrix}^{-1} = \frac{1}{24-(+20)}\begin{bmatrix} 8 & 4 \\ 5 & 3 \end{bmatrix} = \begin{bmatrix} 2 & 1 \\ 5/4 & 3/4 \end{bmatrix}$$

Premultiplying (1) by this matrix gives

$$I\begin{bmatrix} x \\ y \end{bmatrix} = \begin{bmatrix} 2 & 1 \\ 5/4 & 3/4 \end{bmatrix}\begin{bmatrix} -18 \\ 34 \end{bmatrix} = \begin{bmatrix} -2 \\ 3 \end{bmatrix}$$

or $x = -2$, $y = 3$.

2.15 If $[b^{ij}] = [a_{ij}]^{-1}$, solve the $n \times n$ system

$$y_i = a_{ij}x_j \qquad (1)$$

for the x_j in terms of the y_i.

Multiply both sides of (1) by b^{ki} and sum on i:

$$b^{ki}y_i = b^{ki}a_{ij}x_j = \delta^k_j x_j = x_k$$

Therefore, $x_j = b^{ji}y_i$.

2.16 Write the quadratic form in $\mathbf{R}^4$

$$Q = 7x_1^2 - 4x_1x_3 + 3x_1x_4 - x_2^2 + 10x_2x_4 + x_3^2 - 6x_3x_4 + 3x_4^2$$

in the matrix form $\mathbf{x}^T A\mathbf{x}$ with A symmetric.

$$Q = [x_1 \quad x_2 \quad x_3 \quad x_4]\begin{bmatrix} 7 & 0 & -4 & 3 \\ 0 & -1 & 0 & 10 \\ 0 & 0 & 1 & -6 \\ 0 & 0 & 0 & 3 \end{bmatrix}\begin{bmatrix} x_1 \\ x_2 \\ x_3 \\ x_4 \end{bmatrix} = [x_1 \quad x_2 \quad x_3 \quad x_4]\begin{bmatrix} 7 & 0 & -2 & 3/2 \\ 0 & -1 & 0 & 5 \\ -2 & 0 & 1 & -3 \\ 3/2 & 5 & -3 & 3 \end{bmatrix}\begin{bmatrix} x_1 \\ x_2 \\ x_3 \\ x_4 \end{bmatrix}$$

LINEAR TRANSFORMATIONS

2.17 Show that under a change of coordinates $\bar{x}_i = a_{ij}x_j$, the quadric hypersurface $c_{ij}x_ix_j = 1$ transforms to $\bar{c}_{ij}\bar{x}_i\bar{x}_j = 1$, where

$$\bar{c}_{ij} = c_{rs}b_{ri}b_{sj} \qquad \text{with} \qquad (b_{ij}) = (a_{ij})^{-1}$$

This will be worked using matrices, from which the component form can be easily deduced. The hypersurface has the equation $\mathbf{x}^TC\mathbf{x} = 1$ in unbarred coordinates, and $\bar{\mathbf{x}} = A\mathbf{x}$ defines a barred coordinate system. Substituting $\mathbf{x} = B\bar{\mathbf{x}}$ $(B = A^{-1})$ into the equation of the quadric, we have

$$(B\bar{\mathbf{x}})^TC(B\bar{\mathbf{x}}) = 1 \qquad \text{or} \qquad \bar{\mathbf{x}}^TB^TCB\bar{\mathbf{x}} = 1$$

Thus, in the barred coordinate system, the equation of the quadric is $\bar{\mathbf{x}}^T\bar{C}\bar{\mathbf{x}} = 1$, where $\bar{C} = B^TCB$.

DISTANCE IN A BARRED COORDINATE SYSTEM

2.18 Calculate the coefficients g_{ij} in the distance formula (*2.14*) for the barred coordinate system in $\mathbf{R}^2$ defined by $\bar{x}_i = a_{ij}x_j$, where $a_{11} = a_{22} = 1$, $a_{12} = 0$, and $a_{21} = 2$.

We have merely to calculate $G = (AA^T)^{-1}$, where $A = (a_{ij})$:

$$AA^T = \begin{bmatrix} 1 & 0 \\ 2 & 1 \end{bmatrix}\begin{bmatrix} 1 & 2 \\ 0 & 1 \end{bmatrix} = \begin{bmatrix} 1\cdot 1+0 & 1\cdot 2+0 \\ 2\cdot 1+0 & 2\cdot 2+1\cdot 1 \end{bmatrix} = \begin{bmatrix} 1 & 2 \\ 2 & 5 \end{bmatrix}$$

By (*2.11b*),

$$(AA^T)^{-1} = \begin{bmatrix} 1 & 2 \\ 2 & 5 \end{bmatrix}^{-1} = \frac{1}{5-4}\begin{bmatrix} 5 & -2 \\ -2 & 1 \end{bmatrix} = \begin{bmatrix} 5 & -2 \\ -2 & 1 \end{bmatrix}$$

Thus, $g_{11} = 5$, $g_{12} = g_{21} = -2$, $g_{22} = 1$.

2.19 Test the distance formula obtained in Problem 2.18 by finding the distance between the aliases of $(x_i) = (1, -3)$ and $(y_i) = (0, -2)$, which points are a distance $\sqrt{2}$ apart.

The coordinates for the given points in the barred system are found to be

$$\bar{\mathbf{x}} = \begin{bmatrix} 1 & 0 \\ 2 & 1 \end{bmatrix}\begin{bmatrix} 1 \\ -3 \end{bmatrix} = \begin{bmatrix} 1 \\ -1 \end{bmatrix} \qquad \bar{\mathbf{y}} = \begin{bmatrix} 1 & 0 \\ 2 & 1 \end{bmatrix}\begin{bmatrix} 0 \\ -2 \end{bmatrix} = \begin{bmatrix} 0 \\ -2 \end{bmatrix}$$

or $(\bar{x}_i) = (1, -1)$ and $(\bar{y}_i) = (0, -2)$. Using the g_{ij} calculated in Problem 2.18,

$$d(\bar{\mathbf{x}}, \bar{\mathbf{y}}) = \sqrt{5(1-0)^2 - 2\cdot 2(1-0)(-1+2) + 1(-1+2)^2} = \sqrt{2}$$

2.20 Prove formula (*2.14*).

In unbarred coordinates, the distance formula has the matrix form

$$d(\mathbf{x}, \mathbf{y}) = \|\mathbf{x} - \mathbf{y}\| = \sqrt{(\mathbf{x}-\mathbf{y})^T(\mathbf{x}-\mathbf{y})}$$

Now, $\bar{\mathbf{x}} = A\mathbf{x}$ or $\mathbf{x} = B\bar{\mathbf{x}}$, where $B = A^{-1}$; so we have by substitution,

$$\begin{aligned} d(\mathbf{x}, \mathbf{y}) &= \sqrt{(B\bar{\mathbf{x}} - B\bar{\mathbf{y}})^T(B\bar{\mathbf{x}} - B\bar{\mathbf{y}})} = \sqrt{(B(\bar{\mathbf{x}} - \bar{\mathbf{y}}))^T B(\bar{\mathbf{x}} - \bar{\mathbf{y}})} \\ &= \sqrt{(\bar{\mathbf{x}} - \bar{\mathbf{y}})^T B^T B(\bar{\mathbf{x}} - \bar{\mathbf{y}})} = \sqrt{(\bar{\mathbf{x}} - \bar{\mathbf{y}})^T G(\bar{\mathbf{x}} - \bar{\mathbf{y}})} \\ &= d(\bar{\mathbf{x}}, \bar{\mathbf{y}}) \end{aligned}$$

where $G \equiv B^TB = (A^{-1})^TA^{-1} = (A^T)^{-1}A^{-1} = (AA^T)^{-1}$, the last two equalities following from Problem 2.13.

RECTANGULAR COORDINATES

2.21 Suppose that $(x^i) = (x, y, z)$ and $(\bar{x}^i) = (\bar{x}, \bar{y}, \bar{z})$ (the use of superscripts here anticipates future notation) denote two rectangular coordinate systems at O and that the direction angles of the $\bar{x}^i$-axis relative to the x-, y-, and z-axes are $(\alpha_i, \beta_i, \gamma_i)$, $i = 1, 2, 3$. Show that the correspondence between the coordinate systems is given by $\bar{\mathbf{x}} = A\mathbf{x}$, where $\mathbf{x} = (x, y, z)$, $\bar{\mathbf{x}} = (\bar{x}, \bar{y}, \bar{z})$, and where the matrix

$$A = \begin{bmatrix} \cos\alpha_1 & \cos\beta_1 & \cos\gamma_1 \\ \cos\alpha_2 & \cos\beta_2 & \cos\gamma_2 \\ \cos\alpha_3 & \cos\beta_3 & \cos\gamma_3 \end{bmatrix}$$

is orthogonal.

Let the unit vectors along the $\bar{x}$-, $\bar{y}$-, and $\bar{z}$-axes be $\bar{\mathbf{i}} = \overrightarrow{OP}$, $\bar{\mathbf{j}} = \overrightarrow{OQ}$, and $\bar{\mathbf{k}} = \overrightarrow{OR}$, respectively (see Fig. 2-1). If $\bar{\mathbf{x}}$ is the position vector of any point $W(x, y, z)$, then

$$\bar{\mathbf{x}} = \bar{x}\,\bar{\mathbf{i}} + \bar{y}\,\bar{\mathbf{j}} + \bar{z}\,\bar{\mathbf{k}}$$

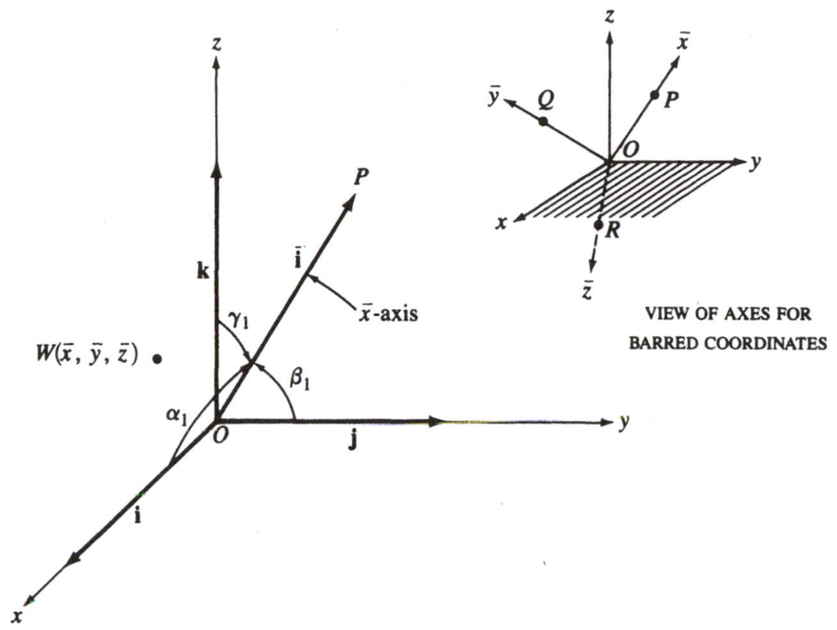

Fig. 2-1

We know that the (x, y, z)-coordinates of P are $(\cos\alpha_1, \cos\beta_1, \cos\gamma_1)$. Similar statements hold for the coordinates of Q and R, respectively. Hence:

$$\begin{aligned}\bar{\mathbf{i}} &= (\cos\alpha_1)\mathbf{i} + (\cos\beta_1)\mathbf{j} + (\cos\gamma_1)\mathbf{k}\\ \bar{\mathbf{j}} &= (\cos\alpha_2)\mathbf{i} + (\cos\beta_2)\mathbf{j} + (\cos\gamma_2)\mathbf{k}\\ \bar{\mathbf{k}} &= (\cos\alpha_3)\mathbf{i} + (\cos\beta_3)\mathbf{j} + (\cos\gamma_3)\mathbf{k}\end{aligned}$$

Substituting these into the expression for $\bar{\mathbf{x}}$ and collecting coefficients of $\mathbf{i}$, $\mathbf{j}$, and $\mathbf{k}$:

$$\begin{aligned}\bar{\mathbf{x}} = {}&(\bar{x}\cos\alpha_1 + \bar{y}\cos\alpha_2 + \bar{z}\cos\alpha_3)\mathbf{i}\\ &+ (\bar{x}\cos\beta_1 + \bar{y}\cos\beta_2 + \bar{z}\cos\beta_3)\mathbf{j}\\ &+ (\bar{x}\cos\gamma_1 + \bar{y}\cos\gamma_2 + \bar{z}\cos\gamma_3)\mathbf{k}\end{aligned}$$

Hence, the x-coordinate of W is the coefficient of $\mathbf{i}$, or

$$x = \bar{x}\cos\alpha_1 + \bar{y}\cos\alpha_2 + \bar{z}\cos\alpha_3$$

Similarly,

$$\begin{aligned}y &= \bar{x}\cos\beta_1 + \bar{y}\cos\beta_2 + \bar{z}\cos\beta_3\\ z &= \bar{x}\cos\gamma_1 + \bar{y}\cos\gamma_2 + \bar{z}\cos\gamma_3\end{aligned}$$

In terms of the matrix A defined above, we can write these three equations in the matrix form

$$\mathbf{x} = A^T\bar{\mathbf{x}} \tag{1}$$

Now, the (i, j)-element of the matrix AA^T is

$$\cos\alpha_i\cos\alpha_j + \cos\beta_i\cos\beta_j + \cos\gamma_i\cos\gamma_j$$

for $i, j = 1, 2, 3$. Note that the diagonal elements,

$$(\cos\alpha_i)^2 + (\cos\beta_i)^2 + (\cos\gamma_i)^2 \qquad (i = 1, 2, 3)$$

are the three quantities $\overrightarrow{OP}\cdot\overrightarrow{OP}$, $\overrightarrow{OQ}\cdot\overrightarrow{OQ}$, $\overrightarrow{OR}\cdot\overrightarrow{OR}$; i.e., they are unity. If $i \neq j$, then the corresponding element of AA^T is either $\overrightarrow{OP}\cdot\overrightarrow{OQ}$, $\overrightarrow{OP}\cdot\overrightarrow{OR}$, or $\overrightarrow{OQ}\cdot\overrightarrow{OR}$, and is therefore zero (since these vectors are mutually orthogonal). Hence, $AA^T = I$ (and also $A^TA = I$), and, from (1),

$$A\mathbf{x} = AA^T\bar{\mathbf{x}} = \bar{\mathbf{x}}$$

CURVILINEAR COORDINATES

2.22 A curvilinear coordinate system $(\bar{x}, \bar{y})$ is defined in terms of rectangular coordinates (x, y) by

$$\begin{aligned}\bar{x} &= x^2 - xy \\ \bar{y} &= xy\end{aligned} \qquad (1)$$

Show that in the barred coordinate system the equation of the line $y = x - 1$ is $\bar{y} = \bar{x}^2 - \bar{x}$. [In the alibi interpretation, (*1*) deforms the straight line into a parabola.]

It helps initially to parameterize the equation of the line as $x = t$, $y = t - 1$. Substitution of $x = t$, $y = t - 1$ in the change-of-coordinates formula gives the parametric equations of the line in the barred coordinate system:

$$\begin{aligned}\bar{x} &= t^2 - t(t-1) = t \\ \bar{y} &= t(t-1) = t^2 - t\end{aligned} \qquad (2)$$

Now t may be eliminated from (*2*) to give $\bar{y} = \bar{x}^2 - \bar{x}$.

CHAIN RULE

2.23 Suppose that under a change of coordinates, $\bar{x}_i = \bar{x}_i(x_1, x_2, \ldots, x_n)$ $(1 \leqq i \leqq n)$, the real-valued vector functions $(\bar{T}_i)$ and (T_i) are related by the formula

$$\bar{T}_i = T_r \frac{\partial x_r}{\partial \bar{x}_i} \qquad (1)$$

Find the transformation rule for the partial derivatives of (T_i)—that is, express the $\partial \bar{T}_i / \partial \bar{x}_j$ in terms of the $\partial T_r / \partial x_s$—given that all second-order partial derivatives are zero.

Begin by taking the partial derivative with respect to $\bar{x}_j$ of both sides of (*1*), using the product rule:

$$\frac{\partial \bar{T}_i}{\partial \bar{x}_j} = \frac{\partial}{\partial \bar{x}_j}\left\{T_r \frac{\partial x_r}{\partial \bar{x}_i}\right\} = \frac{\partial T_r}{\partial \bar{x}_j}\frac{\partial x_r}{\partial \bar{x}_i} + T_r \frac{\partial}{\partial \bar{x}_j}\left\{\frac{\partial x_r}{\partial \bar{x}_i}\right\}$$

By assumption, the second term on the right is zero; and, by the chain rule,

$$\frac{\partial T_r}{\partial \bar{x}_j} = \frac{\partial T_r}{\partial x_s}\frac{\partial x_s}{\partial \bar{x}_j}$$

Consequently, the desired transformation rule is

$$\frac{\partial \bar{T}_i}{\partial \bar{x}_j} = \frac{\partial T_r}{\partial x_s}\frac{\partial x_s}{\partial \bar{x}_j}\frac{\partial x_r}{\partial \bar{x}_i}$$

Supplementary Problems

2.24 Display the matrices (*a*) $[u^{ij}]_{35}$, (*b*) $[u^{ji}]_{35}$, (*c*) $[u^{ij}]_{53}$, (*d*) $[\delta^i_j]_{36}$.

2.25 Carry out the following matrix multiplications:

(*a*) $\begin{bmatrix} 3 & -1 & 2 \\ 0 & 1 & -1 \\ 1 & 2 & 0 \end{bmatrix}\begin{bmatrix} 1 \\ 2 \\ 2 \end{bmatrix}$ (*b*) $\begin{bmatrix} 3 & -1 \\ 2 & 0 \end{bmatrix}\begin{bmatrix} 1 & 1 & -1 \\ 2 & 1 & 1 \end{bmatrix}$

2.26 Prove by the product rule and by use of the summation convention the associative law for matrices:

$$(AB)C = A(BC)$$

where $A \equiv (a_{ij})$, $B \equiv (b_{ij})$, and $C \equiv (c_{ij})$ are arbitrary matrices, but compatible for multiplication.

2.27 Prove: (*a*) if A and B are symmetric matrices and if $AB = BA = C$, then C is symmetric; (*b*) if A and B are skew-symmetric and if $AB = -BA = C$, then C is skew-symmetric.

2.28 Prove that the product of two orthogonal matrices is orthogonal.

2.29 Evaluate the determinants

$$(a)\ \begin{vmatrix} 3 & -2 \\ 1 & 5 \end{vmatrix} \qquad (b)\ \begin{vmatrix} 2 & 1 & -1 \\ 3 & 0 & 1 \\ 1 & -1 & 2 \end{vmatrix} \qquad (c)\ \begin{vmatrix} -1 & 1 & -1 & 1 & 0 \\ 1 & 0 & 1 & 1 & 1 \\ 0 & 1 & 0 & 0 & 0 \\ -1 & 1 & 0 & 1 & 1 \\ 1 & 1 & 0 & 0 & 0 \end{vmatrix}$$

2.30 In the Laplace expansion of the fourth-order determinant $|a_{ij}|$, the six-term summation $e_{2ijk}a_{12}a_{2i}a_{3j}a_{4k}$ appears. (*a*) Write out this sum explicitly, then (*b*) represent it as a third-order determinant.

2.31 Prove that if a matrix has two rows the same, its determinant is zero. (*Hint*: First show that interchanging any two subscripts reverses the sign of the permutation symbol.)

2.32 Calculate the inverse of

$$(a)\ \begin{bmatrix} 3 & 1 \\ 5 & 2 \end{bmatrix} \qquad (b)\ \begin{bmatrix} 0 & 1 & 2 \\ 1 & -1 & 0 \\ 2 & 1 & -1 \end{bmatrix}$$

2.33 (*a*) Verify the following formulas for the permutation symbols e_{ij} and e_{ijk} (for distinct values of the indices only):

$$e_{ij} = \frac{j-i}{|j-i|} \qquad e_{ijk} = \frac{(j-i)(k-i)(k-j)}{|j-i|\,|k-i|\,|k-j|}$$

(*b*) Prove the general formula:

$$e_{i_1 i_2 \ldots i_n} = \frac{(i_2-i_1)(i_3-i_1)\cdots(i_n-i_1)(i_3-i_2)\cdots(i_n-i_2)\cdots(i_n-i_{n-1})}{|i_2-i_1|\,|i_3-i_1|\cdots|i_n-i_1|\,|i_3-i_2|\cdots|i_n-i_2|\cdots|i_n-i_{n-1}|} \equiv \prod_{p>q} \frac{i_p-i_q}{|i_p-i_q|}$$

2.34 Calculate the angle between the $\mathbf{R}^6$-vectors $\mathbf{x} = (3, -1, 0, 1, 2, -3)$ and $\mathbf{y} = (-2, 1, 0, 1, 0, 0)$.

2.35 Find two linearly independent vectors in $\mathbf{R}^3$ which are orthogonal to the vector $(3, -2, 1)$.

2.36 Solve for x and y by use of matrices:

$$\begin{aligned} 3x - 4y &= -23 \\ 5x + 3y &= \ \ 10 \end{aligned}$$

2.37 Write out the quadratic form in $\mathbf{R}^3$ represented by $Q = \mathbf{x}^T A \mathbf{x}$, where

$$A = \begin{bmatrix} 1 & 4 & 3 \\ 4 & 2 & 0 \\ 3 & 0 & -1 \end{bmatrix}$$

2.38 Represent with a symmetric matrix A the quadratic form in $\mathbf{R}^4$

$$Q = -3x_1^2 - x_2^2 + x_3^2 - x_1x_2 - x_1x_3 + 6x_1x_4$$

2.39 Given the hyperplane $c_r x_r = 1$, how do the coefficients c_i transform under a change of coordinates $\bar{x}_i = a_{ij} x_j$?

2.40 Calculate the g_{ij} for the distance formula (*2.14*) in a barred coordinate system defined by $\bar{\mathbf{x}} = A\mathbf{x}$, with

$$A = \begin{bmatrix} 1 & -2 \\ 2 & 3 \end{bmatrix}$$

2.41 Test the distance formula of Problem 2.40 on the pair of points whose unbarred coordinates are (2, −1) and (2, −4).

2.42 (*a*) Show that for independent functions $\bar{x}_i = \bar{x}_i(x_1, x_2, \ldots, x_n)$,

$$\frac{\partial \bar{x}_i}{\partial x_r} \frac{\partial x_r}{\partial \bar{x}_j} = \delta^i_j \qquad (1)$$

(*b*) Take the partial derivative with respect to x_k of (*1*) to establish the formula

$$\frac{\partial^2 \bar{x}_i}{\partial x_k \partial x_r} \frac{\partial x_r}{\partial \bar{x}_j} = -\frac{\partial^2 x_r}{\partial \bar{x}_s \partial \bar{x}_j} \frac{\partial \bar{x}_i}{\partial x_r} \frac{\partial \bar{x}_s}{\partial x_k} \qquad (2)$$

Chapter 3

General Tensors

3.1 COORDINATE TRANSFORMATIONS

At this point the notation for coordinates will be changed to that usual in tensor calculus.

Superscripts for Vector Components

The coordinates of a point (vector) in $\mathbf{R}^n$ will henceforth be denoted as $(x^1, x^2, x^3, \ldots, x^n)$. Thus, the familiar subscripts are now replaced by superscripts, and the upper position is no longer reserved for exponents alone. It will be clear by context whether a character represents a vector component or the power of a scalar.

EXAMPLE 3.1 If a power of some vector component is to be indicated, obviously parentheses are necessary; thus, $(x^3)^2$ and $(x^{n-1})^5$ represent, respectively, the square of the third component, and the $(n-1)$st component raised to the fifth power, of the vector $\mathbf{x}$. If u is introduced as a real number, then u^2 and u^3 are powers of u and not vector components. If $(c)^k$ appears without explanation, the parentheses indicate the use of the superscript k as an exponent and not as the index of a vector component.

Rectangular Coordinates

Coordinates in $\mathbf{R}^n$ are called *rectangular* (also *rectangular cartesian* or *cartesian*) if they are patterned after the usual orthogonal coordinate systems of two- and three-dimensional analytic geometry. A general definition that is workable in this setting is, in effect, an assertion of the converse of the Pythagorean theorem.

Definition 1: A coordinate system (x^i) is *rectangular* if the distance between two arbitrary points $P(x^1, x^2, \ldots, x^n)$ and $Q(y^1, y^2, \ldots, y^n)$ is given by

$$PQ = \sqrt{(x^1 - y^1)^2 + (x^2 - y^2)^2 + \cdots + (x^n - y^n)^2} \equiv \sqrt{\delta_{ij}\,\Delta x^i\,\Delta x^j}$$

where $\Delta x^i \equiv x^i - y^i$.

Under orthogonal coordinate changes, which are isometric, the above formula for distance is invariant (cf. Section 2.5). Hence, all coordinate systems $(\bar{x}^i)$ defined by $\bar{x}^i = a^i_r x^r$, where (a^i_j) is such that $a^r_i a^r_j = \delta_{ij}$, are rectangular. It can be shown that these are the only rectangular coordinate systems whose origin coincides with that of the (x^i)-system.

Curvilinear Coordinates

Suppose that in some region of $\mathbf{R}^n$ two coordinate systems are defined, and that these two systems are connected by equations of the form

$$\mathscr{T} \,:\, \bar{x}^i = \bar{x}^i(x^1, x^2, \ldots, x^n) \qquad (1 \leqq i \leqq n) \tag{3.1}$$

where, for each i, the function, or *scalar field*, $x^i(x^1, x^2, \ldots, x^n)$ maps the given region in $\mathbf{R}^n$ to the reals and has continuous second-partial derivatives at every point in the region (is *class* C^2). The transformation $\mathscr{T}$, if bijective, is called a *coordinate transformation*, as in Section 2.6. If (x^i) are ordinary rectangular coordinates, the $(\bar{x}^i)$ are called *curvilinear coordinates* unless $\mathscr{T}$ is linear, in which case $(\bar{x}^i)$ are called *affine coordinates*.

For convenience, the three most common curvilinear-coordinate systems are presented below. In each case, a "reverse" notation is employed: the two- or three-dimensional curvilinear system (x^i) is defined by the mapping $\mathscr{T}$ that takes it into a rectangular system $(\bar{x}^i)$ of the same dimension.

Polar coordinates (Fig. 3-1). Let $(\bar{x}^1, \bar{x}^2) = (x, y)$ and $(x^1, x^2) = (r, \theta)$, under the restriction $r > 0$.

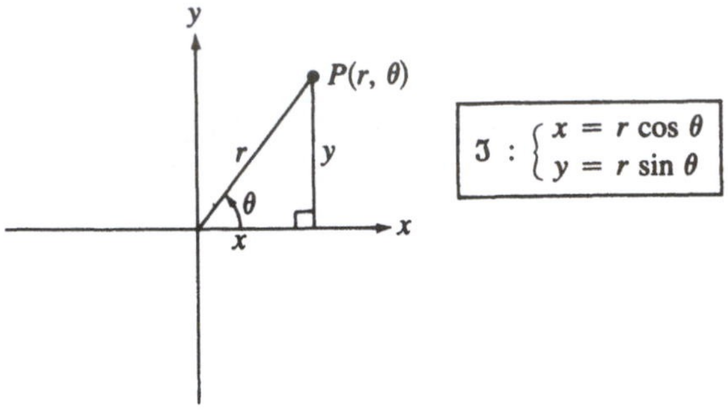

Fig. 3-1

Then,

$$\mathcal{T} : \begin{cases} \bar{x}^1 = x^1 \cos x^2 \\ \bar{x}^2 = x^1 \sin x^2 \end{cases} \qquad \mathcal{T}^{-1} : \begin{cases} x^1 = \sqrt{(\bar{x}^1)^2 + (\bar{x}^2)^2} \\ x^2 = \tan^{-1}(\bar{x}^2/\bar{x}^1) \end{cases} \tag{3.2}$$

(The inverse given here is, in the equation for x^2, valid only in the first and fourth quadrants of the $\bar{x}_1\bar{x}_2$-plane; other solutions must be used over the other two quadrants. Likewise for the θ-coordinate in the cylindrical and spherical systems.)

Cylindrical coordinates (Fig. 3-2). If $(\bar{x}^1, \bar{x}^2, \bar{x}^3) = (x, y, z)$ and $(x^1, x^2, x^3) = (r, \theta, z)$, where $r > 0$,

$$\mathcal{T} : \begin{cases} \bar{x}^1 = x^1 \cos x^2 \\ \bar{x}^2 = x^1 \sin x^2 \\ \bar{x}^3 = x^3 \end{cases} \qquad \mathcal{T}^{-1} : \begin{cases} x^1 = \sqrt{(\bar{x}^1)^2 + (\bar{x}^2)^2} \\ x^2 = \tan^{-1}(\bar{x}^2/\bar{x}^1) \\ x^3 = \bar{x}^3 \end{cases} \tag{3.3}$$

Spherical coordinates (Fig. 3-3). If $(\bar{x}_1, \bar{x}_2, \bar{x}_3) = (x, y, z)$ and $(x^1, x^2, x^3) = (\rho, \varphi, \theta)$, where $\rho > 0$ and $0 \leqq \varphi \leqq \pi$,

$$\mathcal{T} : \begin{cases} \bar{x}^1 = x^1 \sin x^2 \cos x^3 \\ \bar{x}^2 = x^1 \sin x^2 \sin x^3 \\ \bar{x}^3 = x^1 \cos x^2 \end{cases} \qquad \mathcal{T}^{-1} : \begin{cases} x^1 = \sqrt{(\bar{x}^1)^2 + (\bar{x}^2)^2 + (\bar{x}^3)^2} \\ x^2 = \cos^{-1}\left(\bar{x}^3/\sqrt{(\bar{x}^1)^2 + (\bar{x}^2)^2 + (\bar{x}^3)^2}\right) \\ x^3 = \tan^{-1}(\bar{x}^2/\bar{x}^1) \end{cases} \tag{3.4}$$

(*Caution*: In an older but still common notation for spherical coordinates, θ denotes the polar angle and φ the equatorial angle.)

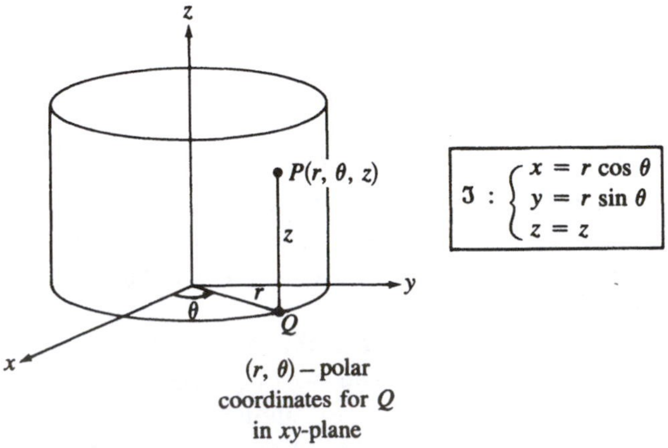

Fig. 3-2

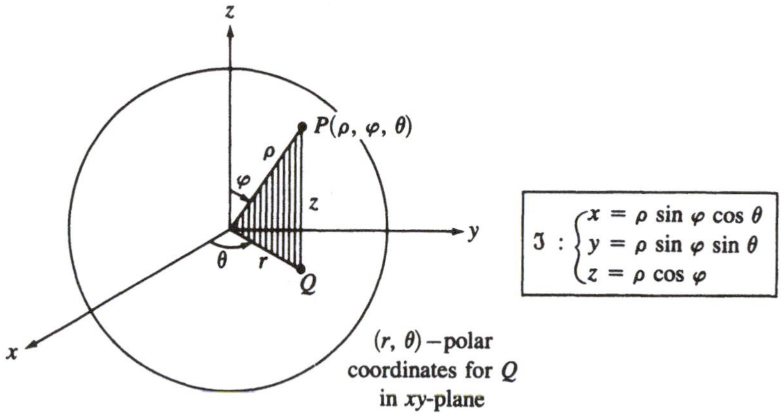

Fig. 3-3

The Jacobian

The n^2 first-order partial derivatives $\partial\bar{x}^i/\partial x^j$ arising from (*3.1*) are normally arranged in an $n \times n$ matrix,

$$J = \begin{bmatrix} \dfrac{\partial\bar{x}^1}{\partial x^1} & \dfrac{\partial\bar{x}^1}{\partial x^2} & \cdots & \dfrac{\partial\bar{x}^1}{\partial x^n} \\ \dfrac{\partial\bar{x}^2}{\partial x^1} & \dfrac{\partial\bar{x}^2}{\partial x^2} & \cdots & \dfrac{\partial\bar{x}^2}{\partial x^n} \\ \cdots & \cdots & \cdots & \cdots \\ \dfrac{\partial\bar{x}^n}{\partial x^1} & \dfrac{\partial\bar{x}^n}{\partial x^2} & \cdots & \dfrac{\partial\bar{x}^n}{\partial x^n} \end{bmatrix} \tag{3.5}$$

Matrix J is the *Jacobian matrix*, and its determinant $\mathcal{J} \equiv \det J$ is the *Jacobian*, of the transformation $\mathcal{T}$.

EXAMPLE 3.2 In $\mathbf{R}^2$ let a curvilinear coordinate system $(\bar{x}^i)$ be defined from rectangular coordinates (x^i) by the equations

$$\mathcal{T} \; : \; \begin{cases} \bar{x}^1 = x^1 x^2 \\ \bar{x}^2 = (x^2)^2 \end{cases}$$

Since $\partial\bar{x}^1/\partial x^1 = x^2$, $\partial\bar{x}^1/\partial x^2 = x^1$, $\partial\bar{x}^2/\partial x^1 = 0$, and $\partial\bar{x}^2/\partial x^2 = 2x^2$, the Jacobian of $\mathcal{T}$ is

$$\mathcal{J} = \begin{vmatrix} x^2 & x^1 \\ 0 & 2x^2 \end{vmatrix} = 2(x^2)^2$$

A well-known theorem from analysis states that $\mathcal{T}$ is locally bijective on an open set $\mathcal{U}$ in $\mathbf{R}^n$ if and only if $\mathcal{J} \neq 0$ at each point of $\mathcal{U}$. When $\mathcal{J} \neq 0$ in $\mathcal{U}$ and $\mathcal{T}$ is class C^2 in $\mathcal{U}$, then (*3.1*) is termed an *admissible change of coordinates* for $\mathcal{U}$.

EXAMPLE 3.3 The curvilinear coordinates of Example 3.2 are admissible for the regions $x^2 > 0$ and $x^2 < 0$ (both open sets in the plane). See Problem 3.1.

In an admissible change of coordinates, the inverse transformation $\mathcal{T}^{-1}$ (the local existence of which is guaranteed by the theorem mentioned above) is also class C^2, on $\bar{\mathcal{U}}$, the image of $\mathcal{U}$ under $\mathcal{T}$. Moreover, if $\mathcal{T}^{-1}$ has the form

$$\mathcal{T}^{-1} \; : \; x^i = x^i(\bar{x}^1, \bar{x}^2, \ldots, \bar{x}^n) \qquad (1 \leq i \leq n) \tag{3.6}$$

on $\bar{\mathscr{U}}$, the Jacobian matrix $\bar{J}$ of $\mathscr{T}^{-1}$ is the inverse of J. Thus, $J\bar{J} = \bar{J}J = I$, or

$$\frac{\partial \bar{x}^i}{\partial x^r}\frac{\partial x^r}{\partial \bar{x}^j} = \frac{\partial x^i}{\partial \bar{x}^r}\frac{\partial \bar{x}^r}{\partial x^j} = \delta^i_j \tag{3.7}$$

[cf. Problem 2.42(*a*)]. It also follows that $\bar{\mathscr{J}} = 1/\mathscr{J}$.

General Coordinate Systems

In later developments it will be necessary to adopt coordinate systems that are not tied to rectangular coordinates in any way [via (*3.1*)] and to define distance in terms of an arc-length formula for arbitrary curves, with points represented abstractly by n-tuples $(x^1, x^2, \ldots, x^n)$. Each such distance functional or *metric* will be invariant under admissible changes of coordinates, and admissible coordinate systems will exist for each separate metric. Under such metrics, $\mathbf{R}^n$ will generally become non-Euclidean; e.g., the angle sum of a triangle will not invariably equal π.

Although the curvilinear coordinate systems presented above are explicitly associated with the Euclidean metric (since they are connected via (*3.1*) with rectangular coordinates and Euclidean space), those same systems could be formally adopted in a non-Euclidean space if some purpose were served by doing so. The point to be made is that *the space metric and the coordinate system used to describe that metric are completely independent of each other*, except in the single instance of rectangular coordinates, whose very definition (see Definition 1) involves the Euclidean metric.

Usefulness of Coordinate Changes

A primary concern in studying tensor analysis is the manner in which a change of coordinates affects the way geometrical objects or physical laws are described. For example, in rectangular coordinates the equation of a circle of radius a centered at the origin is quadratic,

$$(\bar{x}^1)^2 + (\bar{x}^2)^2 = a^2$$

but in polar coordinates, (*3.2*), that same circle has the simple linear equation $x^1 = a$. The reader is no doubt familiar with the sometimes dramatic change that takes place in a differential equation under a change of variables, which is nothing but a change of coordinates. This idea of changing the description of phenomena by changing coordinate systems lies at the heart of not only what a tensor means, but how it is used in practice.

3.2 FIRST-ORDER TENSORS

Consider a *vector field* $\mathbf{V} = (V^i)$ defined on some subset $\mathscr{S}$ of $\mathbf{R}^n$ [that is, for each i, the component $V^i = V^i(\mathbf{x})$ is a scalar field (real-valued function) as $\mathbf{x}$ varies over $\mathscr{S}$]. In each admissible coordinate system of a region $\mathscr{U}$ containing $\mathscr{S}$, let the n components $V^1, V^2, \ldots, V^n$ of $\mathbf{V}$ be expressible as n real-valued functions; say, as

$$T^1, \quad T^2, \quad \ldots, \quad T^n \qquad \text{in the } (x^i)\text{-system}$$

and

$$\bar{T}^1, \quad \bar{T}^2, \quad \ldots, \quad \bar{T}^n \qquad \text{in the } (\bar{x}^i)\text{-system}$$

where (x^i) and $(\bar{x}^i)$ are related by (*3.1*) and (*3.6*).

Definition 2: The vector field $\mathbf{V}$ is a *contravariant tensor of order one* (or *contravariant vector*) provided its components (T^i) and $(\bar{T}^i)$ relative to the respective coordinate systems (x^i) and $(\bar{x}^i)$ obey the law of transformation

$$\textbf{\textit{contravariant vector}} \quad \bar{T}^i = T^r \frac{\partial \bar{x}^i}{\partial x^r} \qquad (1 \leqq i \leqq n) \tag{3.8}$$

EXAMPLE 3.4 Let $\mathscr{C}$ be a curve given parametrically in the (x^i)-system by

$$x^i = x^i(t) \qquad (a \leqq t \leqq b)$$

The tangent vector field $\mathbf{T} = (T^i)$ is defined by the usual differentiation formula

$$T^i = \frac{dx^i}{dt}$$

Under a change of coordinates (3.1), the same curve is given in the $(\bar{x}^i)$-system by

$$\bar{x}^i = \bar{x}^i(t) \equiv \bar{x}^i(x^1(t), x^2(t), \ldots, x^n(t)) \qquad (a \leqq t \leqq b)$$

and the tangent vector for $\mathscr{C}$ in the $(\bar{x}^i)$-system has components

$$\bar{T}^i = \frac{d\bar{x}^i}{dt}$$

But, by the chain rule,

$$\frac{d\bar{x}^i}{dt} = \frac{\partial \bar{x}^i}{\partial x^r}\frac{dx^r}{dt} \qquad \text{or} \qquad \bar{T}^i = T^r \frac{\partial \bar{x}^i}{\partial x^r}$$

proving that $\mathbf{T}$ is a contravariant vector. (Note that because $\mathbf{T}$ is defined only on the curve $\mathscr{C}$, we have $\mathscr{S} = \mathscr{C}$ for this particular vector field.) We conclude in general that under a change of coordinates, *the tangent vector of a smooth curve transforms as a contravariant tensor of order one.*

Remark 1: In some treatments of the subject, tensors are defined to possess certain *weights*, with (3.8) replaced by

$$\textbf{\textit{weighted contravariant vector}} \quad \bar{T}^i = wT^r \frac{\partial \bar{x}^i}{\partial x^r} \qquad (1 \leqq i \leqq n) \tag{3.9}$$

for some real-valued function w (the "weight of $\mathbf{T}$").

In framing the next definition we (arbitrarily) shift to a subscript notation for the components of the vector field.

Definition 3: The vector field $\mathbf{V}$ is a *covariant tensor of order one* (or *covariant vector*) provided its components (T_i) and $(\bar{T}_i)$ relative to an arbitrary pair of coordinate systems (x^i) and $(\bar{x}^i)$, respectively, obey the law of transformation

$$\textbf{\textit{covariant vector}} \quad \bar{T}_i = T_r \frac{\partial x^r}{\partial \bar{x}^i} \qquad (1 \leqq i \leqq n) \tag{3.10}$$

EXAMPLE 3.5 Let $F(\mathbf{x})$ denote a differentiable scalar field defined in a coordinate system (x^i) of $\mathbf{R}^n$. The gradient of F is defined as the vector field

$$\nabla F \equiv \left(\frac{\partial F}{\partial x^1}, \frac{\partial F}{\partial x^2}, \ldots, \frac{\partial F}{\partial x^n}\right)$$

In a barred coordinate system, the gradient is given by $\overline{\nabla F} = (\partial \bar{F}/\partial \bar{x}^i)$, where $\bar{F}(\bar{\mathbf{x}}) \equiv F \circ \mathbf{x}(\bar{\mathbf{x}})$. The chain rule for partial derivatives, together with the functional relations (3.6), gives

$$\frac{\partial \bar{F}}{\partial \bar{x}^i} = \frac{\partial F}{\partial x^r}\frac{\partial x^r}{\partial \bar{x}^i}$$

which is just (3.10) for $T_i = \partial F/\partial x^i$, $\bar{T}_i = \partial \bar{F}/\partial \bar{x}^i$. Thus, *the gradient of an arbitrary differentiable function is a covariant vector.*

Remark 2: Tangent vectors and gradient vectors are really two different kinds of vectors. Tensor calculus is vitally concerned with the distinction between contravariance and covariance, and consistently employs upper indices to indicate the one and lower indices to indicate the other.

Remark 3: From this point on, we shall frequently refer to first-order tensors, contravariant or covariant as the case may be, simply as "vectors"; they are, of course, actually vector *fields*, defined *on* $\mathbf{R}^n$. This usage will coexist with our earlier employment of "vectors" to denote real n-tuples; i.e., elements *of* $\mathbf{R}^n$. There is no conflict here insofar as the n-tuples make up the vector field corresponding to the identity mapping $V^i(\mathbf{x}) = x^i \quad (i = 1, 2, \ldots, n)$. But the vector (x^i) does not enjoy the transformation property of a tensor; so, to emphasize that fact, we shall sometimes refer to it as a *position vector*.

3.3 INVARIANTS

Objects, functions, equations, or formulas that are independent of the coordinate system used to express them have intrinsic value and are of fundamental significance; they are called *invariants*. Roughly speaking, the product of a contravariant vector and a covariant vector always is an invariant. The following is a more precise statement of this fact.

Theorem 3.1: Let S^i and T_i be the components of a contravariant and covariant vector, respectively. If the inner product $E \equiv S^r T_r$ is defined in each coordinate system, then E is an invariant.

EXAMPLE 3.6 In Examples 3.4 and 3.5 it was established that the tangent vector, $(S^i) = (dx^i/dt)$, to a curve $\mathscr{C}$ and the gradient of a function, $(T_i) = (\partial F/\partial x^i)$, are contravariant and covariant vectors, respectively. Let us verify Theorem 3.1 for these two vectors. Define

$$E = S^r T_r \equiv \frac{\partial F}{\partial x^r}\frac{dx^r}{dt}$$

Now, by the chain rule,

$$E = \frac{dF}{dt}$$

so the assertion of Theorem 3.1 is that the value of

$$\frac{d}{dt}\left[F \circ (x^i(t))\right] \equiv \frac{d}{dt}\left[\hat{F}(t)\right]$$

is independent of the particular coordinate system (x^i) used to specify the curve. To visualize this, the reader should study Fig. 3-4, which shows how the composition $\hat{F} = F \circ (x^i(t))$ works out in $\mathbf{R}^3$. It is apparent here that the map $\hat{F}$ entirely bypasses the coordinate system (x^1, x^2, x^3). Thus, $\hat{F}$—and with it, $d\hat{F}/dt$—is an invariant with respect to coordinate changes.

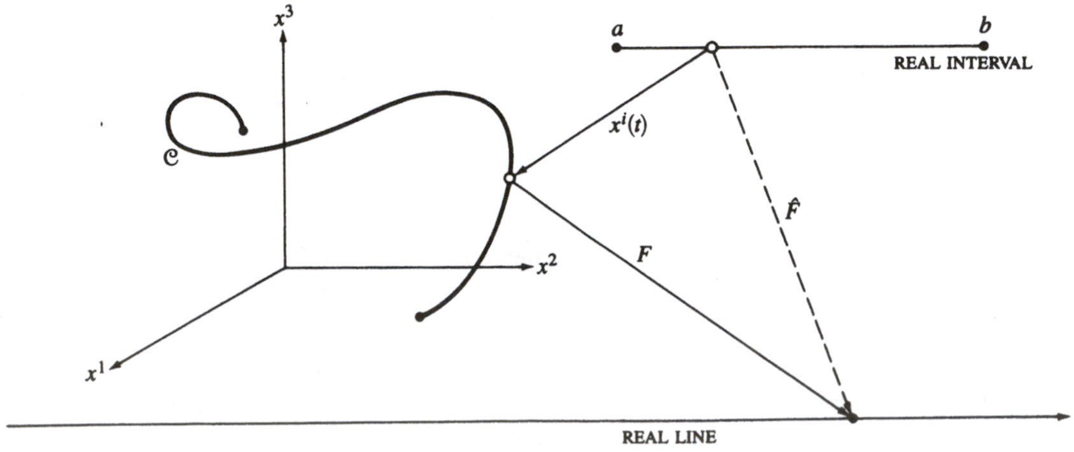

Fig. 3-4

3.4 HIGHER-ORDER TENSORS

Tensors of arbitrary order may be defined. Although most work does not involve tensors of order greater than 4, the general definition will be included here for completeness. We begin with the three types of second-order tensors.

Second-Order Tensors

Let $\mathbf{V} = (V^{ij})$ denote a *matrix field*; that is, (V^{ij}) is an $n \times n$ matrix of scalar fields $V^{ij}(\mathbf{x})$, all defined over the same region $\mathcal{U} = \{\mathbf{x}\}$ in $\mathbf{R}^n$. As before, it will be assumed that $\mathbf{V}$ has a representation (T^{ij}) in (x^i) and $(\bar{T}^{ij})$ in $(\bar{x}^i)$, where (x^i) and $(\bar{x}^i)$ are admissible coordinates related by *(3.1)* and *(3.6)*.

Definition 4: The matrix field $\mathbf{V}$ is a *contravariant tensor of order two* if its components (T^{ij}) in (x^i) and $(\bar{T}^{ij})$ in $(\bar{x}^i)$ obey the law of transformation

$$\textbf{\textit{contravariant tensor}} \qquad \bar{T}^{ij} = T^{rs} \frac{\partial \bar{x}^i}{\partial x^r} \frac{\partial \bar{x}^j}{\partial x^s} \qquad (1 \leqq i, j \leqq n) \tag{3.11}$$

Again going over to subscript notation for the components of the matrix field, we state

Definition 5: The matrix field $\mathbf{V}$ is a *covariant tensor of order two* if its components (T_{ij}) in (x^i) and $(\bar{T}_{ij})$ in $(\bar{x}^i)$ obey the law of transformation

$$\textbf{\textit{covariant tensor}} \qquad \bar{T}_{ij} = T_{rs} \frac{\partial x^r}{\partial \bar{x}^i} \frac{\partial x^s}{\partial \bar{x}^j} \qquad (1 \leqq i, j \leqq n) \tag{3.12}$$

Theorem 3.2: Suppose that (T_{ij}) is a covariant tensor of order two. If the matrix $[T_{ij}]_{nn}$ is invertible on $\mathcal{U}$, with inverse matrix $[T^{ij}]_{nn}$, then (T^{ij}) is a contravariant tensor of order two.

Definition 6: The matrix field $\mathbf{V}$ is a *mixed tensor of order two*, contravariant of order one and covariant of order one, if its components (T^i_j) in (x^i) and $(\bar{T}^i_j)$ in $(\bar{x}^i)$ obey the law of transformation

$$\textbf{\textit{mixed tensor}} \qquad \bar{T}^i_j = T^r_s \frac{\partial \bar{x}^i}{\partial x^r} \frac{\partial x^s}{\partial \bar{x}^j} \qquad (1 \leqq i, j \leqq n) \tag{3.13}$$

Tensors of Arbitrary Order

Vector and matrix fields are inadequate for higher-order tensors. It is necessary to introduce a *generalized vector field* $\mathbf{V}$, which is an ordered array of n^m $(m = p + q)$ scalar fields, $(V^{i_1 i_2 \ldots i_p}_{j_1 j_2 \ldots j_q})$, defined over a region $\mathcal{U}$ in $\mathbf{R}^n$; let $(T^{i_1 i_2 \ldots i_p}_{j_1 j_2 \ldots j_q})$ denote the set of component-functions in various coordinate systems which are defined on $\mathcal{U}$.

Definition 7: The generalized vector field $\mathbf{V}$ is a *tensor of order* $m = p + q$, contravariant of order p and covariant of order q, if its components $(T^{i_1 i_2 \ldots i_p}_{j_1 j_2 \ldots j_q})$ in (x^i) and $(\bar{T}^{i_1 i_2 \ldots i_p}_{j_1 j_2 \ldots j_q})$ in $(\bar{x}^i)$ obey the law of transformation

$$\textbf{\textit{general tensor}} \qquad \bar{T}^{i_1 i_2 \ldots i_p}_{j_1 j_2 \ldots j_q} = T^{r_1 r_2 \ldots r_p}_{s_1 s_2 \ldots s_q} \frac{\partial \bar{x}^{i_1}}{\partial x^{r_1}} \frac{\partial \bar{x}^{i_2}}{\partial x^{r_2}} \cdots \frac{\partial \bar{x}^{i_p}}{\partial x^{r_p}} \frac{\partial x^{s_1}}{\partial \bar{x}^{j_1}} \frac{\partial x^{s_2}}{\partial \bar{x}^{j_2}} \cdots \frac{\partial x^{s_q}}{\partial \bar{x}^{j_q}} \tag{3.14}$$

with the obvious range for free indices.

3.5 THE STRESS TENSOR

It was the concept of *stress* in mechanics that originally led to the invention of tensors (*tenseur*, that which exerts tension, stress). Suppose that the unit cube is in equilibrium under forces applied to three of its faces [Fig. 3-5(*a*)]. Since each face has unit area, each force vector represents the *force*

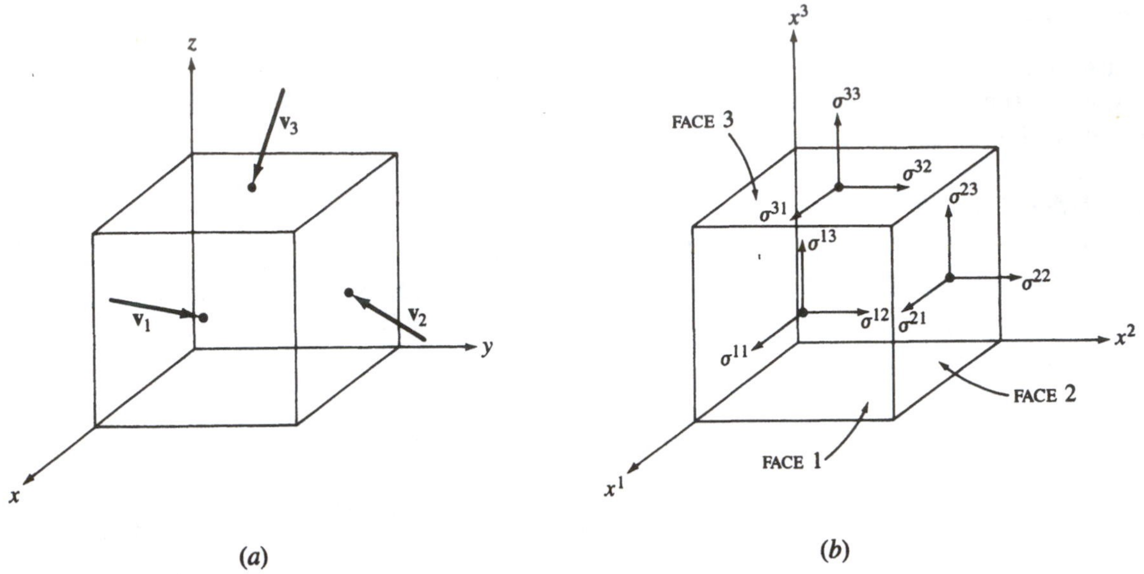

Fig. 3-5

per unit area, or *stress*. Those forces are represented in the component form in Fig. 3-5(*b*). Using the standard basis $\mathbf{e}_1, \mathbf{e}_2, \mathbf{e}_2$, we have

$$\begin{aligned} \mathbf{v}_1 &= \sigma^{1s}\mathbf{e}_s && \text{(stress on face 1)} \\ \mathbf{v}_2 &= \sigma^{2s}\mathbf{e}_s && \text{(stress on face 2)} \\ \mathbf{v}_3 &= \sigma^{3s}\mathbf{e}_s && \text{(stress on face 3)} \end{aligned} \tag{3.15}$$

Stress on a Cube Section

The question arises: What stress $\mathbf{F}$ is transmitted to a planar cross section of the cube that has unit normal $\mathbf{n}$? To answer this, refer to Fig. 3-6, which shows the tetrahedron formed by the cross section and the coordinate planes. Let A be the cross-sectional area. By the assumed equilibrium of the cube, the *stresses* on the x^1x^2-, x^1x^3-, and x^2x^3-bases of the tetrahedron are $-\mathbf{v}_3$, $-\mathbf{v}_2$, and $-\mathbf{v}_1$, respectively, as shown componentwise in Fig. 3-6. Hence, the *forces* on these same bases are $B_1(-\mathbf{v}_3)$, $B_2(-\mathbf{v}_2)$, and $B_3(-\mathbf{v}_1)$, respectively. For the tetrahedron itself to be in equilibrium, the resultant force on it must vanish:

$$A\mathbf{F} + B_1(-\mathbf{v}_3) + B_2(-\mathbf{v}_2) + B_3(-\mathbf{v}_1) = 0$$

or, solving for $\mathbf{F}$,

$$\mathbf{F} = \frac{B_3}{A}\mathbf{v}_1 + \frac{B_2}{A}\mathbf{v}_2 + \frac{B_1}{A}\mathbf{v}_3 \tag{3.16}$$

But B_3 is the projection of A in the x^2x^3-plane: $B_3 = A\mathbf{n}\mathbf{e}_1$ or $B_3/A = \mathbf{n}\mathbf{e}_1$. Similarly, $B_2/A = \mathbf{n}\mathbf{e}_2$ and $B_1/A = \mathbf{n}\mathbf{e}_3$. Substituting these expressions and the expressions (*3.15*) into (*3.16*), we find that

$$\mathbf{F} = \sigma^{rs}(\mathbf{n}\mathbf{e}_r)\mathbf{e}_s \tag{3.17}$$

Contravariance of Stress Under Change of Coordinates

An interesting formula results from (*3.17*) when we change the basis of $\mathbf{R}^3$ by a transformation of the form $\mathbf{e}_i = a_i^j\mathbf{f}_j$ (with $|a_i^j| \neq 0$). In terms of coordinates,

$$x^i\mathbf{e}_i = x^i(a_i^j\mathbf{f}_j) = (a_i^jx^i)\mathbf{f}_j \equiv \bar{x}^j\mathbf{f}_j$$

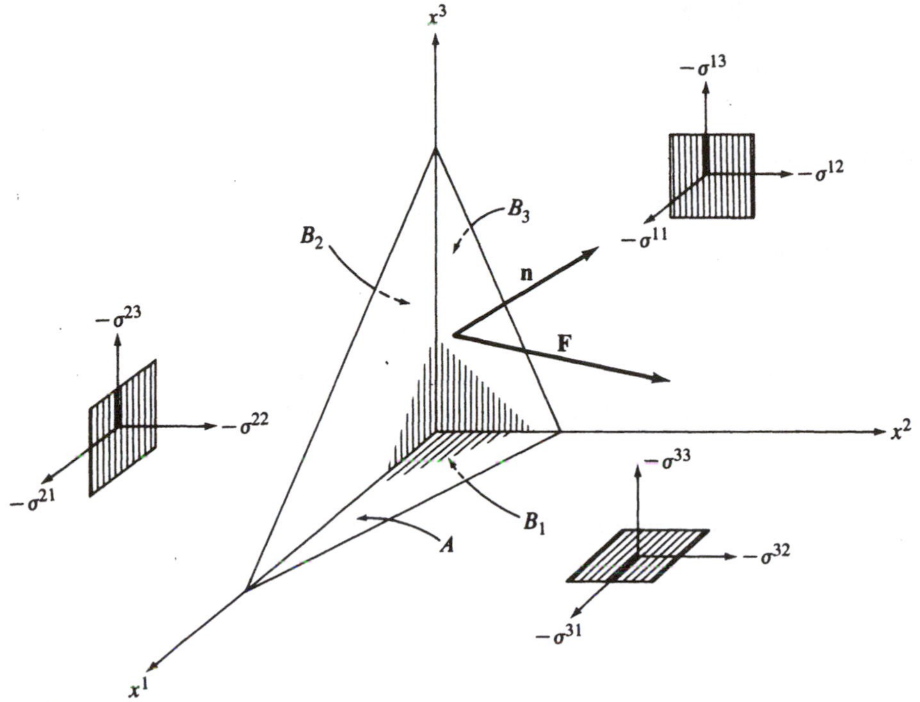

Fig. 3-6

That is, we have a new coordinate system $(\bar{x}^i)$ that is related to (x^i) via

$$\bar{x}^j = a^j_i x^i \tag{3.18}$$

Note that here we have

$$\frac{\partial \bar{x}^j}{\partial x^i} = a^j_i$$

Substituting $\mathbf{e}_r = a^i_r \mathbf{f}_i$ into (3.17) yields the stress components $(\bar{\sigma}^{ij})$ in the new coordinate system, as follows:

$$\mathbf{F} = \sigma^{rs}[\mathbf{n}(a^i_r \mathbf{f}_i)](a^j_s \mathbf{f}_j) = \sigma^{rs} a^i_r a^j_s (\mathbf{n}\mathbf{f}_i)\mathbf{f}_j \equiv \bar{\sigma}^{ij}(\mathbf{n}\mathbf{f}_i)\mathbf{f}_j$$

with

$$\bar{\sigma}^{ij} = \sigma^{rs} a^i_r a^j_s = \sigma^{rs} \frac{\partial \bar{x}^i}{\partial x^r} \frac{\partial \bar{x}^j}{\partial x^s} \tag{3.19}$$

A comparison of (3.19) with the transformation law (3.11) leads to the conclusion that *the stress components σ^{ij} define a second-order contravariant tensor*, at least for linear coordinate changes.

3.6 CARTESIAN TENSORS

Tensors corresponding to admissible *linear* coordinate changes, $\mathscr{T}\ :\ \bar{x}^i = a^i_j x^j \quad (|a^i_j| \neq 0)$, are called *affine tensors*. If (a^i_j) is orthogonal (and $\mathscr{T}$ is distance-preserving), the corresponding tensors are *cartesian tensors*. Now, an object that is a tensor with respect to all one-one linear transformations is necessarily a tensor with respect to all orthogonal linear transformations, but the converse is not true. Hence, *affine tensors are special cartesian tensors*. Likewise, *affine invariants* are particular *cartesian invariants*.

Affine Tensors

A transformation of the form $\mathscr{T}\ :\ \bar{x}^i = a^i_j x^j \quad (|a^i_j| \neq 0)$ takes a rectangular coordinate system (x^i) into a system $(\bar{x}^i)$ having oblique axes; thus affine tensors are defined on the class of all such oblique coordinate systems. Since the Jacobian matrices of $\mathscr{T}$ and $\mathscr{T}^{-1}$ are

$$J = \left[\frac{\partial \bar{x}^i}{\partial x^j}\right]_{nn} = [a^i_j]_{nn} \qquad \text{and} \qquad J^{-1} = \left[\frac{\partial x^i}{\partial \bar{x}^j}\right]_{nn} \equiv [b^i_j]_{nn} \tag{3.20}$$

the transformation laws for affine tensors are:

$$\begin{array}{rl} \textit{contravariant} & \bar{T}^i = a^i_r T^r, \quad \bar{T}^{ij} = a^i_r a^j_s T^{rs}, \quad \bar{T}^{ijk} = a^i_r a^j_s a^k_t T^{rst}, \quad \ldots \\ \textit{covariant} & \bar{T}_i = b^r_i T_r, \quad \bar{T}_{ij} = b^r_i b^s_j T_{rs}, \quad \bar{T}_{ijk} = b^r_i b^s_j b^t_k T_{rst}, \quad \ldots \\ \textit{mixed} & \bar{T}^i_j = a^i_r b^s_j T^r_s, \quad \bar{T}^i_{jk} = a^i_r b^s_j b^t_k T^r_{st}, \quad \ldots \end{array} \tag{3.21}$$

Under the less stringent conditions (*3.21*), more objects can qualify as tensors than before; for instance, an ordinary position vector $\mathbf{x} = (x^i)$ becomes an (affine) tensor (see Problem 3.9), and the partial derivatives of a tensor define an (affine) tensor (as implied by Problem 2.23).

Cartesian Tensors

When the above linear transformation $\mathscr{T}$ is restricted to be orthogonal, then $J^{-1} = J^T$, or

$$b^i_j = a^j_i \qquad (1 \leqq i, j \leqq n)$$

so that the transformation laws for cartesian tensors are, from (*3.21*),

$$\begin{array}{rl} \textit{contravariant} & \bar{T}^i = a^i_r T^r, \quad \bar{T}^{ij} = a^i_r a^j_s T^{rs}, \quad \ldots \\ \textit{covariant} & \bar{T}_i = a^i_r T_r, \quad \bar{T}_{ij} = a^i_r a^j_s T_{rs}, \quad \ldots \\ \textit{mixed} & \bar{T}^i_j = a^i_r a^j_s T^r_s, \quad \ldots \end{array}$$

A striking feature of these forms is that contravariant and covariant behaviors do not distinguish themselves. Consequently, all cartesian tensors are notated the same way—with subscripts:

$$\begin{array}{rl} \textit{allowable coordinate changes} & \bar{x}_i = a_{ij} x_j \quad \text{or} \quad x_i = a_{ji} \bar{x}_j \\ \textit{cartesian tensor laws} & \bar{T}_i = a_{ir} T_r, \quad \bar{T}_{ij} = a_{ir} a_{js} T_{rs}, \quad \ldots \end{array} \tag{3.22}$$

Because an orthogonal transformation takes one rectangular coordinate system into another (having the same origin), cartesian tensors appertain to the rectangular (cartesian) coordinate systems. There are, of course, even more cartesian tensors than affine tensors.

Note that $JJ^T = I$ implies $\mathscr{J}^2 = 1$, or $\mathscr{J} = \pm 1$. Objects that obey the tensor laws (*3.22*) when the allowable coordinate changes are such that

$$\mathscr{J} = |a_{ij}| = +1$$

are called *direct* cartesian tensors.

Solved Problems

CHANGE OF COORDINATES

3.1 For the transformation of Example 3.2, (*a*) obtain the equations for $\mathscr{T}^{-1}$; (*b*) compute $\bar{J}$ from (*a*), and compare with J^{-1}.

(*a*) Solving $\bar{x}^1 = x^1x^2$, $\bar{x}^2 = (x^2)^2$ for x^1 and x^2, we find that

$$\mathscr{T}: \begin{cases} \bar{x}^1 = x^1 x^2 \\ \bar{x}^2 = (x^2)^2 \end{cases} \qquad \mathscr{T}^{-1}: \begin{cases} x^1 = \bar{x}^1/\sqrt{\bar{x}^2} \\ x^2 = \sqrt{\bar{x}^2} \end{cases} \tag{1}$$

is a one-one mapping between the regions $x^2>0$ and $\bar{x}^2>0$, and that

$$\mathcal{T}:\begin{cases}\bar{x}^1 = x^1x^2\\ \bar{x}^2=(x^2)^2\end{cases}\qquad \mathcal{T}^{-1}:\begin{cases}x^1=-\bar{x}^1/\sqrt{\bar{x}^2}\\ x^2=-\sqrt{\bar{x}^2}\end{cases}\tag{2}$$

is one-one between $x^2<0$ and $\bar{x}^2>0$. Note that the two regions of the x^1x^2-plane are separated by the line on which the Jacobian of $\mathcal{T}$ vanishes.

(*b*) From Example 3.2,

$$J=\begin{bmatrix}x^2 & x^1\\ 0 & 2x^2\end{bmatrix}\qquad\text{and so}\qquad J^{-1}=\frac{1}{2(x^2)^2}\begin{bmatrix}2x^2 & -x^1\\ 0 & x^2\end{bmatrix}$$

valid in both regions $x^2>0$ and $x^2<0$. Now, on $\bar{x}^2>0$, differentiation of the inverse transformation (*1*), followed by a change back to unbarred coordinates, yields

$$\bar{J}=\begin{bmatrix}\dfrac{\partial x^1}{\partial\bar{x}^1} & \dfrac{\partial x^1}{\partial\bar{x}^2}\\[2ex] \dfrac{\partial x^2}{\partial\bar{x}^1} & \dfrac{\partial x^2}{\partial\bar{x}^2}\end{bmatrix}=\begin{bmatrix}(\bar{x}^2)^{-1/2} & -\frac{1}{2}\bar{x}^1(\bar{x}^2)^{-3/2}\\[2ex] 0 & \frac{1}{2}(\bar{x}^2)^{-1/2}\end{bmatrix}=\begin{bmatrix}(x^2)^{-1} & -\frac{1}{2}x^1(x^2)^{-2}\\[2ex] 0 & \frac{1}{2}(x^2)^{-1}\end{bmatrix}$$

It is seen that on $x^2>0$, $\bar{J}=J^{-1}$.

Similarly, from (*2*), with $x^2<0$,

$$\bar{J}=\begin{bmatrix}-(\bar{x}^2)^{-1/2} & \frac{1}{2}\bar{x}^1(\bar{x}^2)^{-3/2}\\[2ex] 0 & -\frac{1}{2}(\bar{x}^2)^{-1/2}\end{bmatrix}=\begin{bmatrix}+(x^2)^{-1} & -\frac{1}{2}x^1(x^2)^{-2}\\[2ex] 0 & +\frac{1}{2}(x^2)^{-1}\end{bmatrix}=J^{-1}$$

3.2 For polar coordinates as defined by (*3.2*), (*a*) calculate the Jacobian matrix of $\mathcal{T}$ and infer the region over which $\mathcal{T}$ is bijective; (*b*) calculate the Jacobian matrix of $\mathcal{T}^{-1}$ for the region

$$\{(r,\theta)\mid r>0,\ -\pi/2<\theta<\pi/2\}$$

i.e., the right half-plane, and verify that it is the inverse of the matrix of (*a*).

(*a*)

$$J=\begin{bmatrix}\dfrac{\partial}{\partial x^1}(x^1\cos x^2) & \dfrac{\partial}{\partial x^2}(x^1\cos x^2)\\[2ex] \dfrac{\partial}{\partial x^1}(x^1\sin x^2) & \dfrac{\partial}{\partial x^2}(x^1\sin x^2)\end{bmatrix}=\begin{bmatrix}\cos x^2 & -x^1\sin x^2\\[2ex] \sin x^2 & x^1\cos x^2\end{bmatrix}$$

whence $\mathcal{J}=x^1\equiv r$. Therefore, $\mathcal{T}$ is bijective on the open set $r>0$, which is the entire plane punctured at the origin.

(*b*) For $\mathcal{T}^{-1}$ we have, over the right half-plane,

$$\frac{\partial x^1}{\partial\bar{x}^1}=\frac{\bar{x}^1}{\sqrt{(\bar{x}^1)^2+(\bar{x}^2)^2}}\qquad\frac{\partial x^1}{\partial\bar{x}^2}=\frac{\bar{x}^2}{\sqrt{(\bar{x}^1)^2+(\bar{x}^2)^2}}$$

$$\frac{\partial x^2}{\partial\bar{x}^1}=\frac{1}{1+(\bar{x}^2/\bar{x}^1)^2}\left[-\frac{\bar{x}^2}{(\bar{x}^1)^2}\right]=\frac{-\bar{x}^2}{(\bar{x}^1)^2+(\bar{x}^2)^2}\qquad\frac{\partial x^2}{\partial\bar{x}^2}=\frac{\bar{x}^1}{(\bar{x}^1)^2+(\bar{x}^2)^2}$$

and so

$$\bar{J}=\begin{bmatrix}\dfrac{\bar{x}^1}{\sqrt{(\bar{x}^1)^2+(\bar{x}^2)^2}} & \dfrac{\bar{x}^2}{\sqrt{(\bar{x}^1)^2+(\bar{x}^2)^2}}\\[2ex] \dfrac{-\bar{x}^2}{(\bar{x}^1)^2+(\bar{x}^2)^2} & \dfrac{\bar{x}^1}{(\bar{x}^1)^2+(\bar{x}^2)^2}\end{bmatrix}=\begin{bmatrix}\cos x^2 & \sin x^2\\[2ex] -\dfrac{\sin x^2}{x^1} & \dfrac{\cos x^2}{x^1}\end{bmatrix}$$

Now compute J^{-1}:

$$J^{-1}=\frac{1}{x^1}\begin{bmatrix}x^1\cos x^2 & x^1\sin x^2\\[2ex] -\sin x^2 & \cos x^2\end{bmatrix}=\begin{bmatrix}\cos x^2 & \sin x^2\\[2ex] -\dfrac{\sin x^2}{x^1} & \dfrac{\cos x^2}{x^1}\end{bmatrix}=\bar{J}$$

CONTRAVARIANT VECTORS

3.3 If $\mathbf{V} = (T^i)$ is a contravariant vector, show that the partial derivatives $T^i_{,j} \equiv \partial T^i/\partial x^j$, defined in each coordinate system, transform according to the rule

$$\bar{T}^i_{,j} = T^r_{,s} \frac{\partial \bar{x}^i}{\partial x^r} \frac{\partial x^s}{\partial \bar{x}^j} + T^r \frac{\partial^2 \bar{x}^i}{\partial x^r \partial x^s} \frac{\partial x^s}{\partial \bar{x}^j}$$

Differentiate both sides of

$$\bar{T}^i = T^r \frac{\partial \bar{x}^i}{\partial x^r}$$

with respect to $\bar{x}^j$, using the product rule:

$$\bar{T}^i_{,j} \equiv \frac{\partial \bar{T}^i}{\partial \bar{x}^j} = \frac{\partial}{\partial \bar{x}^j}\left(T^r \frac{\partial \bar{x}^i}{\partial x^r}\right) = \frac{\partial T^r}{\partial \bar{x}^j} \frac{\partial \bar{x}^i}{\partial x^r} + T^r \frac{\partial}{\partial \bar{x}^j}\left(\frac{\partial \bar{x}^i}{\partial x^r}\right) \tag{1}$$

By the chain rule for partial derivatives, (2.15),

$$\frac{\partial T^r}{\partial \bar{x}^j} = \frac{\partial T^r}{\partial x^s} \frac{\partial x^s}{\partial \bar{x}^j} \equiv T^r_{,s} \frac{\partial x^s}{\partial \bar{x}^j} \qquad \text{and} \qquad \frac{\partial}{\partial \bar{x}^j}\left(\frac{\partial \bar{x}^i}{\partial x^r}\right) = \left[\frac{\partial}{\partial x^s}\left(\frac{\partial \bar{x}^i}{\partial x^r}\right)\right] \frac{\partial x^s}{\partial \bar{x}^j}$$

Substituting these expressions into (1) yields the desired formula.

3.4 Suppose that (T^i) is a contravariant vector on $\mathbf{R}^2$ and that $(T^i) = (x^2, x^1)$ in the (x^i)-system. Calculate $(\bar{T}^i)$ in the $(\bar{x}^i)$-system, under the change of coordinates

$$\begin{aligned} \bar{x}^1 &= (x^2)^2 \neq 0 \\ \bar{x}^2 &= x^1 x^2 \end{aligned}$$

By definition of contravariance,

$$\bar{T}^i = T^r \frac{\partial \bar{x}^i}{\partial x^r} = T^1 \frac{\partial \bar{x}^i}{\partial x^1} + T^2 \frac{\partial \bar{x}^i}{\partial x^2}$$

Note that the top row of the Jacobian matrix J is needed for the case $i = 1$, and the bottom row is needed for $i = 2$.

$$J = \begin{bmatrix} \dfrac{\partial \bar{x}^1}{\partial x^1} & \dfrac{\partial \bar{x}^1}{\partial x^2} \\[2ex] \dfrac{\partial \bar{x}^2}{\partial x^1} & \dfrac{\partial \bar{x}^2}{\partial x^2} \end{bmatrix} = \begin{bmatrix} 0 & 2x^2 \\ x^2 & x^1 \end{bmatrix}$$

Thus,

$$\bar{T}^1 = T^1(0) + T^2(2x^2) = 2x^1x^2 \qquad \bar{T}^2 = T^1(x^2) + T^2(x^1) = (x^2)^2 + (x^1)^2$$

which, in terms of barred coordinates, are

$$\bar{T}^1 = 2\bar{x}^2 \qquad \bar{T}^2 = \bar{x}^1 + \frac{(\bar{x}^2)^2}{\bar{x}^1}$$

3.5 Show that a contravariant vector can be constructed the components of which take on a given set of values $(a, b, c, \ldots)$ in some particular coordinate system. (The prescribed values may be point functions.)

Let $(a, b, c, \ldots) \equiv (a^i)$ be the given values to be assigned in the coordinate system (x^i). Set $V^i = a^i$ for the values in (x^i), and for any other admissible coordinate system $(\bar{x}^i)$, set $\bar{V}^i = a^r(\partial \bar{x}^i/\partial x^r)$. To show that (V^i) is a contravariant tensor, let (y^i) and $(\bar{y}^i)$ be any two admissible coordinate systems. Then, $y^i = f^i(x^1, x^2, \ldots, x^n)$ and $\bar{y}^i = g^i(x^1, x^2, \ldots, x^n)$, and, by definition, the values of (V^i) in (y^i) and $(\bar{y}^i)$ are, respectively, $T^i = a^r(\partial y^i/\partial x^r)$ and $\bar{T}^i = a^r(\partial \bar{y}^i/\partial x^r)$. But, by the chain rule,

$$\bar{T}^i = a^r \frac{\partial \bar{y}^i}{\partial x^r} = a^r \frac{\partial \bar{y}^i}{\partial y^s} \frac{\partial y^s}{\partial x^r} = T^s \frac{\partial \bar{y}^i}{\partial y^s} \qquad \text{QED}$$

COVARIANT VECTORS

3.6 Calculate $(\bar{T}_i)$ in the $(\bar{x}^i)$-system if $\mathbf{V} = (T_i) \equiv (x^2, x^1 + 2x^2)$ is a covariant vector under the coordinate transformation of Problem 3.4.

To avoid radicals, compute J^{-1} in terms of (x^i):

$$J^{-1} = \begin{bmatrix} \dfrac{-x^1}{2(x^2)^2} & \dfrac{1}{x^2} \\ \dfrac{1}{2x^2} & 0 \end{bmatrix}$$

By covariance,

$$\bar{T}_i = T_r \frac{\partial x^r}{\partial \bar{x}^i} = T_1 \frac{\partial x^1}{\partial \bar{x}^i} + T_2 \frac{\partial x^2}{\partial \bar{x}^i} \qquad (i = 1, 2)$$

For $i = 1$, read off the partials from the first column of J^{-1}:

$$\bar{T}_1 = T_1(-x^1/2(x^2)^2) + T_2(1/2x^2) = -x^1/2x^2 + x^1/2x^2 + 1 = 1$$

Similarly, for $i = 2$, use the second column of J^{-1}:

$$\bar{T}_2 = T_1(1/x^2) + T_2(0) = x^2(1/x^2) = 1$$

Hence, $(\bar{T}_i) = (1, 1)$ at all points in the $(\bar{x}^i)$-system ($\bar{x}^1 = 0$ excluded).

3.7 Use the fact that ∇f is a covariant vector (Example 3.5) to bring the partial differential equation

$$x \frac{\partial f}{\partial x} = y \frac{\partial f}{\partial y} \qquad (1)$$

into simpler form by the change of variables $\bar{x} = xy$, $\bar{y} = (y)^2$; then solve.

Write $\nabla f = (\partial f/\partial x, \partial f/\partial y) \equiv (T_i)$, $(x^1, x^2) = (x, y)$, $(\bar{x}^1, \bar{x}^2) = (\bar{x}, \bar{y})$, and

$$\bar{T}_i \equiv \frac{\partial \bar{f}}{\partial \bar{x}^i} = T_r \frac{\partial x^r}{\partial \bar{x}^i}$$

Again calculating J first, then its inverse, we have

$$\left(\frac{\partial x^i}{\partial \bar{x}^j}\right) \equiv J^{-1} = \begin{bmatrix} y & x \\ 0 & 2y \end{bmatrix}^{-1} = \begin{bmatrix} \dfrac{1}{y} & \dfrac{-x}{2(y)^2} \\ 0 & \dfrac{1}{2y} \end{bmatrix}$$

so that

$$\frac{\partial \bar{f}}{\partial \bar{x}} \equiv \bar{T}_1 = T_r \frac{\partial x^r}{\partial \bar{x}^1} = T_1 \cdot \frac{1}{y} + T_2 \cdot 0 = \frac{1}{y} \frac{\partial f}{\partial x}$$

$$\frac{\partial \bar{f}}{\partial \bar{y}} \equiv \bar{T}_2 = T_r \frac{\partial x^r}{\partial \bar{x}^2} = T_1 \cdot \frac{-x}{2(y)^2} + T_2 \cdot \frac{1}{2y} = -\frac{x}{2(y)^2} \frac{\partial f}{\partial x} + \frac{1}{2y} \frac{\partial f}{\partial y}$$

But, by (*1*),

$$\frac{\partial \bar{f}}{\partial \bar{y}} = \frac{1}{2(y)^2} \left(-x \frac{\partial f}{\partial x} + y \frac{\partial f}{\partial y}\right) = 0$$

which implies that $\bar{f} = F(\bar{x})$, a function of $\bar{x}$ alone; therefore, $f = F(xy)$ is the general solution to (*1*).

INVARIANTS

3.8 Prove Theorem 3.1.

We must show that if (S^i) and (T_i) are tensors of the indicated types and order, then the quantity $E \equiv S^i T_i$ is invariant with respect to coordinate changes; that is, $\bar{E} = E$, where $\bar{E} = \bar{S}^i \bar{T}_i$. But observe that

$$\bar{S}^i = S^r \frac{\partial \bar{x}^i}{\partial x^r} \qquad \text{and} \qquad \bar{T}_i = T_s \frac{\partial x^s}{\partial \bar{x}^i}$$

so that, in view of (*3.7*),

$$\bar{E} = \bar{S}^i \bar{T}_i = S^r \frac{\partial \bar{x}^i}{\partial x^r} \cdot T_s \frac{\partial x^s}{\partial \bar{x}^i} = S^r T_s \frac{\partial \bar{x}^i}{\partial x^r} \frac{\partial x^s}{\partial \bar{x}^i} = S^r T_s \delta^s_r = S^r T_r = E$$

3.9 Show that under linear coordinate changes of $\mathbf{R}^n$, $\bar{x}^i = a^i_j x^j$ $(|a^i_j| \neq 0)$, the equation of a hyperplane $A_i x^i = 1$ is invariant provided the normal vector (A_i) is covariant.

In view of Theorem 3.1, it suffices to show that $(T^i) = (x^i)$ is a contravariant affine tensor. But this is immediate:

$$\bar{T}^i \equiv \bar{x}^i = a^i_j x^j \equiv a^i_j T^j$$

which is the transformation law (*3.21*).

SECOND-ORDER CONTRAVARIANT TENSORS

3.10 Suppose that the components of a contravariant tensor $\mathbf{T}$ of order 2 in a coordinate system (x^i) of $\mathbf{R}^2$ are $T^{11} = 1$, $T^{12} = 1$, $T^{21} = -1$, and $T^{22} = 2$. (*a*) Find the components $\bar{T}^{ij}$ of $\mathbf{T}$ in the $(\bar{x}^i)$-system, connected to the (x^i)-system via

$$\bar{x}^1 = (x^1)^2 \neq 0$$
$$\bar{x}^2 = x^1 x^2$$

(*b*) Compute the values of the $\bar{T}^{ij}$ at the point which corresponds to $x^1 = 1, x^2 = -2$.

For economy of effort, the problem will be worked using matrices.

(*a*) Writing

$$J^i_j \equiv J'^{\,j}_i \equiv \frac{\partial \bar{x}^i}{\partial x^j}$$

we have from (*2.1b*),

$$\bar{T}^{ij} = T^{rs} \frac{\partial \bar{x}^i}{\partial x^r} \frac{\partial \bar{x}^j}{\partial x^s} = J^i_r T^{rs} J'^{\,s}_j$$

That is,

$$\bar{T} = JTJ^T$$
$$= \begin{bmatrix} 2x^1 & 0 \\ x^2 & x^1 \end{bmatrix} \begin{bmatrix} 1 & 1 \\ -1 & 2 \end{bmatrix} \begin{bmatrix} 2x^1 & x^2 \\ 0 & x^1 \end{bmatrix} = \begin{bmatrix} 4(x^1)^2 & 2x^1x^2 + 2(x^1)^2 \\ 2x^1x^2 - 2(x^1)^2 & 2(x^1)^2 + (x^2)^2 \end{bmatrix}$$

(*b*) At the point $(1, -2)$,

$$\bar{T}^{11} = 4(1)^2 = 4 \qquad \bar{T}^{12} = 2(1)(-2) + 2(1)^2 = -2$$
$$\bar{T}^{21} = 2(1)(-2) - 2(1)^2 = -6 \qquad \bar{T}^{22} = 2(1)^2 + (-2)^2 = 6$$

3.11 Show that if (S^i) and (T^i) are contravariant vectors on $\mathbf{R}^n$, the matrix $[U^{ij}] \equiv [S^i T^j]_{nn}$, defined in this manner for all coordinate systems, represents a contravariant tensor of order 2.

Multiply

$$\bar{S}^i = S^r \frac{\partial \bar{x}^i}{\partial x^r} \qquad \text{and} \qquad \bar{T}^j = T^s \frac{\partial \bar{x}^j}{\partial x^s}$$

to obtain

$$\bar{U}^{ij} = \bar{S}^i \bar{T}^j = S^r \frac{\partial \bar{x}^i}{\partial x^r} \cdot T^s \frac{\partial \bar{x}^j}{\partial x^s} = U^{rs} \frac{\partial \bar{x}^i}{\partial x^r} \frac{\partial \bar{x}^j}{\partial x^s}$$

which is the tensor law. (The notion of the "outer product" of two tensors will be further developed in Chapter 4.)

SECOND-ORDER COVARIANT TENSORS

3.12 Show that if T_i are the components of covariant vector **T**, then $S_{ij} \equiv T_i T_j - T_j T_i$ are the components of a skew-symmetric covariant tensor **S**.

The skew-symmetry is obvious. From the transformation law for **T**,

$$\bar{T}_i \bar{T}_j - \bar{T}_j \bar{T}_i = T_r \frac{\partial x^r}{\partial \bar{x}^i} \cdot T_s \frac{\partial x^s}{\partial \bar{x}^j} - T_s \frac{\partial x^s}{\partial \bar{x}^j} \cdot T_r \frac{\partial x^r}{\partial \bar{x}^i}$$

$$= T_r T_s \frac{\partial x^r}{\partial \bar{x}^i} \frac{\partial x^s}{\partial \bar{x}^j} - T_s T_r \frac{\partial x^r}{\partial \bar{x}^i} \frac{\partial x^s}{\partial \bar{x}^j} = (T_r T_s - T_s T_r) \frac{\partial x^r}{\partial \bar{x}^i} \frac{\partial x^s}{\partial \bar{x}^j}$$

or

$$\bar{S}_{ij} = S_{rs} \frac{\partial x^r}{\partial \bar{x}^i} \frac{\partial x^s}{\partial \bar{x}^j}$$

which establishes the covariant tensor character of **S**.

3.13 If a symmetric array (T_{ij}) transforms according to

$$\bar{T}_{ij} = T_{rt} \frac{\partial x^k}{\partial \bar{x}^s} \frac{\partial x^s}{\partial \bar{x}^j} \frac{\partial x^t}{\partial \bar{x}^i} \frac{\partial \bar{x}^r}{\partial x^k}$$

show that it defines a second-order covariant tensor.

$$\bar{T}_{ij} = T_{rt} \left(\frac{\partial \bar{x}^r}{\partial x^k} \frac{\partial x^k}{\partial \bar{x}^s} \right) \frac{\partial x^s}{\partial \bar{x}^j} \frac{\partial x^t}{\partial \bar{x}^i} = T_{rt} \delta^r_s \frac{\partial x^s}{\partial \bar{x}^j} \frac{\partial x^t}{\partial \bar{x}^i}$$

$$= T_{st} \frac{\partial x^s}{\partial \bar{x}^j} \frac{\partial x^t}{\partial \bar{x}^i} = T_{ts} \frac{\partial x^t}{\partial \bar{x}^i} \frac{\partial x^s}{\partial \bar{x}^j}$$

3.14 Let $\mathbf{U} = (U_{ij})$ be a covariant tensor of order 2. Under the same coordinate change as in Problem 3.10, (*a*) calculate the components $\bar{U}_{ij}$, if $U_{11} = x^2$, $U_{12} = U_{21} = 0$, $U_{22} = x^1$; (*b*) verify that the quantity $T^{ij} U_{ij} = E$ is an invariant, where the T^{ij} and $\bar{T}^{ij}$ are obtained from Problem 3.10.

(*a*) In terms of the inverse Jacobian matrix, the covariant transformation law is

$$\bar{U}_{ij} = \frac{\partial x^r}{\partial \bar{x}^i} U_{rs} \frac{\partial x^s}{\partial \bar{x}^j} = \bar{J}^r_i U_{rs} \bar{J}^s_j = \bar{J}'^i_r U_{rs} \bar{J}^s_j \qquad \text{or} \qquad \bar{U} = \bar{J}^T U \bar{J}$$

Substituting

$$\bar{J} = \begin{bmatrix} 2x^1 & 0 \\ x^2 & x^1 \end{bmatrix}^{-1} = \begin{bmatrix} \dfrac{1}{2x^1} & 0 \\ -\dfrac{x^2}{2(x^1)^2} & \dfrac{1}{x^1} \end{bmatrix} \qquad U = \begin{bmatrix} x^2 & 0 \\ 0 & x^1 \end{bmatrix}$$

we find

$$\bar{U} = \begin{bmatrix} \dfrac{1}{2x^1} & -\dfrac{x^2}{2(x^1)^2} \\ 0 & \dfrac{1}{x^1} \end{bmatrix} \begin{bmatrix} x^2 & 0 \\ 0 & x^1 \end{bmatrix} \begin{bmatrix} \dfrac{1}{2x^1} & 0 \\ -\dfrac{x^2}{2(x^1)^2} & \dfrac{1}{x^1} \end{bmatrix} = \begin{bmatrix} \dfrac{x^1x^2 + (x^2)^2}{4(x^1)^3} & -\dfrac{x^2}{2(x^1)^2} \\ -\dfrac{x^2}{2(x^1)^2} & \dfrac{1}{x^1} \end{bmatrix}$$

from which the $\bar{U}_{ij}$ may be read off.

(*b*) Continuing in the matrix approach, we note that E is the *trace* (sum of diagonal elements) of the matrix TU^T.

$$TU^T = \begin{bmatrix} 1 & 1 \\ -1 & 2 \end{bmatrix} \begin{bmatrix} x^2 & 0 \\ 0 & x^1 \end{bmatrix} = \begin{bmatrix} x^2 & x^1 \\ -x^2 & 2x^1 \end{bmatrix}$$

$$E = x^2 + 2x^1$$

and

$$\bar{T}\bar{U}^T = \begin{bmatrix} 4(x^1)^2 & 2x^1x^2 + 2(x^1)^2 \\ 2x^1x^2 - 2(x^1)^2 & 2(x^1)^2 + (x^2)^2 \end{bmatrix} \begin{bmatrix} \dfrac{x^1x^2 + (x^2)^2}{4(x^1)^3} & -\dfrac{x^2}{2(x^1)^2} \\ -\dfrac{x^2}{2(x^1)^2} & \dfrac{1}{x^1} \end{bmatrix}$$

$$= \begin{bmatrix} 0 & 2x^1 \\ -\dfrac{3x^2}{2} & x^2 + 2x^1 \end{bmatrix}$$

$$\bar{E} = x^2 + 2x^1 = E$$

3.15 Prove Theorem 3.2.

Observe first of all that if a covariant matrix (second-order tensor) U has inverse V in unbarred coordinates, then $\bar{U}$ has inverse $\bar{V}$ in barred coordinates; i.e., $(\bar{U})^{-1} = \overline{U^{-1}}$. Now, by Problem 3.14(*a*),

$$\bar{U} = \bar{J}^T U \bar{J}$$

Inverting both sides of this matrix equation, applying Problem 2.13, and recalling that $J\bar{J} = I$, we obtain

$$\overline{U^{-1}} = \bar{J}^{-1}U^{-1}(\bar{J}^T)^{-1} = JU^{-1}J^T$$

which is the contravariant law for U^{-1} [see Problem 3.10(*a*)].

MIXED TENSORS

3.16 Compute the formulas for the tensor components $(\bar{T}^i_j)$ in polar coordinates in terms of (T^i_j) in rectangular coordinates, if the tensor is symmetric in rectangular coordinates. (In contrast to Section 3.1, it is now the curvilinear coordinates that are barred.)

The general formula calls for the calculations

$$\bar{T}^i_j = T^r_s \frac{\partial \bar{x}^i}{\partial x^r} \frac{\partial x^s}{\partial \bar{x}^j} = \frac{\partial \bar{x}^i}{\partial x^r} T^r_s \frac{\partial x^s}{\partial \bar{x}^j} \qquad (T^i_j = T^j_i)$$

Using (*2.1b*), this may be written in matrix form as

$$\bar{T} = JTJ^{-1} = \bar{J}^{-1}T\bar{J} \tag{1}$$

where $T = [T^i_j]_{22}$ and where

$$\bar{J} = \begin{bmatrix} \cos\theta & -r\sin\theta \\ \sin\theta & r\cos\theta \end{bmatrix}$$

is the Jacobian matrix of the transformation from (r, θ) to (x, y). Thus,

$$\bar{T} = \begin{bmatrix} \cos\theta & \sin\theta \\ -\dfrac{\sin\theta}{r} & \dfrac{\cos\theta}{r} \end{bmatrix} \begin{bmatrix} T_1^1 & T_2^1 \\ T_2^1 & T_2^2 \end{bmatrix} \begin{bmatrix} \cos\theta & -r\sin\theta \\ \sin\theta & r\cos\theta \end{bmatrix}$$

$$= \begin{bmatrix} \cos\theta & \sin\theta \\ -\dfrac{\sin\theta}{r} & \dfrac{\cos\theta}{r} \end{bmatrix} \begin{bmatrix} T_1^1\cos\theta + T_2^1\sin\theta & -rT_1^1\sin\theta + rT_2^1\cos\theta \\ T_2^1\cos\theta + T_2^2\sin\theta & -rT_2^1\sin\theta + rT_2^2\cos\theta \end{bmatrix}$$

The final matrix multiplication can be carried out routinely, simplifying by means of trigonometric identities:

$$\bar{T} = \begin{bmatrix} T_1^1\cos^2\theta + T_2^1\sin 2\theta + T_2^2\sin^2\theta & -\dfrac{r}{2}T_1^1\sin 2\theta + rT_2^1\cos 2\theta + \dfrac{r}{2}T_2^2\sin 2\theta \\ -T_1^1\dfrac{\sin 2\theta}{2r} + T_2^1\dfrac{\cos 2\theta}{r} + T_2^2\dfrac{\sin 2\theta}{2r} & T_1^1\sin^2\theta - T_2^1\sin 2\theta + T_2^2\cos^2\theta \end{bmatrix}$$

Observe that $\bar{T}$ does not share the symmetry of T: $\bar{T}_1^2 = r^{-2}\bar{T}_2^1$.

3.17 Prove that the determinant of a mixed tensor of order two is invariant.

By (*1*) of Problem 3.16, we have—whether or not T is symmetric—

$$|\bar{T}| = |JTJ^{-1}| = |J|\,|T|\,|J^{-1}| = \mathcal{J}\,|T|\,\mathcal{J}^{-1} = |T|$$

GENERAL TENSORS

3.18 Display the transformation law for a third-order tensor that is contravariant of order two and covariant of order one.

Take $p = 2$ and $q = 1$ in Definition 7 and, to avoid unnecessary subscripts, write i, j, k, r, s, t in place of $i_1, i_2, j_1, r_1, r_2, s_1$. Then (*3.14*) gives

$$\bar{T}_k^{ij} = T_t^{rs}\frac{\partial \bar{x}^i}{\partial x^r}\frac{\partial \bar{x}^j}{\partial x^s}\frac{\partial x^t}{\partial \bar{x}^k}$$

3.19 Let $\mathbf{T} = (T_{klm}^{ij})$ denote a tensor of the order and type indicated by the indices. Prove that $\mathbf{S} = (T_k) \equiv (T_{kij}^{ij})$ is a covariant vector.

The transformation law (*3.14*) for $\mathbf{T}$ is

$$\bar{T}_{klm}^{ij} = T_{tuv}^{rs}\frac{\partial \bar{x}^i}{\partial x^r}\frac{\partial \bar{x}^j}{\partial x^s}\frac{\partial x^t}{\partial \bar{x}^k}\frac{\partial x^u}{\partial \bar{x}^l}\frac{\partial x^v}{\partial \bar{x}^m}$$

Set $l = i$, $m = j$ and sum:

$$\bar{T}_k \equiv \bar{T}_{kij}^{ij} = T_{tuv}^{rs}\frac{\partial \bar{x}^i}{\partial x^r}\frac{\partial \bar{x}^j}{\partial x^s}\frac{\partial x^t}{\partial \bar{x}^k}\frac{\partial x^u}{\partial \bar{x}^i}\frac{\partial x^v}{\partial \bar{x}^j} = T_{tuv}^{rs}\left(\frac{\partial \bar{x}^i}{\partial x^r}\frac{\partial x^u}{\partial \bar{x}^i}\right)\left(\frac{\partial \bar{x}^j}{\partial x^s}\frac{\partial x^v}{\partial \bar{x}^j}\right)\frac{\partial x^t}{\partial \bar{x}^k}$$

$$= T_{tuv}^{rs}\delta_r^u\delta_s^v\frac{\partial x^t}{\partial \bar{x}^k} = T_{trs}^{rs}\frac{\partial x^t}{\partial \bar{x}^k} \equiv T_t\frac{\partial x^t}{\partial \bar{x}^k}$$

CARTESIAN TENSORS

3.20 Show that the permutation symbol (e_{ij}) defines a direct cartesian tensor over $\mathbf{R}^2$. Assume that e_{ij} is defined the same way for all rectangular coordinate systems.

If the coordinate change is $\bar{x}_i = a_{ij}x_j$, where $(a_{ij})^T(a_{kl}) = (\delta_{pq})$ and

$$|a_{ij}| = a_{11}a_{22} - a_{12}a_{21} = 1$$

we must establish the cartesian tensor law (*3.22*):

$$\bar{e}_{ij} = e_{rs}a_{ir}a_{js} \qquad (n=2)$$

We examine separately the four possible cases:

$$\begin{aligned} \boldsymbol{i=j=1} &\quad e_{rs}a_{1r}a_{1s} = a_{11}a_{12} - a_{12}a_{11} = 0 = \bar{e}_{11} \\ \boldsymbol{i=1, j=2} &\quad e_{rs}a_{1r}a_{2s} = a_{11}a_{22} - a_{12}a_{21} = 1 = \bar{e}_{12} \\ \boldsymbol{i=2, j=1} &\quad e_{rs}a_{2r}a_{1s} = a_{21}a_{12} - a_{22}a_{11} = -1 = \bar{e}_{21} \\ \boldsymbol{i=j=2} &\quad e_{rs}a_{2r}a_{2s} = a_{21}a_{22} - a_{22}a_{21} = 0 = \bar{e}_{22} \end{aligned}$$

3.21 Prove that (*a*) the coefficients c_{ij} of the quadratic form $c_{ij}x^ix^j = 1$ transform as an affine tensor and (*b*) the trace c_{ii} of (c_{ij}) is a cartesian invariant.

(*a*) If $\bar{x}^i = a^i_j x^j$ and $x^i = b^i_j \bar{x}^j$, where $(b^i_j) = (a^i_j)^{-1}$, the quadratic form goes over into

$$1 = c_{ij}(b^i_r \bar{x}^r)(b^j_s \bar{x}^s) \equiv \bar{c}_{rs}\bar{x}^r\bar{x}^s$$

with $\bar{c}_{rs} = b^i_r b^j_s c_{ij}$. But this formula is just (*3.21*) for a covariant affine tensor of order two.

(*b*) Assuming an orthogonal transformation, $(b^i_j) = (a^i_j)^T$, we have

$$\bar{c}_{rs} = b^i_r a^s_j c_{ij}$$

Hence, $\bar{c}_{rr} = (b^i_r a^r_j)c_{ij} = \delta^i_j c_{ij} = c_{ii}$.

3.22 Establish the identity between the permutation symbol and the Kronecker delta:

$$e_{rij}e_{rkl} \equiv \delta_{ik}\delta_{jl} - \delta_{il}\delta_{jk} \tag{3.23}$$

The identity implies $n = 3$, so that there are potentially $3^4 = 81$ separate cases to consider. However, this number can be quickly reduced to only 4 cases by the following reasoning: If either $i = j$ or $k = l$, then both sides vanish. For example, if $i = j$, then on the left $e_{rij} = 0$, and on the right,

$$\delta_{ik}\delta_{jl} - \delta_{jl}\delta_{ik} = 0$$

Hence, we need only consider the cases in which both $i \neq j$ and $k \neq l$. Upon writing out the sum on the left, two of the terms drop out, since $i \neq j$:

$$e_{1ij}e_{1kl} + e_{2ij}e_{2kl} + e_{3ij}e_{3kl} = e_{1'2'3'}e_{1'kl} \qquad (i = 2',\ j = 3')$$

where (1′2′3′) denotes some permutation of (123). Thus, there are left only two cases, each with two subcases.

Case 1: $e_{1'2'3'}e_{1'kl} \neq 0$ (with $i = 2'$, $j = 3'$). Here, either $k = 2'$ and $l = 3'$ or $k = 3'$ and $l = 2'$. If the former, then the left member of (*3.23*) is $+1$, while the right member equals

$$\delta_{2'2'}\delta_{3'3'} - \delta_{2'3'}\delta_{3'2'} = 1 - 0 = 1$$

If the latter, then both members equal -1, as can be easily verified.

Case 2: $e_{1'2'3'}e_{1'kl} = 0$ (with $i = 2'$, $j = 3'$). Since $k \neq l$, either $k = 1'$ or $l = 1'$. If $k = 1'$, then the right member of (*3.23*) equals

$$\delta_{2'1'}\delta_{3'l} - \delta_{2'l}\delta_{3'1'} = 0 - 0 = 0$$

If $l = 1'$, we have $\delta_{2'k}\delta_{3'1'} - \delta_{2'1'}\delta_{3'k} = 0 - 0 = 0$.

This completes the examination of all cases, and the identity is established.

Supplementary Problems

3.23 Suppose that the following transformation connects the (x^i) and $(\bar{x}^i)$ coordinate systems:

$$\mathcal{T} : \begin{cases} \bar{x}^1 = \exp(x^1 + x^2) \\ \bar{x}^2 = \exp(x^1 - x^2) \end{cases}$$

(*a*) Calculate the Jacobian matrix J and the Jacobian $\mathcal{J}$. Show that $\mathcal{J} \neq 0$ over all of $\mathbf{R}^2$. (*b*) Give equations for $\mathcal{T}^{-1}$. (*c*) Calculate the Jacobian matrix $\bar{J}$ of $\mathcal{T}^{-1}$ and compare with J^{-1}.

3.24 Prove that if (T_i) defines a covariant vector, and if the components $S_{ij} \equiv T_i T_j + T_j T_i$ are defined in each coordinate system, then (S_{ij}) is a symmetric covariant tensor. (Compare Problem 3.12.)

3.25 Prove that if (T_i) defines a covariant vector and, in each coordinate system, we define

$$\frac{\partial T_i}{\partial x^j} - \frac{\partial T_j}{\partial x^i} \equiv T_{ij}$$

then (T_{ij}) is a skew-symmetric covariant tensor of the second order. [*Hint*: Model the proof on Problem 3.3.]

3.26 Convert the partial differential equation

$$y \frac{\partial f}{\partial x} = x \frac{\partial f}{\partial y}$$

to polar form (making use of the fact that ∇f is a covariant vector), and solve for $f(x, y)$.

3.27 Show that the quadratic form $Q = g_{ij}x^i x^j$ is an affine invariant provided (g_{ij}) is a covariant affine tensor. [Converse of Problem 3.21(*a*).]

3.28 Prove that the partial derivatives of a contravariant vector (T^i) define a mixed affine tensor of order two. [*Hint*: Compare Problem 2.23.]

3.29 Prove that the Kronecker delta (δ^i_j), uniformly defined in all coordinate systems, is a mixed tensor of order two.

3.30 Show that the permutation symbol (e_{ij}) of order two, uniformly defined in all coordinate systems, is not—Problem 3.20 notwithstanding—covariant under arbitrary coordinate changes. [*Hint*: Use $x^1 = \bar{x}^1\bar{x}^2$, $x^2 = \bar{x}^2$, at the point $(\bar{x}^i) = (1, 2)$.]

3.31 By use of *(3.23)*, establish the familiar identity for the vector product of three vectors,

$$\mathbf{u} \times (\mathbf{v} \times \mathbf{w}) = (\mathbf{uw})\mathbf{v} - (\mathbf{uv})\mathbf{w}$$

or, in coordinate form,

$$e_{ijk}u_j(e_{krs}v_r w_s) = (u_j w_j)v_i - (u_j v_j)w_i$$

3.32 (*a*) Show that if (T^i_j) is a mixed tensor, then $(T^i_j + T^j_i)$ is not generally a tensor. (*b*) Show that a mixed tensor of order two, symmetric in a given coordinate system, will transform as a *symmetric* tensor if the Jacobian matrix is orthogonal.

3.33 Prove: (*a*) If (T^i_j) is a mixed tensor of order two, T^i_i is an invariant; (*b*) if (S^i_{jk}) and (T^i) are tensors of the type and order indicated, $S^r_{jr}T^j$ is an invariant.

3.34 If $\mathbf{T} \equiv (T^{ijk}_{ml})$ is a tensor, contravariant of order 3 and covariant of order 2, show that $\mathbf{S} \equiv (T^{ijk}_{kj})$ is a contravariant vector.

3.35 Show that the derivative, $d\mathbf{T}/dt$, of the tangent vector $\mathbf{T} \equiv (T^i) = (dx^i/dt)$ to a curve $x^i = x^i(t)$ is a contravariant affine tensor. Is it a cartesian tensor?

3.36 (*a*) Use the theory of tensors to prove that the scalar product $\mathbf{uv} \equiv u_i v_i$ of two vectors $\mathbf{u} = (u_i)$ and $\mathbf{v} = (v_i)$ is a cartesian invariant. (*b*) Is $\mathbf{uv}$ an affine invariant?

Chapter 4

Tensor Operations; Tests for Tensor Character

4.1 FUNDAMENTAL OPERATIONS

From two given tensors,

$$\mathbf{S} = (S^{i_1 i_2 \ldots i_p}_{j_1 j_2 \ldots j_q}) \qquad \mathbf{T} = (T^{k_1 k_2 \ldots k_r}_{l_1 l_2 \ldots l_s}) \tag{4.1}$$

certain operations, to be described, will produce a third tensor.

Sums, Linear Combinations

Let $p = r$ and $q = s$ in (*4.1*). Since the transformation law (*3.14*) is linear in the tensor components, it is clear that

$$\mathbf{S} + \mathbf{T} \equiv (S^{i_1 i_2 \ldots i_r}_{j_1 j_2 \ldots j_s} + T^{i_1 i_2 \ldots i_r}_{j_1 j_2 \ldots j_s}) \tag{4.2a}$$

is a tensor of the same type and order as the original two tensors. More generally, if $\mathbf{T}_1, \mathbf{T}_2, \ldots, \mathbf{T}_\mu$ are tensors of the same type and order and if $\lambda_1, \lambda_2, \ldots, \lambda_\mu$ are invariant scalars, then

$$\lambda_1 \mathbf{T}_1 + \lambda_2 \mathbf{T}_2 + \cdots + \lambda_\mu \mathbf{T}_\mu \tag{4.2b}$$

is a tensor of that same type and order.

Outer Product

The *outer product* of the tensors **S** and **T** of (*4.1*) is the tensor

$$[\mathbf{ST}] \equiv (S^{i_1 i_2 \ldots i_p}_{j_1 j_2 \ldots j_q} \cdot T^{k_1 k_2 \ldots k_r}_{l_1 l_2 \ldots l_s}) \tag{4.3}$$

which is of order $m = p + q + r + s$ (the sum of the orders of **S** and **T**), contravariant of order $p + r$ and covariant of order $q + s$. Note that $[\mathbf{ST}] = [\mathbf{TS}]$.

EXAMPLE 4.1 Given two tensors, $\mathbf{S} = (S^i_j)$ and $\mathbf{T} = (T_k)$, the outer product $[\mathbf{ST}] = (S^i_j T_k) \equiv (P^i_{jk})$ is a tensor because

$$\bar{P}^i_{jk} \equiv \bar{S}^i_j \bar{T}_k = \left(S^r_s \frac{\partial \bar{x}^i}{\partial x^r} \frac{\partial x^s}{\partial \bar{x}^j}\right)\left(T_u \frac{\partial x^u}{\partial \bar{x}^k}\right) = P^r_{su} \frac{\partial \bar{x}^i}{\partial x^r} \frac{\partial x^s}{\partial \bar{x}^j} \frac{\partial x^u}{\partial \bar{x}^k}$$

Inner Product

To take the inner product of two tensors, one equates an upper (contravariant) index of one tensor to a lower (covariant) index of the other, and sums products of components over the repeated index. In effect, the contravariant and covariant behaviors cancel out, which lowers the total order of the two tensors.

To state this more formally, set $i_\alpha = u = l_\beta$ in (*4.1*). Then the inner product corresponding to this pair of indices is

$$\mathbf{ST} \equiv (S^{i_1 \ldots u \ldots i_p}_{j_1 j_2 \ldots j_q} T^{k_1 k_2 \ldots k_r}_{l_1 \ldots u \ldots l_s}) \tag{4.4}$$

It is seen that there will exist $ps + rq$ inner products **ST** and **TS**; in general, all of these will be distinct. Each will be a tensor of order

$$m = p + q + r + s - 2$$

EXAMPLE 4.2 From the tensors $\mathbf{S} = (S^{ij})$ and $\mathbf{T} = (T_{klm})$, form the inner product $\mathbf{U} = (U^j_{km}) \equiv (S^{uj}T_{kum})$. We have:

$$\begin{aligned}\bar{U}^j_{km} &= \left(S^{pr}\frac{\partial \bar{x}^u}{\partial x^p}\frac{\partial \bar{x}^j}{\partial x^r}\right)\left(T_{sqt}\frac{\partial x^s}{\partial \bar{x}^k}\frac{\partial x^q}{\partial \bar{x}^u}\frac{\partial x^t}{\partial \bar{x}^m}\right) \\ &= S^{pr}T_{sqt}\left(\frac{\partial \bar{x}^u}{\partial x^p}\frac{\partial x^q}{\partial \bar{x}^u}\right)\frac{\partial \bar{x}^j}{\partial x^r}\frac{\partial x^s}{\partial \bar{x}^k}\frac{\partial x^t}{\partial \bar{x}^m} = S^{pr}T_{sqt}\,\delta^q_p\,\frac{\partial \bar{x}^j}{\partial x^r}\frac{\partial x^s}{\partial \bar{x}^k}\frac{\partial x^t}{\partial \bar{x}^m} \\ &= S^{pr}T_{spt}\frac{\partial \bar{x}^j}{\partial x^r}\frac{\partial x^s}{\partial \bar{x}^k}\frac{\partial x^t}{\partial \bar{x}^m} \equiv U^r_{st}\frac{\partial \bar{x}^j}{\partial x^r}\frac{\partial x^s}{\partial \bar{x}^k}\frac{\partial x^t}{\partial \bar{x}^m}\end{aligned}$$

which verifies that $\mathbf{U}$ is a tensor of order 3, contravariant of order 1 and covariant of order 2.

EXAMPLE 4.3 With (T_{ij}) and (T^{ij}) as in Theorem 3.2,

$$T^{iu}T_{uj} = \delta^i_j$$

As an inner product, the left side defines a second-order tensor that is contravariant of order one and covariant of order one. This constitutes a new proof (cf. Problem 3.29) of the tensor nature of the Kronecker delta.

In the special case when $\mathbf{S}$ is a contravariant vector and $\mathbf{T}$ is a covariant vector, the inner product $\mathbf{ST}$ is of the form S^iT_i, which is an invariant (Theorem 3.1). Because the tensor $\mathbf{ST}$ is of order

$$m = p + q + r + s - 2 = 1 + 0 + 0 + 1 - 2 = 0$$

an invariant is regarded as a tensor of order zero.

Contraction

Another order-reducing operation, like the inner product but applying to single tensors, is that of contracting a tensor on a pair of indices. In tensor $\mathbf{S}$ of *(4.1)* set $i_\alpha = u = j_\beta$ and sum on u; the resulting tensor (Problem 4.7),

$$\mathbf{S}' = (S^{i_1 \ldots u \ldots i_p}_{j_1 \ldots u \ldots j_q}) \tag{4.5}$$

is called a *contraction* of $\mathbf{S}$, with *contraction indices* i_α and j_β. $\mathbf{S}'$ is contravariant of order $p-1$ and covariant of order $q-1$.

Combined Operations

It is clear that one may form new tensors from old in a variety of ways by performing a sequence of the tensor operations discussed above. For example, one might form the outer product of two tensors, then take an inner product of this with a third tensor; or contract on one or more pairs of indices, either before or after taking a product. It is noteworthy that an inner product of two tensors may be characterized as a contraction of their outer product: $\mathbf{ST} = [\mathbf{ST}]'$. See Fig. 4-1.

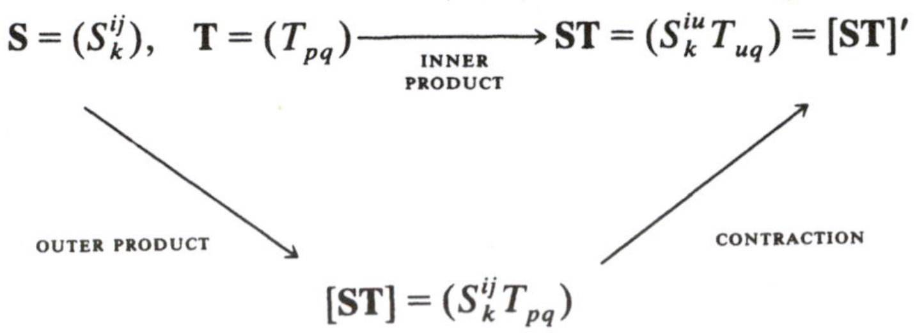

Fig. 4-1

4.2 TESTS FOR TENSOR CHARACTER

It is useful to have an alternative method for verifying tensor character that does not directly appeal to the tensor transformation laws. Roughly stated, the principle is this: If it can be shown that the inner product **TV** is a tensor for all vectors **V**, then **T** is a tensor. This idea is often referred to as the Quotient Rule for tensors; the official Quotient Theorem is our Theorem 4.2 below.

The following statements are useful criteria or "tests" for tensor character; they may all be derived as special cases of the Quotient Theorem.

(1) If $T_iV^i \equiv E$ is invariant for all contravariant vectors (V^i), then (T_i) is a covariant vector (tensor of order 1).

(2) If $T_{ij}V^i \equiv U_j$ are components of a covariant vector for all contravariant vectors (V^i), then (T_{ij}) is a covariant tensor of order 2.

(3) If $T_{ij}U^iV^j \equiv E$ is invariant for all contravariant vectors (U^i) and (V^i), then (T_{ij}) is a covariant tensor of order 2.

(4) If (T_{ij}) is symmetric and $T_{ij}V^iV^j \equiv E$ is invariant for all contravariant vectors (V^i), then (T_{ij}) is a covariant tensor of order 2.

EXAMPLE 4.4 Establish criterion (1).

Since E is invariant, $\bar{E} = E$, or $\bar{T}_i\bar{V}^i = T_iV^i$. Substitute in this equation the transformation law for (V^i) and change the dummy index on the right:

$$\bar{T}_i\left(V^j \frac{\partial \bar{x}^i}{\partial x^j}\right) = T_jV^j \qquad \text{or} \qquad \left(T_j - \bar{T}_i \frac{\partial \bar{x}^i}{\partial x^j}\right)V^j = 0$$

The latter equation must hold when (V^i) is any of the contravariant vectors represented in (x^i) by $(\delta^i_1), (\delta^i_2), \ldots, (\delta^i_n)$; their existence is guaranteed by Problem 3.5. Thus, for the kth of these vectors $(1 \leqq k \leqq n)$,

$$\left(T_k - \bar{T}_i \frac{\partial \bar{x}^i}{\partial x^k}\right)\cdot 1 = 0 \qquad \text{or} \qquad T_k = \bar{T}_i \frac{\partial \bar{x}^i}{\partial x^k}$$

which is the law of transformation—from $(\bar{x}^i)$ to (x^i)—of a covariant vector.

The method of Example 4.4 may be easily extended to establish the following result, which in turn implies the Quotient Theorem.

Lemma 4.1: If $T^{i_1i_2\ldots i_p}_{j_1j_2\ldots j_q}U^{(1)}_{i_1}U^{(2)}_{i_2}\cdots U^{(p)}_{i_p}V^{j_1}_{(1)}V^{j_2}_{(2)}\cdots V^{j_q}_{(q)} \equiv E$ is an invariant for arbitrary covariant vectors $(U^{(\alpha)}_{i_\alpha}) \equiv \mathbf{U}^{(\alpha)}$ $(\alpha = 1, 2, \ldots, p)$ and arbitrary contravariant vectors $(V^{j_\beta}_{(\beta)}) \equiv \mathbf{V}_{(\beta)}$ $(\beta = 1, 2, \ldots, q)$, then $(T^{i_1i_2\ldots i_p}_{j_1j_2\ldots j_q})$ is a tensor of the type indicated by its indices.

Theorem 4.2 (*Quotient Theorem*): If $T^{i_1i_2\ldots i_p}_{j_1j_2\ldots j_qk}V^k \equiv S^{i_1i_2\ldots i_p}_{j_1j_2\ldots j_q}$ are components of a tensor for an arbitrary contravariant vector (V^k), then $(T^{i_1i_2\ldots i_p}_{j_1j_2\ldots j_qj_{q+1}})$ is a tensor of the type and order indicated.

4.3 TENSOR EQUATIONS

Much of the importance of tensors in mathematical physics and engineering resides in the fact that *if a tensor equation or identity is true in one coordinate system, then it is true in all coordinate systems*.

EXAMPLE 4.5 Suppose that in some special coordinate system, (x^i), the covariant tensor $\mathbf{T} = (T_{ij})$ vanishes. The components of $\mathbf{T}$ in any other coordinate system, $(\bar{x}^i)$, are given by

$$\bar{T}_{ij} = T_{rs}\frac{\partial x^r}{\partial \bar{x}^i}\frac{\partial x^s}{\partial \bar{x}^j} = 0 + 0 + \cdots + 0 = 0$$

Therefore, $\mathbf{T} = \mathbf{0}$ in every coordinate system.

EXAMPLE 4.6 Consider a putative equation

$$R_{ijk}U^k = AW_i^{kl}M_{jk}I_l \tag{1}$$

connecting six entities that may or may not be tensors. If it can be shown that (i) $\mathbf{T} = (T_{ij}) \equiv (R_{ijk}U^k - AW_i^{kl}M_{jk}U_l)$ *is* a tensor, and (ii) a special coordinate system exists in which all T_{ij} are zero, then (*1*) is valid in every coordinate system.

EXAMPLE 4.7 A second-order covariant tensor, or a second-order contravariant tensor, that is known to be symmetric in one coordinate system must be symmetric in every coordinate system. (This statement does not extend to a second-order mixed tensor; see Problem 3.16.)

Another application of the principle yields (Problem 4.15) a useful fact in tensor analysis, often taken for granted:

Theorem 4.3: If (T_{ij}) is a covariant tensor of order two whose determinant vanishes in one particular coordinate system, then its determinant vanishes in all coordinate systems.

Corollary 4.4: A covariant tensor of order two that is invertible in one coordinate system is invertible in all coordinate systems.

Solved Problems

TENSOR SUMS

4.1 Show that if λ and μ are invariants and S^i and T^i are components of contravariant vectors, the vector defined in all coordinate systems by $(\lambda S^i + \mu T^i)$ is a contravariant vector.

Since $\bar{\lambda} = \lambda$ and $\bar{\mu} = \mu$,

$$\bar{\lambda}\bar{S}^i + \bar{\mu}\bar{T}^i = \lambda\left(S^r \frac{\partial \bar{x}^i}{\partial x^r}\right) + \mu\left(T^r \frac{\partial \bar{x}^i}{\partial x^r}\right) = (\lambda S^r + \mu T^r)\frac{\partial \bar{x}^i}{\partial x^r}$$

as desired.

4.2 Prove that (*a*) the array defined in each coordinate system by $(T_{ij} - T_{ji})$, where (T_{ij}) is a given covariant tensor, is a covariant tensor; (*b*) the array defined in each coordinate system by $(T^i_j - T^j_i)$, where (T^i_j) is a given mixed tensor, is not generally a tensor, but is a cartesian tensor.

(*a*) By (*4.2b*), the array is a tensor if and only if $(T^*_{ij}) \equiv (T_{ji})$ is a covariant tensor. But the transformation law for (T_{ij}) gives

$$\bar{T}_{ji} = T_{rs}\frac{\partial x^r}{\partial \bar{x}^j}\frac{\partial x^s}{\partial \bar{x}^i} \qquad \text{or} \qquad \bar{T}^*_{ij} = T^*_{sr}\frac{\partial x^s}{\partial \bar{x}^i}\frac{\partial x^r}{\partial \bar{x}^j}$$

which shows that (T^*_{ij}) is indeed a covariant tensor.

(*b*) We give a second proof [recall Problem 3.32(*a*)], based on (*4.2b*). The question is whether $(U^i_j) \equiv (T^j_i)$ is a tensor. From the transformation law for (T^i_j),

$$\bar{T}^j_i = T^r_s\frac{\partial \bar{x}^j}{\partial x^r}\frac{\partial x^s}{\partial \bar{x}^i} \qquad \text{or} \qquad \bar{U}^i_j = U^s_r\frac{\partial x^s}{\partial \bar{x}^i}\frac{\partial \bar{x}^j}{\partial x^r}$$

Thus, (U^i_j) does not obey a tensor law, unless, for all p, q,

$$\frac{\partial \bar{x}^p}{\partial x^q} = \frac{\partial x^q}{\partial \bar{x}^p} \qquad \text{or} \qquad J = (J^{-1})^T$$

i.e., unless the Jacobian matrix is orthogonal—as it is for orthogonal linear transformations (cartesian tensors).

OUTER PRODUCT

4.3 Show that the outer product of two contravariant vectors is a contravariant tensor of order two.

With (S^i) and (T^i) as the given vectors,

$$\bar{S}^i\bar{T}^j = \left(S^r \frac{\partial \bar{x}^i}{\partial x^r}\right)\left(T^s \frac{\partial \bar{x}^j}{\partial x^s}\right) = S^r T^s \frac{\partial \bar{x}^i}{\partial x^r}\frac{\partial \bar{x}^j}{\partial x^s}$$

which is the correct transformation law for the outer product to be a contravariant tensor of order two.

INNER PRODUCT

4.4 Prove that the inner product $(T^r U_{ir})$ is a tensor if (T^i) and (U_{ij}) are tensors of the types indicated.

With $V_j \equiv T^i U_{ji}$,

$$\bar{V}_j = \left(T^r \frac{\partial \bar{x}^i}{\partial x^r}\right)\left(U_{st} \frac{\partial x^s}{\partial \bar{x}^j}\frac{\partial x^t}{\partial \bar{x}^i}\right) = (T^r U_{st}\delta^t_r)\frac{\partial x^s}{\partial \bar{x}^j} = V_s \frac{\partial x^s}{\partial \bar{x}^j}$$

which is the desired transformation law.

4.5 Prove that if $\mathbf{g} = (g_{ij})$ is a covariant tensor of order two, and $\mathbf{U} = (U^i)$ and $\mathbf{V} = (V^i)$ are contravariant vectors, then the double inner product $\mathbf{gUV} = g_{ij}U^iV^j$ is an invariant.

The transformation laws are

$$\bar{g}_{ij} = g_{rs}\frac{\partial x^r}{\partial \bar{x}^i}\frac{\partial x^s}{\partial \bar{x}^j} \qquad \bar{U}^i = U^t \frac{\partial \bar{x}^i}{\partial x^t} \qquad \bar{V}^j = V^u \frac{\partial \bar{x}^j}{\partial x^u}$$

Multiply, and sum over i and j:

$$\bar{\mathbf{g}}\bar{\mathbf{U}}\bar{\mathbf{V}} = \bar{g}_{ij}\bar{U}^i\bar{V}^j = g_{rs}U^tV^u \frac{\partial x^r}{\partial \bar{x}^i}\frac{\partial x^s}{\partial \bar{x}^j}\frac{\partial \bar{x}^i}{\partial x^t}\frac{\partial \bar{x}^j}{\partial x^u} = g_{rs}U^tV^u\delta^r_t\delta^s_u = g_{rs}U^rV^s = \mathbf{gUV}$$

CONTRACTION

4.6 Assuming that contraction of a tensor yields a tensor, how many tensors may be created by repeated contraction of the tensor $\mathbf{T} = (T^{ij}_{kl})$?

Single contraction produces the four mixed tensors

$$(T^{uj}_{ul}) \qquad (T^{uj}_{ku}) \qquad (T^{iu}_{ul}) \qquad (T^{iu}_{ku})$$

and double contraction produces the two zero-order tensors (invariants) T^{uv}_{uv} and T^{uv}_{vu}. Thus there are six tensors, in general all distinct.

4.7 Show that any contraction of the tensor $\mathbf{T} = (T^i_{jk})$ results in a covariant vector.

We may contract on either $i = j$ or $i = k$. For $(S_k) \equiv (T^i_{ik})$, we have the transformation law

$$\bar{S}_k \equiv \bar{T}^i_{ik} = T^r_{st}\frac{\partial \bar{x}^i}{\partial x^r}\frac{\partial x^s}{\partial \bar{x}^i}\frac{\partial x^t}{\partial \bar{x}^k} = T^r_{st}\delta^s_r\frac{\partial x^t}{\partial \bar{x}^k} = T^r_{rt}\frac{\partial x^t}{\partial \bar{x}^k} = S_t\frac{\partial x^t}{\partial \bar{x}^k}$$

and, for $(U_j) \equiv (T^i_{ji})$,

$$\bar{U}_j \equiv \bar{T}^i_{ji} = T^r_{st}\frac{\partial \bar{x}^i}{\partial x^r}\frac{\partial x^s}{\partial \bar{x}^j}\frac{\partial x^t}{\partial \bar{x}^i} = T^r_{st}\delta^t_r\frac{\partial x^s}{\partial \bar{x}^j} = T^r_{sr}\frac{\partial x^s}{\partial \bar{x}^j} = U_s\frac{\partial x^s}{\partial \bar{x}^j}$$

In either case, the transformation law is that of a covariant vector.

COMBINED OPERATIONS

4.8 Suppose that $\mathbf{S} = (S^{ij}_k)$ and $\mathbf{T} = (T^i_j)$ are tensors from which a contravariant vector $\mathbf{V} = (V^i)$ is to be constructed using a combination of outer/inner products and contractions. (*a*) Show that there are six possibilities for **V**, which can all be distinct. (*b*) Verify that each possible **V** is obtainable as a contraction of an inner product **ST**.

(*a*) Writing $[\mathbf{ST}] \equiv \mathbf{U} = (U^{ijk}_{lm})$, we obtain the contravariant vectors as the double contractions of **U**:

$$(U^{uvk}_{uv}) \qquad (U^{uvk}_{vu}) \qquad (U^{ujv}_{uv}) \qquad (U^{ujv}_{vu}) \qquad (U^{iuv}_{uv}) \qquad (U^{iuv}_{vu})$$

(*b*) The vector $(U^{uvk}_{uv}) \equiv (S^{uv}_u T^k_v)$ may be obtained by first taking the inner product $(S^{iv}_l T^k_v)$ and then contracting on $i = u = l$. Likewise for the other five vectors of (*a*).

TESTS FOR TENSOR CHARACTER

4.9 Prove criterion (2) of Section 4.2 without invoking the Quotient Theorem.

We are to verify that (T_{ij}) is a covariant tensor of order two if it is given that for every contravariant vector (V^i), $T_{ij}V^i \equiv U_j$ are components of a covariant vector. Start out with the transformation law for (U_j) [from (x^i) to $(\bar{x}^i)$]:

$$\bar{U}_j = U_s \frac{\partial x^s}{\partial \bar{x}^j} \qquad \text{or} \qquad \bar{T}_{ij}\bar{V}^i = T_{is}V^i \frac{\partial x^s}{\partial \bar{x}^j}$$

Now substitute the transformation law for $\bar{V}^i$ [from $(\bar{x}^i)$ to (x^i)]:

$$\bar{T}_{ij}\bar{V}^i = T_{is}\left(\bar{V}^p \frac{\partial x^i}{\partial \bar{x}^p}\right) \frac{\partial x^s}{\partial \bar{x}^j}$$

Replace the dummy index i by p on the left and by r on the right:

$$\bar{T}_{pj}\bar{V}^p = T_{rs}\bar{V}^p \frac{\partial x^r}{\partial \bar{x}^p} \frac{\partial x^s}{\partial \bar{x}^j} \qquad \text{or} \qquad \left(\bar{T}_{pj} - T_{rs} \frac{\partial x^r}{\partial \bar{x}^p} \frac{\partial x^s}{\partial \bar{x}^j}\right)\bar{V}^p = 0$$

The proof is concluded as in Example 4.4.

4.10 Prove criterion (3) of Section 4.2.

Here we must show that (T_{ij}) is a covariant tensor, assuming that $T_{ij}U^iV^j$ is invariant. Using criterion (1), we conclude that $(T_{ij}U^i)$ is a covariant vector. Using criterion (2), it follows that since (U^i) is arbitrary, (T_{ij}) is a covariant tensor of order two, the desired conclusion.

4.11 Prove criterion (4) of Section 4.2.

We wish to show that if (T_{ij}) is a symmetric array such that $T_{ij}V^iV^j$ is an invariant for every contravariant vector (V^i), then (T_{ij}) is a (symmetric) covariant tensor of order two.

Let (U^i) and (V^i) denote arbitrary contravariant vectors and let $(W^i) \equiv (U^i + V^i)$, a contravariant vector by (*4.2a*). Then,

$$\begin{aligned} T_{ij}W^iW^j &\equiv T_{ij}(U^i + V^i)(U^j + V^j) \\ &= T_{ij}U^iU^j + T_{ij}V^iU^j + T_{ij}U^iV^j + T_{ij}V^iV^j \\ &= T_{ij}U^iU^j + T_{ij}V^iV^j + 2T_{ij}U^iV^j \end{aligned}$$

where the symmetry of (T_{ij}) has been used in the last step. Now, by hypothesis, the left-hand side and the first two terms of the right-hand side of the above identity are invariants. Therefore, $T_{ij}U^iV^j$ must be an invariant, and the desired conclusion follows from criterion (3).

4.12 Use Lemma 4.1 to write a proof of the Quotient Theorem, Theorem 4.2.

In the notation of the theorem and lemma, $S^{i_1 i_2 \cdots i_p}_{j_1 j_2 \cdots j_q} \cdot U^{(1)}_{i_1} U^{(2)}_{i_2} \cdots U^{(p)}_{i_p} V^{j_1}_{(1)} V^{j_2}_{(2)} \cdots V^{j_q}_{(q)}$ is a tensor of order zero, or an invariant, for arbitrary $\mathbf{U}^{(\alpha)}$ and $\mathbf{V}_{(\beta)}$; that is,

$$T^{i_1 i_2 \cdots i_p}_{j_1 j_2 \cdots j_q k} U^{(1)}_{i_1} U^{(2)}_{i_2} \cdots U^{(p)}_{i_p} V^{j_1}_{(1)} V^{j_2}_{(2)} \cdots V^{j_q}_{(q)} V^k$$

is an invariant, with (V^k) also arbitrary. It then follows from Lemma 4.1 (with q replaced by $q+1$) that $(T^{i_1 i_2 \cdots i_p}_{j_1 j_2 \cdots j_q k})$ is a tensor, contravariant of order p and covariant of order $q+1$.

From the above method of proof, it is clear that the Quotient Theorem is equally valid when the "divisor" is an arbitrary covariant vector. This form of the theorem will be used in Problem 4.13.

4.13 Use the Quotient Theorem to prove Theorem 3.2.

If $\mathbf{U} = (U^i)$ is a contravariant vector, the inner product

$$\mathbf{V} = \mathbf{T}\mathbf{U} \equiv (T_{ij} U^j)$$

is a covariant vector. Moreover, because $[T_{ij}]_{nn}$ has an inverse, it follows that as $\mathbf{U}$ runs through all contravariant vectors, $\mathbf{V}$ runs through all covariant vectors. Thus,

$$\mathbf{U} = \mathbf{T}^{-1}\mathbf{V} \equiv (T^{ij} V_j)$$

is a tensor for an arbitrary (V_i), making (T^{ij}) a contravariant tensor of order two.

TENSOR EQUATIONS

4.14 Prove that if (T^i_{jkl}) is a tensor such that, in the (x^i)-system, $T^i_{jkl} = 3T^i_{ljk}$, then $T^i_{jkl} = 3T^i_{ljk}$ in all coordinate systems.

We must prove that $\bar{T}^i_{jkl} = 3\bar{T}^i_{ljk}$ in $(\bar{x}^i)$. But

$$\begin{aligned} \bar{T}^i_{jkl} - 3\bar{T}^i_{ljk} &= T^p_{rst} \frac{\partial \bar{x}^i}{\partial x^p} \frac{\partial x^r}{\partial \bar{x}^j} \frac{\partial x^s}{\partial \bar{x}^k} \frac{\partial x^t}{\partial \bar{x}^l} - 3T^p_{rst} \frac{\partial \bar{x}^i}{\partial x^p} \frac{\partial x^r}{\partial \bar{x}^l} \frac{\partial x^s}{\partial \bar{x}^j} \frac{\partial x^t}{\partial \bar{x}^k} \\ &= T^p_{rst} \frac{\partial \bar{x}^i}{\partial x^p} \frac{\partial x^r}{\partial \bar{x}^j} \frac{\partial x^s}{\partial \bar{x}^k} \frac{\partial x^t}{\partial \bar{x}^l} - 3T^p_{trs} \frac{\partial \bar{x}^i}{\partial x^p} \frac{\partial x^t}{\partial \bar{x}^l} \frac{\partial x^r}{\partial \bar{x}^j} \frac{\partial x^s}{\partial \bar{x}^k} \\ &= (T^p_{rst} - 3T^p_{trs}) \frac{\partial \bar{x}^i}{\partial x^p} \frac{\partial x^r}{\partial \bar{x}^j} \frac{\partial x^s}{\partial \bar{x}^k} \frac{\partial x^t}{\partial \bar{x}^l} = 0 \end{aligned}$$

as desired.

4.15 Prove Theorem 4.3.

By Problem 3.14(*a*), the covariant transformation law has the matrix expression

$$\bar{T} = \bar{J}^T T \bar{J} \qquad \text{whence} \qquad |\bar{T}| = \bar{\mathscr{J}}^2 |T|$$

Thus, $|T| = 0$ implies $|\bar{T}| = 0$.

4.16 Prove that if a mixed tensor (T^i_j) can be expressed as the outer product of contravariant and covariant vectors (U^i) and (V_j) in one coordinate system, then (T^i_j) is the outer product of those vectors in general.

We must prove that $\bar{T}^i_j = \bar{U}^i \bar{V}_j$ for any admissible coordinate system $(\bar{x}^i)$. But, by hypothesis,

$$\bar{T}^i_j - \bar{U}^i \bar{V}_j = T^r_s \frac{\partial \bar{x}^i}{\partial x^r} \frac{\partial x^s}{\partial \bar{x}^j} - \left(U^r \frac{\partial \bar{x}^i}{\partial x^r}\right)\left(V_s \frac{\partial x^s}{\partial \bar{x}^j}\right) = (T^r_s - U^r V_s) \frac{\partial \bar{x}^i}{\partial x^r} \frac{\partial x^s}{\partial \bar{x}^j} = 0$$

Supplementary Problems

4.17 If (U^i) and (V^i) are contravariant vectors, verify that $(2U^i + 3V^i)$ is also a contravariant vector.

4.18 Verify that the outer product of a contravariant vector and a covariant vector is a mixed tensor of order two.

4.19 How many potentially different mixed tensors of order two can be defined by taking the outer product of $\mathbf{S} = (S^{ij}_k)$ and $\mathbf{T} = (T^i_{jk})$, then contracting twice?

4.20 Show that if T^{ij}_{kl} are tensor components, T^{ij}_{ij} is an invariant.

4.21 Prove that if $T^i_{jkl}U^j \equiv S^i_{kl}$ are components of a tensor for any contravariant vector (U^j), then (T^i_{jkl}) is a tensor of the indicated type. [*Hint*: Apply the Quotient Theorem to $(M^i_{klj}) \equiv (T^i_{jkl})$. More generally, the Quotient Theorem is valid for all choices of the inner product.]

4.22 Prove that if $T^i_{jkl}S^{kl} \equiv U^i_j$ are tensor components for arbitrary contravariant tensors (S^{kl}), then (T^i_{jkl}) is a tensor of the indicated type. [*Hint*: Follow Problem 4.9.]

4.23 Prove that if $T^i_{jkl}U^kU^l \equiv V^i_j$ are components of a tensor for an arbitrary contravariant vector (U^i), and if (T^i_{jkl}) is symmetric in the last two lower indices in all coordinate systems, then (T^i_{jkl}) is a tensor of the type indicated.

4.24 Show that Theorem 4.3 and Corollary 4.4 are equivalent.

4.25 Prove the assertion of Example 4.7.

4.26 Prove that if an invariant E can be expressed as the inner product of vectors (U_i) and (V^i) in one coordinate system, then E has that representation in any coordinate system.

Chapter 5

The Metric Tensor

5.1 INTRODUCTION

The notion of distance (or *metric*) is fundamental in applied mathematics. Frequently, the distance concept most useful in a particular application is non-Euclidean (under which the Pythagorean relation for geodesic right triangles is not valid). Tensor calculus provides a natural tool for the investigation of general formulations of distance; it studies not only non-Euclidean metrics but also the forms assumed by the Euclidean metric in particular coordinate systems.

Calculus texts often contain derivations of arc-length formulas for polar coordinates that apparently apply only to that one coordinate system; here we develop a concise method for obtaining the arc-length formula for any admissible coordinate system. The theory culminates in later chapters with a method for distinguishing between a metric that is genuinely non-Euclidean and one that is Euclidean but disguised by the peculiarities of a particular system of coordinates.

5.2 ARC LENGTH IN EUCLIDEAN SPACE

The classical expressions from calculus for arc length in various coordinate systems lead to a general formula of the type

$$L = \int_a^b \sqrt{\left| g_{ij} \frac{dx^i}{dt} \frac{dx^j}{dt} \right|}\, dt \tag{5.1a}$$

where $g_{ij} = g_{ij}(x^1, x^2, \ldots, x^n) = g_{ji}$ are functions of the coordinates and L gives the length of the arc $a \leqq t \leqq b$ of the curve $x^i = x^i(t) \quad (1 \leqq i \leqq n)$.

EXAMPLE 5.1 The arc-length formula for Euclidean three-space in a rectangular coordinate system (x^1, x^2, x^3) may be recalled:

$$L = \int_a^b \sqrt{\left(\frac{dx^1}{dt}\right)^2 + \left(\frac{dx^2}{dt}\right)^2 + \left(\frac{dx^3}{dt}\right)^2}\, dt = \int_a^b \sqrt{\delta_{ij} \frac{dx^i}{dt} \frac{dx^j}{dt}}\, dt$$

This is (*5.1a*), with $g_{ij} = \delta_{ij}$.

The formula in Example 5.1 has the equally informative differential form

$$ds^2 = (dx^1)^2 + (dx^2)^2 + (dx^3)^2 = \delta_{ij}\, dx^i\, dx^j$$

More generally, (*5.1a*) is equivalent to

$$\pm ds^2 = g_{ij}\, dx^i\, dx^j \tag{5.1b}$$

EXAMPLE 5.2 For convenient reference, formulas for the Euclidean metric in the nonrectangular coordinate systems heretofore considered are collected below.

Polar coordinates: $(x^1, x^2) = (r, \theta)$; Fig. 3-1.

$$ds^2 = (dx^1)^2 + (x^1)^2(dx^2)^2 \tag{5.2}$$

Cylindrical coordinates: $(x^1, x^2, x^3) = (r, \theta, z)$; Fig. 3-2.

$$ds^2 = (dx^1)^2 + (x^1)^2(dx^2)^2 + (dx^3)^2 \tag{5.3}$$

Spherical coordinates: $(x^1, x^2, x^3) = (\rho, \varphi, \theta)$; Fig. 3-3.

$$ds^2 = (dx^1)^2 + (x^1)^2(dx^2)^2 + (x^1 \sin x^2)^2(dx^3)^2 \tag{5.4}$$

Affine coordinates: (see Fig. 5-1).

$$ds^2 = (dx^1)^2 + (dx^2)^2 + (dx^3)^2 + 2\cos\alpha\, dx^1\, dx^2 + 2\cos\beta\, dx^1\, dx^3 + 2\cos\gamma\, dx^2\, dx^3 \tag{5.5}$$

Formula (*5.5*) is derived in Problem 5.9. Note that the matrix (g_{ij}) defining the Euclidean metric is nondiagonal in affine coordinates.

Although formulated for Euclidean space, (*5.1*) is extended, in the section that follows, to provide the distance concept for non-Euclidean spaces as well.

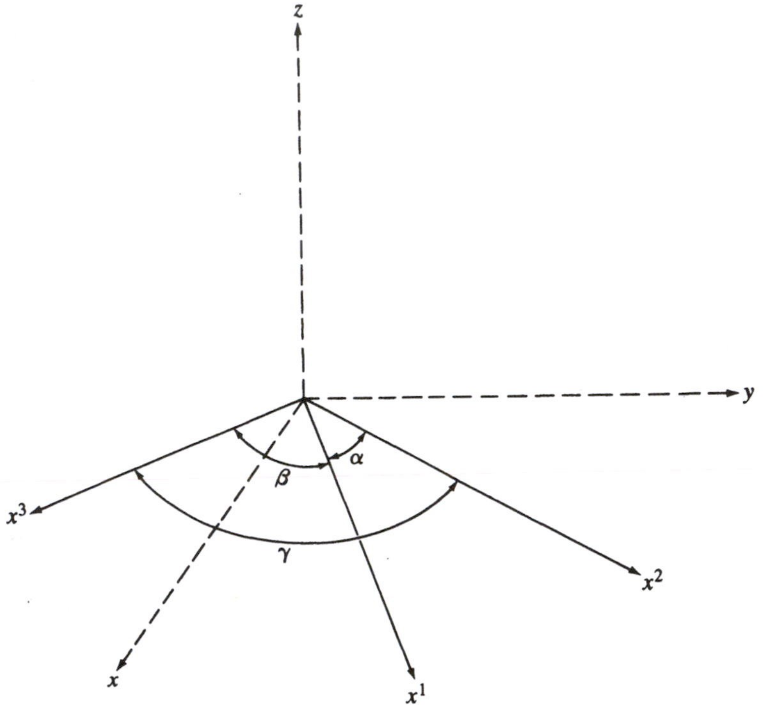

Fig. 5-1

5.3 GENERALIZED METRICS; THE METRIC TENSOR

Assume that a matrix field $\mathbf{g} = (g_{ij})$ exists satisfying in all (admissible) coordinate systems (x^i) and in some (open) region of space:

A. $\mathbf{g}$ is of differentiability class C^2 (i.e., all second-order partial derivatives of the g_{ij} exist and are continuous).

B. $\mathbf{g}$ is symmetric (i.e., $g_{ij} = g_{ji}$).

C. $\mathbf{g}$ is nonsingular (i.e., $|g_{ij}| \neq 0$).

D. The differential form (*5.1b*), and hence the distance concept generated by $\mathbf{g}$, is invariant with respect to a change of coordinates.

Sometimes, particularly in geometric applications of tensors, a property stronger than C above is assumed:

C′. $\mathbf{g}$ is *positive definite* [i.e., $g_{ij}v^iv^j > 0$ for all nonzero vectors $\mathbf{v} = (v^1, v^2, \ldots, v^n)$].

Under property C′, $|g_{ij}|$ and $g_{11}, g_{22}, \ldots, g_{nn}$ are all positive. Furthermore, the inverse matrix field $\mathbf{g}^{-1}$ is also positive definite.

For later use we define the *arc-length parameter* for a curve $\mathscr{C} \ : \ x^i = x^i(t) \quad (a \leqq t \leqq b)$:

$$s(t) = \int_a^t \sqrt{\varepsilon g_{ij} \frac{dx^i}{du} \frac{dx^j}{du}}\, du \tag{5.6a}$$

where $\varepsilon = +1$ or -1 according as

$$g_{ij} \frac{dx^i}{du} \frac{dx^j}{du} \geqq 0 \qquad \text{or} \qquad g_{ij} \frac{dx^i}{du} \frac{dx^j}{du} < 0$$

The functional ε is called the *indicator* of the vector (dx^i/du) relative to the metric (g_{ij}). One can, of course, use absolute value signs instead of the indicator, but the latter notation works better in algebraic manipulations. In terms of the arc-length parameter, the length of $\mathscr{C}$ is $L = s(b)$.

Differentiating (*5.6a*) and squaring yields the equivalent formula

$$\left(\frac{ds}{dt}\right)^2 = \varepsilon g_{ij} \frac{dx^i}{dt} \frac{dx^j}{dt} \tag{5.6b}$$

Finally, introducing the differentials

$$dx^i \equiv \frac{dx^i(t)}{dt}\, dt$$

the values of which are independent of the choice of curve parameter, we retrieve (*5.1b*) as

$$\varepsilon\, ds^2 = g_{ij}\, dx^i\, dx^j \tag{5.6c}$$

EXAMPLE 5.3 Suppose that on $\mathbf{R}^3$ a matrix field is given in (x^i) by

$$(g_{ij}) = \begin{bmatrix} (x^1)^2 - 1 & 1 & 0 \\ 1 & (x^2)^2 & 0 \\ 0 & 0 & \dfrac{64}{9} \end{bmatrix} \qquad \text{where} \qquad [(x^1)^2 - 1](x^2)^2 \neq 1$$

(*a*) Show that, if extended to all admissible coordinate systems according to the transformation law for covariant tensors, this matrix field is a metric; i.e., it satisfies properties A–D above. (*b*) For this metric, compute the arc-length parameter and the length of the curve

$$\mathscr{C} \ : \ \begin{cases} x^1 = 2t - 1 \\ x^2 = 2t^2 \\ x^3 = t^3 \end{cases} \qquad (0 \leqq t \leqq 1)$$

(*a*) Property A obtains since g_{ij} is a polynomial in x^1 and x^2 for each i, j. Since the matrix (g_{ij}) is symmetric, property B holds. Since

$$|g_{ij}| = \frac{64}{9} \begin{vmatrix} (x^1)^2 - 1 & 1 \\ 1 & (x^2)^2 \end{vmatrix} = \frac{64}{9} \{(x^2)^2[(x^1)^2 - 1] - 1\} \neq 0$$

property C obtains. Property D follows from Problem 4.5.

(*b*) It is convenient, here and later, to rewrite (*5.6b*) as the matrix product

$$\varepsilon \left(\frac{ds}{dt}\right)^2 = \left(\frac{dx^i}{dt}\right)^T (g_{ij}) \left(\frac{dx^i}{dt}\right) \tag{5.6d}$$

Along the given curve, this becomes

$$\varepsilon \left(\frac{ds}{dt}\right)^2 = [2 \quad 4t \quad 3t^2] \begin{bmatrix} (2t-1)^2 - 1 & 1 & 0 \\ 1 & (2t^2)^2 & 0 \\ 0 & 0 & \dfrac{64}{9} \end{bmatrix} \begin{bmatrix} 2 \\ 4t \\ 3t^2 \end{bmatrix}$$

$$= 64t^6 + 64t^4 + 16t^2 = (8t^3 + 4t)^2$$

Hence, $\varepsilon = 1$ and

$$s(t) = \int_0^t (8u^3 + 4u)\,du = [2u^4 + 2u^2]_0^t = 2t^4 + 2t^2$$

from which $L = 2(1)^4 + 2(1)^2 = 4$.

The properties postulated of **g** make it a tensor, the so-called *fundamental* or *metric tensor*. In fact, property D ensures that

$$g_{ij}V^iV^j \equiv E$$

is an invariant for every contravariant vector $(V^i) = (dx^i/dt)$. (By solving an ordinary differential equation, one can exhibit the curve that possesses a given tangent vector.) Then, in view of property B, criterion (4) of Section 4.2 implies

Theorem 5.1: The metric $\mathbf{g} = (g_{ij})$ is a covariant tensor of the second order.

In Problem 3.14(*a*), the matrix equation $U = J^T\bar{U}J$ was found for the transformation of a second-order covariant tensor **U**. If $(\bar{x}^i)$ is a rectangular system and $\mathbf{U} = \mathbf{g}$ is the Euclidean metric tensor, then in (x^i), $U = G$, and in $(\bar{x}^i)$, $\bar{U} = \bar{G} = I$; thus we have proved

Theorem 5.2: If the Jacobian matrix of the transformation from a given coordinate system (x^i) to a rectangular system $(\bar{x}^i)$ is $J = (\partial\bar{x}^i/\partial x^j)$, then the matrix $G \equiv (g_{ij})$ of the Euclidean metric tensor in the (x^i)-system is given by

$$G = J^TJ \tag{5.7}$$

Remark 1: Equation (*5.7*) illustrates the following well-known result of matrix theory: Any symmetric, positive definite matrix A has a nonsingular "square root" C such that $A = C^TC$.

It should be emphasized that *only* the Euclidean metric admits of a representation of the form (*5.7*). For, by very definition, if **g** is non-Euclidean, there exists no coordinate system $(\bar{x}^i)$ in which $\bar{G} = I$.

EXAMPLE 5.4 Cylindrical coordinates (x^i) and rectangular coordinates $(\bar{x}^i)$ are connected through

$$\bar{x}^1 = x^1\cos x^2 \qquad \bar{x}^2 = x^1\sin x^2 \qquad \bar{x}^3 = x^3$$

Thus

$$J = \begin{bmatrix} \cos x^2 & -x^1\sin x^2 & 0 \\ \sin x^2 & x^1\cos x^2 & 0 \\ 0 & 0 & 1 \end{bmatrix}$$

and the (Euclidean) metric for cylindrical coordinates is given by

$$G = J^TJ = \begin{bmatrix} \cos x^2 & \sin x^2 & 0 \\ -x^1\sin x^2 & x^1\cos x^2 & 0 \\ 0 & 0 & 1 \end{bmatrix}\begin{bmatrix} \cos x^2 & -x^1\sin x^2 & 0 \\ \sin x^2 & x^1\cos x^2 & 0 \\ 0 & 0 & 1 \end{bmatrix}$$

$$= \begin{bmatrix} 1 & 0 & 0 \\ 0 & (x^1)^2 & 0 \\ 0 & 0 & 1 \end{bmatrix}$$

or $g_{11} = g_{33} = 1$, $g_{22} = (x^1)^2$, and $g_{ij} = 0$ for $i \neq j$. These results verify (*5.3*).

In spite of the apparent restriction to the Euclidean distance concept, in connection with such results as Theorem 5.2, the reader should keep in mind that one is free to choose as the metric tensor for $\mathbf{R}^n$ any **g** that obeys properties A–D above. For instance, it can be shown by methods to be developed later that the metric chosen in Example 5.3 is non-Euclidean.

5.4 CONJUGATE METRIC TENSOR; RAISING AND LOWERING INDICES

One of the fundamental concepts of tensor calculus resides in the "raising" or "lowering" of indices in tensors. If we are given a contravariant vector (T^i) and if, for the moment, (g_{ij}) represents any covariant tensor of the second order, then we know (Problem 4.4) that the inner product $(S_i) = (g_{ij}T^j)$ is a covariant vector. Now, if (g_{ij}) is in fact the metric tensor whereby distance in $\mathbf{R}^n$ is defined, it will prove useful in many contexts to consider (S_i) and (T^i) as covariant and contravariant aspects of a single notion. Thus, we write T_i instead of S_i:

$$T_i = g_{ij}T^j$$

and say that taking the inner product with the metric tensor has *lowered* a contravariant index to a covariant index.

Because the matrix (g_{ij}) is invertible (property C of Section 5.3), the above relation is equivalent to

$$T^i = g^{ij}T_j$$

where $(g^{ij}) = (g_{ij})^{-1}$; now we say that a covariant index has been *raised* to a contravariant index.

Definition 1: The inverse of the fundamental matrix field (metric tensor),

$$[g^{ij}]_{nn} = [g_{ij}]_{nn}^{-1}$$

is called the *conjugate metric tensor*.

Both metric tensors are freely applied to create new, more covariant ($\mathbf{g}$) or more contravariant ($\mathbf{g}^{-1}$) counterparts to given tensors. Thus, starting with the mixed tensor (T^{ij}_k),

$$T^{ijk} \equiv g^{ir}T_r^{jk}$$
$$T^j_{ik} \equiv g_{ir}T^{jr}_k$$

and

$$T_{ijk} \equiv g_{is}T^s_{jk} \equiv g_{is}g_{jr}T^{sr}_k.$$

5.5 GENERALIZED INNER-PRODUCT SPACES

Suppose that a metric $\mathbf{g}$ has been imposed on $\mathbf{R}^n$ and that $\mathbf{U}$ and $\mathbf{V}$ are two vectors on the metric space. It is essential to the definition of a geometrically significant *inner product* $\mathbf{UV}$ that its value depend only on the vectors $\mathbf{U}$ and $\mathbf{V}$, and not on the particular coordinate system used to specify these vectors. (There are other requirements on an inner product, but they are secondary.) This fact motivates

Definition 2: To each pair of contravariant vectors $\mathbf{U} = (U^i)$ and $\mathbf{V} = (V^i)$ is associated the real number

$$\mathbf{UV} \equiv g_{ij}U^iV^j \equiv U^iV_i \equiv U_iV^i \tag{5.8}$$

called the (*generalized*) *inner product* of $\mathbf{U}$ and $\mathbf{V}$.

In similar fashion, the inner product of two covariant vectors is defined as

$$\mathbf{UV} \equiv g^{ij}U_iV_j \equiv U^iV_i \equiv U_iV^i \tag{5.9}$$

consistent with (*5.8*). We therefore have the rule: *To obtain the inner product of two vectors of the same type, convert one vector to the opposite type and then take the tensor inner product.*

Remark 2: It follows from Problem 4.5—or, more fundamentally, from property D of $\mathbf{g}$—that the inner product (*5.8*) or (*5.9*) is an invariant, as required.

According to (*4.2*), the set of all contravariant vectors on $\mathbf{R}^n$ is a vector space, as is the set of all covariant vectors on $\mathbf{R}^n$. With an inner product as defined above, these vector spaces become (*generalized*) *inner-product spaces*.

5.6 CONCEPTS OF LENGTH AND ANGLE

Expressions (*2.7*) and (*2.8*) readily extend to a generalized inner-product space, *provided the metric is positive definite*. The *norm* (or *length*) of an arbitrary vector $\mathbf{V}=(V^i)$ or $\mathbf{V}=(V_i)$ is the nonnegative real number

$$\|\mathbf{V}\| \equiv \sqrt{\mathbf{V}^2} = \sqrt{V_iV^i} \tag{5.10}$$

Remark 3: The norm of a vector—and thus the notion of a *normed linear space*—can be defined abstractly (see Problem 5.14), without reference to an inner product.

EXAMPLE 5.5 Show that under the Euclidean metric (*5.2*) for polar coordinates, the vectors

$$(U^i)=(3/5, 4/5x^1) \qquad \text{and} \qquad (V^i)=(-4/5, 3/5x^1)$$

are orthonormal.

Using matrices, we have:

$$\|\mathbf{U}\|^2 = U^iU_i = g_{ij}U^iU^j = \begin{bmatrix} \frac{3}{5} & \frac{4}{5x^1} \end{bmatrix} \begin{bmatrix} 1 & 0 \\ 0 & (x^1)^2 \end{bmatrix} \begin{bmatrix} \frac{3}{5} \\ \frac{4}{5x^1} \end{bmatrix}$$

$$= \begin{bmatrix} \frac{3}{5} & \frac{4}{5x^1} \end{bmatrix} \begin{bmatrix} \frac{3}{5} \\ \frac{4x^1}{5} \end{bmatrix}$$

$$= \frac{9}{25} + \frac{16x^1}{25x^1} = 1$$

or $\|\mathbf{U}\|=1$; likewise, $\|\mathbf{V}\|=1$. Now we verify that the vectors are orthogonal:

$$\mathbf{UV} = g_{ij}U^iV^j = \begin{bmatrix} \frac{3}{5} & \frac{4}{5x^1} \end{bmatrix} \begin{bmatrix} 1 & 0 \\ 0 & (x^1)^2 \end{bmatrix} \begin{bmatrix} -\frac{4}{5} \\ \frac{3}{5x^1} \end{bmatrix}$$

$$= \begin{bmatrix} \frac{3}{5} & \frac{4}{5x^1} \end{bmatrix} \begin{bmatrix} -\frac{4}{5} \\ \frac{3x^1}{5} \end{bmatrix}$$

$$= -\frac{12}{25} + \frac{12x^1}{25x^1} = 0$$

Both normality and orthogonality depend, of course, on the metric alone, and not on the (polar) coordinate system.

The *angle θ between two non-null contravariant vectors* **U** and **V** is defined by

$$\cos\theta \equiv \frac{\mathbf{UV}}{\|\mathbf{U}\|\,\|\mathbf{V}\|} = \frac{g_{ij}U^iV^j}{\sqrt{g_{pq}U^pU^q}\,\sqrt{g_{rs}V^rV^s}} \qquad (0 \leqq \theta \leqq \pi) \tag{5.11}$$

That θ is well-defined follows from the *Cauchy–Schwarz inequality*, which may be written in the form

$$-1 \leqq \frac{\mathbf{UV}}{\|\mathbf{U}\|\,\|\mathbf{V}\|} \leqq 1$$

(see Problem 5.13).

The tangent field to a family of smooth curves is a contravariant vector (Example 3.4), so that (*5.11*) yields the geometrical

Theorem 5.3: In a general coordinate system, if (U^i) and (V^i) are the tangent vectors to two families of curves, then the families are mutually orthogonal if and only if $g_{ij}U^iV^j = 0$.

EXAMPLE 5.6 Show that each member of the family of curves given in polar coordinates by

$$e^{1/r} = a\,(\sec\,\theta + \tan\,\theta) \qquad (a \geqq 0) \tag{1}$$

is orthogonal to each of the curves (limaçons of Pascal)

$$r = \sin\theta + c \qquad (c \geqq 0) \tag{2}$$

(Figure 5-2 indicates the orthogonality of the curve $a = 1$ to the family of limaçons.)

In polar coordinates $x^1 = r$, $x^2 = \theta$, and with curve parameter t, (*1*) becomes—after taking logarithms—

$$\frac{1}{x^1} = \ln a + \ln|\sec t + \tan t| \qquad x^2 = t \tag{1'}$$

With curve parameter u, (*2*) becomes

$$x^1 = \sin u + c \qquad x^2 = u \tag{2'}$$

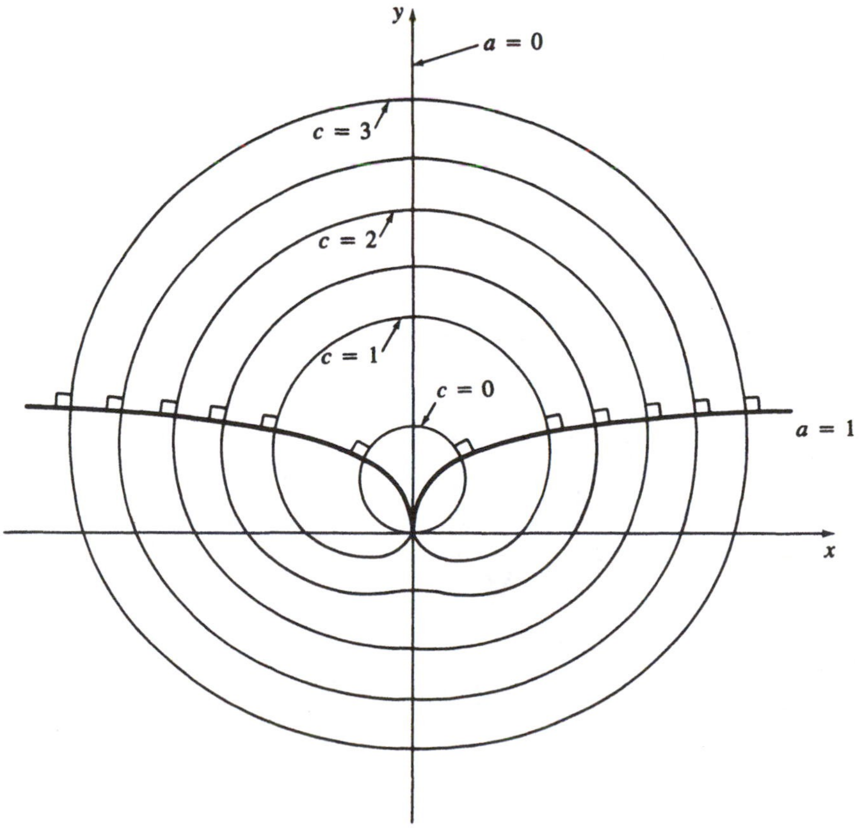

Fig. 5-2

Differentiation of (*1′*) with respect to t yields

$$-\frac{1}{(x^1)^2}\frac{dx^1}{dt} = \sec t = \sec x^2 \qquad \frac{dx^2}{dt} = 1$$

so that the tangent vector to family (*1′*) is

$$(U^1, U^2) = (-(x^1)^2 \sec x^2, 1)$$

Similarly, the tangent vector to family (*2′*) is

$$(V^1, V^2) = (\cos u, 1) = (\cos x^2, 1)$$

Applying Theorem 5.3, with the Euclidean metric tensor in polar coordinates,

$$\begin{aligned} g_{ij}U^iV^j &= g_{11}U^1V^1 + g_{22}U^2V^2 + 0 \\ &= (1)[-(x^1)^2 \sec x^2](\cos x^2) + (x^1)^2(1)(1) \\ &= -(x^1)^2 + (x^1)^2 = 0 \end{aligned}$$

Observe that nonparametric forms of the tangent vectors are used in the orthogonality condition. This is necessary because the metric tensor at the intersection *point* (x^1, x^2) of a curve (*1*) and a curve (*2*) depends on neither the parameter t *along* (*1*) nor the parameter u *along* (*2*).

Solved Problems

ARC LENGTH

5.1 A curve is given in spherical coordinates (x^i) by

$$x^1 = t \qquad x^2 = \arcsin\frac{1}{t} \qquad x^3 = \sqrt{t^2 - 1}$$

Find the length of the arc $1 \leqq t \leqq 2$.

By (*5.4*),

$$\left(\frac{ds}{dt}\right)^2 = \left(\frac{dx^1}{dt}\right)^2 + (x^1)^2\left(\frac{dx^2}{dt}\right)^2 + (x^1 \sin x^2)^2\left(\frac{dx^3}{dt}\right)^2$$

so we first calculate the $(dx^i/dt)^2$:

$$\left(\frac{dx^1}{dt}\right)^2 = 1 \qquad \left(\frac{dx^2}{dt}\right)^2 = \left(\frac{-1/t^2}{\sqrt{1-(1/t)^2}}\right)^2 = \frac{1}{t^2(t^2-1)} \qquad \left(\frac{dx^3}{dt}\right)^2 = \left(\frac{1}{2}\frac{2t}{\sqrt{t^2-1}}\right)^2 = \frac{t^2}{t^2-1}$$

Then

$$\left(\frac{ds}{dt}\right)^2 = 1 + t^2 \cdot \frac{1}{t^2(t^2-1)} + \left(t \cdot \frac{1}{t}\right)^2 \cdot \frac{t^2}{t^2-1} = \frac{2t^2}{t^2-1}$$

and (*5.1a*) gives

$$L = \int_1^2 \frac{\sqrt{2}\,t}{\sqrt{t^2-1}}\,dt = \sqrt{2(t^2-1)}\,\Big]_1^2 = \sqrt{6}$$

5.2 Find the length of the curve

$$\mathscr{C} : \begin{cases} x^1 = 1 \\ x^2 = t \end{cases} \qquad (1 \leqq t \leqq 2)$$

if the metric is that of the hyperbolic plane $(x^2 > 0)$:

$$g_{11} = g_{22} = \frac{1}{(x^2)^2} \qquad g_{12} = g_{21} = 0$$

Since $(dx^i/dt) = (0, 1)$, (*5.6d*) yields ($\varepsilon = 1$)

$$\left(\frac{ds}{dt}\right)^2 = [0 \quad 1]\begin{bmatrix} \frac{1}{t^2} & 0 \\ 0 & \frac{1}{t^2} \end{bmatrix}\begin{bmatrix} 0 \\ 1 \end{bmatrix} = \frac{1}{t^2}$$

and

$$L = \int_1^2 \frac{1}{t}\, dt = \ln 2$$

GENERALIZED METRICS

5.3 Is the form $dx^2 + 3\, dx\, dy + 4\, dy^2 + dz^2$ positive definite?

It must be determined whether the polynomial $Q \equiv (u^1)^2 + 3u^1u^2 + 4(u^2)^2 + (u^3)^2$ is positive unless $u^1 = u^2 = u^3 = 0$. By completing the square,

$$Q = (u^1)^2 + 3u^1u^2 + \frac{9}{4}(u^2)^2 + \frac{7}{4}(u^2)^2 + (u^3)^2 = \left(u^1 + \frac{3}{2}u^2\right)^2 + \frac{7}{4}(u^2)^2 + (u^3)^2$$

All terms are perfect squares with positive coefficients; hence the form is indeed positive definite.

5.4 Show that the formula (*5.1a*) for arc length does not depend on the particular parameterization of the curve.

Given a curve $\mathscr{C}$: $x^i = x^i(t)$ ($a \leqq t \leqq b$), suppose that $x^i = x^i(\bar{t})$ ($\bar{a} \leqq \bar{t} \leqq \bar{b}$) is a different parameterization, where $\bar{t} = \phi(t)$, with $\phi'(t) > 0$ and $\bar{a} = \phi(a)$, $\bar{b} = \phi(b)$. Then, by the chain rule and substitution rule for integrals,

$$L = \int_a^b \sqrt{\left| g_{ij}\frac{dx^i}{dt}\frac{dx^j}{dt}\right|}\, dt = \int_a^b \sqrt{\left| g_{ij}\frac{dx^i}{d\bar{t}}\frac{dx^j}{d\bar{t}}(\phi'(t))^2\right|}\, dt$$

$$= \int_a^b \sqrt{\left| g_{ij}\frac{dx^i}{d\bar{t}}\frac{dx^j}{d\bar{t}}\right|}\, \phi'(t)\, dt = \int_{\bar{a}}^{\bar{b}} \sqrt{\left| g_{ij}\frac{dx^i}{d\bar{t}}\frac{dx^j}{d\bar{t}}\right|}\, d\bar{t} = \bar{L}$$

TENSOR PROPERTY OF THE METRIC

5.5 Find the Euclidean metric tensor (in matrix form) for spherical coordinates, using Theorem 5.2.

Since spherical coordinates (x^i) are connected to rectangular coordinates $(\bar{x}^i)$ via

$$\bar{x}^1 = x^1 \sin x^2 \cos x^3 \qquad \bar{x}^2 = x^1 \sin x^2 \sin x^3 \qquad \bar{x}^3 = x^1 \cos x^2$$

we have

$$J^TJ = \begin{bmatrix} \sin x^2 \cos x^3 & \sin x^2 \sin x^3 & \cos x^2 \\ x^1 \cos x^2 \cos x^3 & x^1 \cos x^2 \sin x^3 & -x^1 \sin x^2 \\ -x^1 \sin x^2 \sin x^3 & x^1 \sin x^2 \cos x^3 & 0 \end{bmatrix}\begin{bmatrix} \sin x^2 \cos x^3 & x^1 \cos x^2 \cos x^3 & -x^1 \sin x^2 \sin x^3 \\ \sin x^2 \sin x^3 & x^1 \cos x^2 \sin x^3 & x^1 \sin x^2 \cos x^3 \\ \cos x^2 & -x^1 \sin x^2 & 0 \end{bmatrix}$$

Since $G = J^TJ$ is known to be symmetric (see Problem 2.4), we need only compute the elements on or above the main diagonal:

$$G = \begin{bmatrix} (\sin^2 x^2)(1) + \cos^2 x^2 & (x^1 \sin x^2 \cos x^2)(1) - x^1 \sin x^2 \cos x^2 & g_{13} \\ g_{21} & ((x^1)^2 \cos^2 x^2)(1) + (x^1)^2 \sin^2 x^2 & g_{23} \\ g_{31} & g_{32} & ((x^1)^2 \sin^2 x^2)(1) \end{bmatrix}$$

where

$$g_{13} = (x^1 \sin^2 x^2)(-\sin x^3 \cos x^3 + \cos x^3 \sin x^3) = 0$$
$$g_{23} = ((x^1)^2 \sin x^2 \cos x^2)(-\cos x^3 \sin x^3 + \sin x^3 \cos x^3) = 0$$

Hence
$$G = \begin{bmatrix} 1 & 0 & 0 \\ 0 & (x^1)^2 & 0 \\ 0 & 0 & (x^1 \sin x^2)^2 \end{bmatrix}$$

5.6 Find the components g_{ij} of the Euclidean metric tensor in the special coordinate system (x^i) defined from rectangular coordinates $(\bar{x}^i)$ by $x^1 = \bar{x}^1$, $x^2 = \exp(\bar{x}^2 - \bar{x}^1)$.

We must compute $J^T J$, where J is the Jacobian matrix of the transformation $\bar{x}^i = \bar{x}^i(x^1, x^2)$. Thus, we solve the above equations for the $\bar{x}^i$:

$$\bar{x}^1 = x^1 \qquad \bar{x}^2 = x^1 + \ln x^2$$

Hence
$$J = \begin{bmatrix} 1 & 0 \\ 1 & (x^2)^{-1} \end{bmatrix}$$

and
$$G = \begin{bmatrix} 1 & 1 \\ 0 & (x^2)^{-1} \end{bmatrix}\begin{bmatrix} 1 & 0 \\ 1 & (x^2)^{-1} \end{bmatrix} = \begin{bmatrix} 2 & (x^2)^{-1} \\ (x^2)^{-1} & (x^2)^{-2} \end{bmatrix}$$

or $g_{11} = 2$, $g_{12} = g_{21} = (x^2)^{-1}$, $g_{22} = (x^2)^{-2}$.

5.7 (*a*) Using the metric of Problem 5.6, calculate the length of the curve

$$\mathscr{C} : x^1 = 3t, \quad x^2 = e^t \qquad (0 \leqq t \leqq 2)$$

(*b*) Interpret geometrically.

(*a*) First calculate the dx^i/dt:

$$\frac{dx^1}{dt} = 3 \qquad \frac{dx^2}{dt} = e^t$$

Then
$$\left(\frac{ds}{dt}\right)^2 = 2\left(\frac{dx^1}{dt}\right)^2 + 2(x^2)^{-1}\left(\frac{dx^1}{dt}\right)\left(\frac{dx^2}{dt}\right) + (x^2)^{-2}\left(\frac{dx^2}{dt}\right)^2$$
$$= 2(9) + 2e^{-t}(3)(e^t) + e^{-2t}(e^{2t}) = 25$$

and
$$L = \int_0^2 5\,dt = 10$$

(*b*) From the transformation equations of Problem 5.6, the curve is described in rectangular coordinates by $\bar{x}^2 = \frac{4}{3}\bar{x}^1$; it is therefore a straight line joining the points which correspond to $t = 0$ and $t = 2$, or $(0, 0)$ and $(6, 8)$. The distance from $(0, 0)$ to $(6, 8)$ is

$$\sqrt{6^2 + 8^2} = 10$$

as found in (*a*).

5.8 Making use of the Euclidean metric for cylindrical coordinates, (*5.3*), calculate the length of arc along the circular helix

$$\bar{x}^1 = a \cos t \qquad \bar{x}^2 = a \sin t \qquad \bar{x}^3 = bt$$

with a and b positive constants, from $t = 0$ to $t = c > 0$. See Fig. 5-3.

In cylindrical coordinates (x^i), where

$$\bar{x}^1 = x^1 \cos x^2 \qquad \bar{x}^2 = x^1 \sin x^2 \qquad \bar{x}^3 = x^3$$

the helical arc is represented by the linear equations

$$x^1 = a \qquad x^2 = t \qquad x^3 = bt \qquad (0 \leqq t \leqq c)$$

$$\left(\frac{ds}{dt}\right)^2 = [0 \quad 1 \quad b]\begin{bmatrix} 1 & 0 & 0 \\ 0 & a^2 & 0 \\ 0 & 0 & 1 \end{bmatrix}\begin{bmatrix} 0 \\ 1 \\ b \end{bmatrix} = [0 \quad 1 \quad b]\begin{bmatrix} 0 \\ a^2 \\ b \end{bmatrix} = a^2 + b^2$$

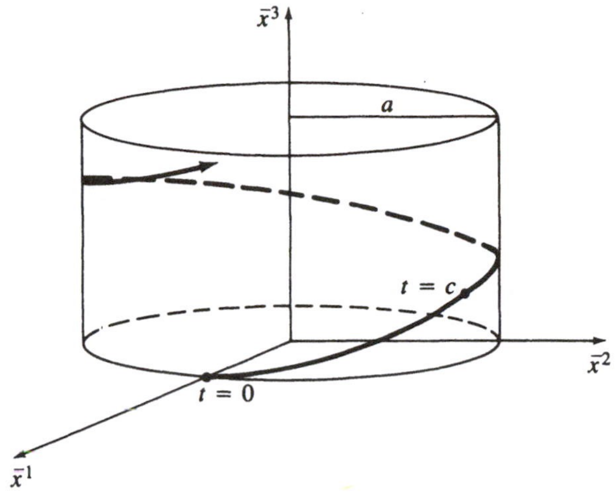

Fig. 5-3

whence

$$L = \int_0^c \sqrt{a^2 + b^2}\, dt = c\sqrt{a^2 + b^2}$$

5.9 (*Affine Coordinates in* $\mathbf{R}^3$) Carpenters taking measurements in a room notice that at the corner they had used as reference point the angles were not true. If the actual measures of the angles are as given in Fig. 5-4, what correction in the usual metric formula,

$$\overline{P_1P_2} = \sqrt{\sum_{i=1}^{3} (x_1^i - x_2^i)^2}$$

should be made to compensate for the errors?

We are asked, in effect, to display $\mathbf{g} = (g_{ij})$ for three-dimensional affine coordinates (x^i). Instead of applying Theorem 5.2, it is much simpler to recall from Problem 3.9 that position vectors are contravariant affine vectors—in particular, the unit vectors

$$\mathbf{u} = (\delta_1^i) \qquad \mathbf{v} = (\delta_2^i) \qquad \mathbf{w} = (\delta_3^i)$$

along the oblique axes (Fig. 5-4). We can now use (*5.11*) in inverse fashion, to obtain:

$$\cos\alpha = \frac{g_{ij}\delta_1^i\delta_2^j}{\sqrt{g_{pq}\delta_1^p\delta_1^q}\sqrt{g_{rs}\delta_2^r\delta_2^s}} = \frac{g_{12}}{\sqrt{g_{11}}\sqrt{g_{22}}} = g_{12}$$

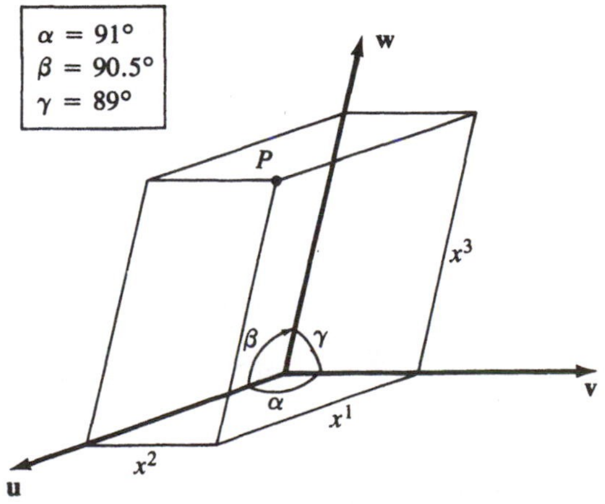

Fig. 5-4

since, obviously, $g_{11} = g_{22} = g_{33} = 1$ $(\pm ds = dx^1$ for motion parallel to **u**; etc.). Likewise,

$$\cos\beta = g_{13} \qquad \cos\gamma = g_{23}$$

and the complete symmetric matrix is

$$G = \begin{bmatrix} 1 & \cos\alpha & \cos\beta \\ \cos\alpha & 1 & \cos\gamma \\ \cos\beta & \cos\gamma & 1 \end{bmatrix} = \begin{bmatrix} 1 & -0.01745 & -0.00873 \\ -0.01745 & 1 & 0.01745 \\ -0.00873 & 0.01745 & 1 \end{bmatrix}$$

It follows that the carpenters must use as the corrected distance formula

$$\overline{P_1P_2} = \sqrt{g_{ij}(x_1^i - x_2^i)(x_1^j - x_2^j)}$$

where the g_{ij} have the numerical values given above.

RAISING AND LOWERING INDICES

5.10 Given that (V^i) is a contravariant vector on $\mathbf{R}^3$, find its associated covariant vector (V_i) in cylindrical coordinates (x^i) under the Euclidean metric.

Since

$$[g_{ij}]_{33} = \begin{bmatrix} 1 & 0 & 0 \\ 0 & (x^1)^2 & 0 \\ 0 & 0 & 1 \end{bmatrix}$$

and $V_i = g_{ir}V^r$, we have in matrix form,

$$\begin{bmatrix} V_1 \\ V_2 \\ V_3 \end{bmatrix} = \begin{bmatrix} 1 & 0 & 0 \\ 0 & (x^1)^2 & 0 \\ 0 & 0 & 1 \end{bmatrix} \begin{bmatrix} V^1 \\ V^2 \\ V^3 \end{bmatrix} = \begin{bmatrix} V^1 \\ (x^1)^2V^2 \\ V^3 \end{bmatrix}$$

5.11 Show that under orthogonal coordinate changes, starting with any particular system of rectangular coordinates, the raising and lowering of indices has no effect on tensors, consistent with the fact (Section 3.6) that there is no distinction between contravariant and covariant cartesian tensors.

It suffices to show merely that $g_{ij} = \delta_{ij} = g^{ij}$ for any admissible coordinate system (x^i), for then it will follow that

$$T^i = \delta^{ij}T_j = T_i \qquad T_{ijk} = \delta_{ir}T^r_{jk} = T^i_{jk} \qquad T^i_{jk} = \delta_{jr}T^{ir}_k = T^{ij}_k$$

and so on. To that end, simply use formula (5.7), with $J = (a^i_j)$ an orthogonal matrix. Because $J^T = J^{-1}$, we have $G = J^{-1}J = I$, or $g_{ij} = \delta_{ij}$, as desired. Since $G^{-1} = I^{-1} = I$, it is also the case that $g^{ij} = \delta_{ij}$.

GENERALIZED NORM

5.12 Show that the length of any contravariant vector (V^i) equals the length of its associated covariant vector (V_i).

By definition,

$$\|(V^i)\| = \sqrt{g_{ij}V^iV^j} \qquad \text{and} \qquad \|(V_i)\| = \sqrt{g^{ij}V_iV_j}$$

But, since $V^i = g^{ir}V_r$ and $g_{ij} = g_{ji}$,

$$g_{ij}V^iV^j = g_{ij}(g^{ir}V_r)(g^{js}V_s) = g_{ji}g^{ir}g^{js}V_rV_s = \delta^r_jg^{js}V_rV_s = g^{rs}V_rV_s$$

and the two lengths are equal.

5.13 Assuming a positive definite metric, show that the basic properties of the cartesian inner product $\mathbf{U}\cdot\mathbf{V}$ are shared by the generalized inner product $\mathbf{UV}$ of contravariant vectors.

(*a*) $\mathbf{UV}=\mathbf{VU}$ (*commutative property*). Follows from symmetry of (g_{ij}).

(*b*) $\mathbf{U}(\mathbf{V}+\mathbf{W})=\mathbf{UV}+\mathbf{UW}$ (*distributive property*). Follows from (*1.2*).

(*c*) $(\lambda\mathbf{U})\mathbf{V}=\mathbf{U}(\lambda\mathbf{V})=\lambda(\mathbf{UV})$ (*associative property*). Follows from $\lambda U_iV^i=U_i(\lambda V^i)=\lambda U_iV^i$.

(*d*) $\mathbf{U}^2\geqq 0$ with equality only if $\mathbf{U}=\mathbf{0}$ (*positive-definiteness*). Follows from the assumed positive-definiteness of (g_{ij}).

(*e*) $(\mathbf{UV})^2\leqq(\mathbf{U}^2)(\mathbf{V}^2)$ (*Cauchy–Schwarz inequality*). This may be derived from the other properties, as follows. If $\mathbf{U}=\mathbf{0}$, the inequality clearly holds. If $\mathbf{U}\neq\mathbf{0}$, property (*d*) ensures that the quadratic polynomial

$$Q(\lambda)\equiv(\lambda\mathbf{U}+\mathbf{V})^2=\mathbf{U}^2\lambda^2+2\mathbf{UV}\lambda+\mathbf{V}^2$$

vanishes for at most one real value of λ. Thus, the discriminant of Q cannot be positive:

$$(\mathbf{UV})^2-(\mathbf{U}^2)(\mathbf{V}^2)\leqq 0$$

and this is the desired inequality.

5.14 A *generalized norm* on a vector space is any real-valued functional $\phi[\quad]$ that satisfies

(i) $\phi[\mathbf{V}]\geqq 0$, with equality only if $\mathbf{V}=\mathbf{0}$;

(ii) $\phi[\lambda\mathbf{V}]=|\lambda|\,\phi[\mathbf{V}]$;

(iii) $\phi[\mathbf{U}+\mathbf{V}]\leqq\phi[\mathbf{U}]+\phi[\mathbf{V}]$ (*triangle inequality*).

Verify these conditions for $\phi[\mathbf{V}]=\|\mathbf{V}\|$, the inner-product norm under a positive definite metric.

(i) and (ii) for $\|\mathbf{V}\|$ are evident. As for (iii), the Cauchy–Schwarz inequality gives

$$\begin{aligned}\|\mathbf{U}+\mathbf{V}\|^2&=(\mathbf{U}+\mathbf{V})^2=\mathbf{U}^2+\mathbf{V}^2+2\mathbf{UV}\\&\leqq\|\mathbf{U}\|^2+\|\mathbf{V}\|^2+2\,\|\mathbf{U}\|\,\|\mathbf{V}\|=(\|\mathbf{U}\|+\|\mathbf{V}\|)^2\end{aligned}$$

from which (iii) follows at once.

ANGLE BETWEEN CONTRAVARIANT VECTORS

5.15 Show that the angle between contravariant vectors is an invariant under a change of coordinate systems.

The defining expression (*5.11*) involves only inner products, which are invariants.

5.16 In $\mathbf{R}^2$ the family of curves $x^2=x^1-c$ (parameterized as $x^1=t$, $x^2=t-c$), has as its system of tangent vectors the vector field $\mathbf{U}=(1,1)$, constant throughout $\mathbf{R}^2$. If (x^i) represent polar coordinates, find the family of orthogonal trajectories, and interpret geometrically.

The metric is given by

$$g=\begin{bmatrix}1&0\\0&(x^1)^2\end{bmatrix}$$

so, by Theorem 5.3, the orthogonality condition becomes

$$g_{ij}U^i\frac{dx^j}{du}=(1)(1)\frac{dx^1}{du}+(x^1)^2(1)\frac{dx^2}{du}=0$$

or, eliminating the differential du,

$$dx^1+(x^1)^2\,dx^2=0$$

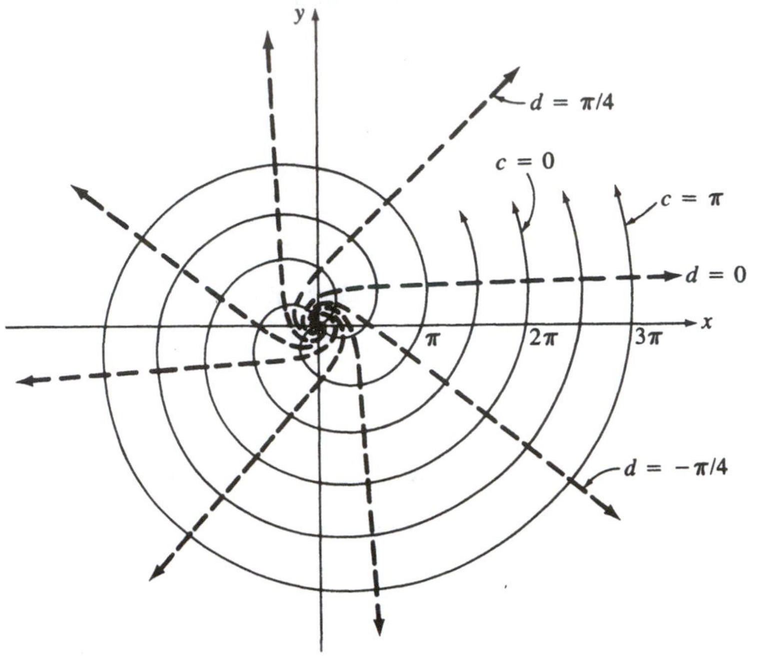

Fig. 5-5

This is a variables-separable differential equation, whose solution is

$$x^1 = \frac{1}{x^2 + d}$$

The given family of curves in the usual polar-coordinate notation is $r = \theta + c$, which is a family of concentric spirals (solid curves in Fig. 5-5). The orthogonal trajectories,

$$r = \frac{1}{\theta + d}$$

are also spirals, each having an asymptote parallel to the line $\theta = -d$; these are the dashed curves in Fig. 5-5.

Note: To solve this problem in rectangular coordinates—that is, to find the orthogonal trajectories of the family

$$\frac{y}{x} = \tan\left(\sqrt{x^2 + y^2} - c\right)$$

under the metric $(g_{ij}) = (\delta_{ij})$—would be difficult or impossible. Quite often, the complication of the metric involved in going over to a specialized curvilinear coordinate system is vastly outweighed by the degree to which the problem is simplified.

5.17 Find the condition for two curves on a sphere of radius a to be orthogonal, if the curves are represented in spherical coordinates by

$$\mathscr{C}_1 \ : \ \theta = f(\varphi) \qquad \text{and} \qquad \mathscr{C}_2 \ : \ \theta = g(\varphi)$$

The two curves can be parameterized in spherical coordinates $(x^i) \equiv (\rho, \varphi, \theta)$ by

$$\mathscr{C}_1 \ : \ \begin{cases} \rho = a \\ \varphi = t \\ \theta = f(t) \end{cases} \qquad \mathscr{C}_2 \ : \ \begin{cases} \rho = a \\ \varphi = u \\ \theta = g(u) \end{cases}$$

At an intersection point (a, φ_0, θ_0) the tangent vectors of $\mathscr{C}_1$ and $\mathscr{C}_2$ are, respectively,

$$\mathbf{U} = (0, 1, f'(\varphi_0)) \qquad \text{and} \qquad \mathbf{V} = (0, 1, g'(\varphi_0))$$

These are orthogonal if and only if $g_{ij}U^iV^j = 0$, or

$$0 = [0 \quad 1 \quad f'(\varphi_0)]\begin{bmatrix} 1 & 0 & 0 \\ 0 & a^2 & 0 \\ 0 & 0 & (a\sin\varphi_0)^2 \end{bmatrix}\begin{bmatrix} 0 \\ 1 \\ g'(\varphi_0) \end{bmatrix}$$

$$= 0 + a^2 + (a\sin\varphi_0)^2 f'(\varphi_0)g'(\varphi_0) = (a^2\sin^2\varphi_0)[\csc^2\varphi_0 + f'(\varphi_0)g'(\varphi_0)]$$

Hence, the desired criterion is that $f'(\varphi_0)g'(\varphi_0) = -\csc^2\varphi_0$ at any intersection point (a, φ_0, θ_0).

5.18 Show that the contravariant vectors $\mathbf{U} = (0, 1, 2bx^2)$ and $\mathbf{V} = (0, -2bx^2, (x^1)^2)$ are orthogonal under the Euclidean metric for cylindrical coordinates. Interpret geometrically along $x^1 = a$, $x^2 = t$, $x^3 = bt^2$.

$$g_{ij}U^iV^j = [0 \quad 1 \quad 2bx^2]\begin{bmatrix} 1 & 0 & 0 \\ 0 & (x^1)^2 & 0 \\ 0 & 0 & 1 \end{bmatrix}\begin{bmatrix} 0 \\ -2bx^2 \\ (x^1)^2 \end{bmatrix} = [0 \quad 1 \quad 2bx^2]\begin{bmatrix} 0 \\ -2bx^2(x^1)^2 \\ (x^1)^2 \end{bmatrix}$$

$$= 0 - 2bx^2(x^1)^2 + 2bx^2(x^1)^2 = 0$$

The geometric interpretation is that $x^1 = a$, $x^2 = t$, $x^3 = bt^2$, for real t, represents a sort of variable-pitch helix on the right circular cylinder $r = a$, having tangent field $\mathbf{U}$. Therefore, any solution of

$$\underbrace{\frac{dx^1}{du} = V^1 = 0}_{\text{or} \quad x^1 = a} \qquad \frac{dx^2}{du} = V^2 = -2bx^2 \qquad \frac{dx^3}{du} = V^3 = a^2 \tag{1}$$

will represent a curve on that cylinder that is orthogonal to this pseudo-helix. See Problem 5.28.

5.19 Show that in any coordinate system (x^i) the contravariant vector (recall Problem 3.5) $\mathbf{V} \equiv (g^{i\alpha})$ is normal to the surface $x^\alpha = \text{const.}$ $(\alpha = 1, 2, \ldots, n)$.

Being "normal to a surface at the surface point P" means being orthogonal, at P, to the tangent vector of any curve lying in the surface and passing through P. Now, for the surface $x^\alpha = \text{const.}$, any such tangent vector $\mathbf{T}$ has as its αth component

$$T^\alpha = \frac{dx^\alpha}{dt} = 0$$

We then have:

$$\mathbf{VT} \equiv g_{ij}V^iT^j = g_{ij}g^{i\alpha}\,T^j = g_{ji}g^{i\alpha}\,T^j = \delta^\alpha_j\,T^j = T^\alpha = 0$$

and the proof is complete.

5.20 Show that in any coordinate system (x^i), the angle θ between the normals to the surfaces $x^\alpha = \text{const.}$ and $x^\beta = \text{const.}$ is given by

$$\cos\theta = \frac{g^{\alpha\beta}}{\sqrt{g^{\alpha\alpha}}\sqrt{g^{\beta\beta}}} \qquad \text{(no sum)} \tag{1}$$

By Problem 5.19, $\mathbf{U} = (g^{i\alpha})$ and $\mathbf{V} = (g^{i\beta})$ are the respective normals to $x^\alpha = \text{const.}$ and $x^\beta = \text{const.}$ Therefore, by the definition (*5.11*)

$$\cos\theta = \frac{UV}{\|U\|\,\|V\|} = \frac{g_{ij}g^{i\alpha}g^{j\beta}}{\sqrt{g_{pq}g^{p\alpha}g^{q\alpha}}\sqrt{g_{rs}g^{r\beta}g^{s\beta}}} = \frac{\delta^\alpha_j g^{j\beta}}{\sqrt{\delta^\alpha_q g^{q\alpha}}\sqrt{\delta^\beta_s g^{s\beta}}} = \frac{g^{\alpha\beta}}{\sqrt{g^{\alpha\alpha}}\sqrt{g^{\beta\beta}}}$$

In consequence of (*1*), *orthogonal coordinates* are defined as those coordinate systems (x^i) relative to which, at all points, $g^{ij} = 0$ $(i \neq j)$, or, equivalently, $g_{ij} = 0$ $(i \neq j)$. Obviously, orthogonal coordinates need not be rectangular: witness polar, cylindrical, and spherical coordinates.

Supplementary Problems

5.21 Using the Euclidean metric for polar coordinates, compute the length of arc for the curve

$$\mathscr{C} \; : \; x^1 = 2a \cos t, \quad x^2 = t \qquad (0 \leqq t \leqq \pi/2)$$

and interpret geometrically.

5.22 Is the form $Q(u^1, u^2, u^3) \equiv 8(u^1)^2 + (u^2)^2 - 6u^1u^3 + (u^3)^2$ positive definite?

5.23 Using the metric

$$G = \begin{bmatrix} 12 & 4 & 0 \\ 4 & 1 & 1 \\ 0 & 1 & (x^1)^2 \end{bmatrix}$$

calculate the length of the curve given by $x^1 = 3 - t$, $x^2 = 6t + 3$, $x^3 = \ln t$, where $1 \leqq t \leqq e$.

5.24 A draftsman calculated several distances between points on his drawing using a set of vertical lines and his T-square. He obtained the distance from (1,2) to (4,6) the usual way:

$$\sqrt{(4-1)^2 + (6-2)^2} = 5$$

Then he noticed his T-square was out several degrees, throwing off all measurements. An accurate reading showed his T-square measured 95.8°. Find, to three decimal places, the error committed in his calculations for the answer 5 obtained above. [*Hint*: Use Problem 5.9 in the special case $x_1^3 = x_2^3 = 0$, with $\alpha = 95.8°$.]

5.25 In curvilinear coordinates (x^i), show that the contravariant vectors

$$\mathbf{U} = (-x^1/x^2, 1, 0) \qquad \mathbf{V} = (1/x^2, 0, 0)$$

are an orthonormal pair, if (x^i) is related to rectangular coordinates $(\bar{x}^i)$ through

$$\bar{x}^1 = x^2 \qquad \bar{x}^2 = x^3 \qquad \bar{x}^3 = x^1x^2$$

where $x^2 \neq 0$.

5.26 Express in (x^i) the covariant vectors associated with **U** and **V** of Problem 5.25.

5.27 Even though (g_{ij}) may define a non-Euclidean metric, prove that the norm (*5.10*) still obeys the following "Euclidean" laws: (*a*) the law of cosines, (*b*) the Pythagorean theorem.

5.28 (*a*) Solve system (*1*) of Problem 5.18. (*b*) Does the solution found in (*a*) include *all* curves orthogonal to the given pseudo-helix? Explain.

5.29 Find the family of orthogonal trajectories in polar coordinates for the family of spirals $x^1 = cx^2$ ($c =$ const.). [*Hint*: Parameterize the family as $x^1 = ce^t$, $x^2 = e^t$.]

5.30 Find the condition for two curves, $z = f(\theta)$ and $z = g(\theta)$, on a right circular cylinder of radius a to be orthogonal.

5.31 Let (x^i) be any coordinate system and (g_{ij}) any positive definite metric tensor realized in that system. Define the *coordinate axes* as the curves $\mathscr{C}_\alpha \; : \; x^i = t\delta^i_\alpha$ $(\alpha = 1, 2, \ldots n)$. Show that the angle ϕ between the coordinate axes $\mathscr{C}_\alpha$ and $\mathscr{C}_\beta$ satisfies the relation

$$\cos \phi = \frac{g_{\alpha\beta}}{\sqrt{g_{\alpha\alpha}}\sqrt{g_{\beta\beta}}} \qquad \text{(no sum)}$$

and is thus distinct, in general, from the angle θ of Problem 5.20.

5.32 Refer to Problems 5.20 and 5.31. (*a*) What property must the metric tensor (g_{ij}) possess in (x^i) for the coordinate axis $\mathscr{C}_\alpha$ to be normal to the surface $x^\alpha = \text{const.}$ (in which case $\theta = \phi$)? (*b*) Show that the property of (*a*) is equivalent to the mutual orthogonality of the coordinate axes.

5.33 Under the metric

$$G = \begin{bmatrix} 1 & \cos 2x^2 \\ \cos 2x^2 & 1 \end{bmatrix} \qquad (2x^2/\pi \text{ nonintegral})$$

compute the norm of the vector $\mathbf{V} = (dx^i/dt)$ evaluated along the curve $x^1 = -\sin 2t$, $x^2 = t$, and use it to find the arc length between $t = 0$ and $t = \pi/2$.

5.34 Under the Euclidean metric for spherical coordinates, (*5.4*), determine a particular family of curves that intersect

$$x^1 = a \qquad x^2 = bt \qquad x^3 = t$$

orthogonally. (Cf. Problem 5.28.)

Chapter 6

The Derivative of a Tensor

6.1 INADEQUACY OF ORDINARY DIFFERENTIATION

Consider a contravariant tensor $\mathbf{T} = (T^i(\mathbf{x}(t)))$ defined on the curve $\mathscr{C}\ :\ \mathbf{x} = \mathbf{x}(t)$. Differentiating the transformation law

$$\bar{T}^i = T^r \frac{\partial \bar{x}^i}{\partial x^r}$$

with respect to t gives

$$\frac{d\bar{T}^i}{dt} = \frac{dT^r}{dt}\frac{\partial \bar{x}^i}{\partial x^r} + T^r \frac{\partial^2 \bar{x}^i}{\partial x^s \partial x^r}\frac{dx^s}{dt}$$

which shows that the ordinary derivative of $\mathbf{T}$ along the curve is a contravariant tensor when and only when the $\bar{x}^i$ are linear functions of the x^r.

Theorem 6.1: The derivative of a tensor is a tensor if and only if coordinate changes are restricted to linear transformations.

EXAMPLE 6.1 With $\mathbf{T} = d\mathbf{x}/dt$ the tangent field along $\mathscr{C}$ (under the choice $t = s =$ arc length), the classical formula for the curvature of $\mathscr{C}$,

$$\kappa = \left\| \frac{d\mathbf{T}}{dt} \right\|$$

will hold in affine coordinates but will fail to define an invariant in curvilinear coordinates, since $d\mathbf{T}/dt$ is not a general tensor. Thus, if we want the curvature of a curve to be an intrinsic concept, independent of coordinate systems, we must invent a more general concept of tensor differentiation. This requires the objects defined next.

6.2 CHRISTOFFEL SYMBOLS OF THE FIRST KIND

Definition and Basic Properties

The n^3 functions

$$\Gamma_{ijk} \equiv \frac{1}{2}\left[\frac{\partial}{\partial x^i}(g_{jk}) + \frac{\partial}{\partial x^j}(g_{ki}) - \frac{\partial}{\partial x^k}(g_{ij})\right] \tag{6.1a}$$

are the *Christoffel symbols of the first kind*. In order to simplify the notation here and elsewhere, we shall adopt the following convention: The partial derivative of a tensor with respect to x^k will be indicated by a final subscript k. Thus,

$$\Gamma_{ijk} \equiv \tfrac{1}{2}(-g_{ijk} + g_{jki} + g_{kij}) \tag{6.1b}$$

EXAMPLE 6.2 Compute the Christoffel symbols corresponding to the Euclidean metric for spherical coordinates:

$$G = \begin{bmatrix} 1 & 0 & 0 \\ 0 & (x^1)^2 & 0 \\ 0 & 0 & (x^1)^2 \sin^2 x^2 \end{bmatrix}$$

Here, $g_{221} = 2x^1$, $g_{331} = 2x^1 \sin^2 x^2$, $g_{332} = 2(x^1)^2 \sin x^2 \cos x^2$, and all other g_{ijk} are zero. Hence, $\Gamma_{ijk} = 0$ unless the triplet ijk includes precisely two 2s (six cases) or precisely two 3s (six cases):

$\bullet\ \Gamma_{221}=\frac{1}{2}(-g_{221}+g_{212}+g_{122})=-x^1$ $\quad\bullet\ \Gamma_{212}=\frac{1}{2}(-g_{212}+g_{122}+g_{221})=x^1$ $\quad\bullet\ \Gamma_{122}=\frac{1}{2}(-g_{122}+g_{221}+g_{212})=x^1$

$\Gamma_{223}=\frac{1}{2}(-g_{223}+g_{232}+g_{322})=0$ $\quad\Gamma_{232}=\frac{1}{2}(-g_{232}+g_{322}+g_{223})=0$ $\quad\Gamma_{322}=\frac{1}{2}(-g_{322}+g_{223}+g_{232})=0$

and

$\bullet\ \Gamma_{331}=\frac{1}{2}(-g_{331}+g_{313}+g_{133})=-x^1\sin^2 x^2$ $\quad\bullet\ \Gamma_{323}=\frac{1}{2}(-g_{323}+g_{233}+g_{332})=(x^1)^2\sin x^2\cos x^2$

$\bullet\ \Gamma_{332}=\frac{1}{2}(-g_{332}+g_{323}+g_{233})=-(x^1)^2\sin x^2\cos x^2$ $\quad\bullet\ \Gamma_{133}=\frac{1}{2}(-g_{133}+g_{331}+g_{313})=x^1\sin^2 x^2$

$\bullet\ \Gamma_{313}=\frac{1}{2}(-g_{313}+g_{133}+g_{331})=x^1\sin^2 x^2$ $\quad\bullet\ \Gamma_{233}=\frac{1}{2}(-g_{233}+g_{332}+g_{323})=(x^1)^2\sin x^2\cos x^2$

(The nine nonzero zymbols are marked with bullets for later reference.)

The two basic properties of the Christoffel symbols of the first kind are:

(i) $\Gamma_{ijk}=\Gamma_{jik}$ (symmetry in the first two indices)

(ii) all Γ_{ijk} vanish if all g_{ij} are constant

A useful formula results from simply permuting the subscripts in (*6.1b*) and summing:

$$\frac{\partial g_{ik}}{\partial x^j}=\Gamma_{ijk}+\Gamma_{jki} \tag{6.2}$$

The converse of property (ii) follows at once from (*6.2*); thus:

Lemma 6.2: In any particular coordinate system, the Christoffel symbols uniformly vanish if and only if the metric tensor has constant components in that system.

Transformation Law

The transformation law for the Γ_{ijk} can be inferred from that for the g_{ij}. By differentiation,

$$\bar{g}_{ijk}=\frac{\partial}{\partial\bar{x}^k}\left(g_{rs}\frac{\partial x^r}{\partial\bar{x}^i}\frac{\partial x^s}{\partial\bar{x}^j}\right)=\frac{\partial g_{rs}}{\partial\bar{x}^k}\frac{\partial x^r}{\partial\bar{x}^i}\frac{\partial x^s}{\partial\bar{x}^j}+g_{rs}\frac{\partial^2 x^r}{\partial\bar{x}^k\partial\bar{x}^i}\frac{\partial x^s}{\partial\bar{x}^j}+g_{rs}\frac{\partial x^r}{\partial\bar{x}^i}\frac{\partial^2 x^s}{\partial\bar{x}^k\partial\bar{x}^j}$$

Use the chain rule on $\partial g_{rs}/\partial\bar{x}^k$:

$$\frac{\partial g_{rs}}{\partial\bar{x}^k}=\frac{\partial g_{rs}}{\partial x^t}\frac{\partial x^t}{\partial\bar{x}^k}\equiv g_{rst}\frac{\partial x^t}{\partial\bar{x}^k}$$

Then rewrite the expression with subscripts permuted cyclically, sum the three expressions (arrows couple terms which cancel out), and divide by 2:

$$-\bar{g}_{ijk}=-g_{rst}\frac{\partial x^r}{\partial\bar{x}^i}\frac{\partial x^s}{\partial\bar{x}^j}\frac{\partial x^t}{\partial\bar{x}^k}+g_{rs}\left(-\frac{\partial^2 x^r}{\partial\bar{x}^k\partial\bar{x}^i}\frac{\partial x^s}{\partial\bar{x}^j}-\frac{\partial^2 x^s}{\partial\bar{x}^k\partial\bar{x}^j}\frac{\partial x^r}{\partial\bar{x}^i}\right)$$

$$\bar{g}_{jki}=g_{str}\frac{\partial x^s}{\partial\bar{x}^j}\frac{\partial x^t}{\partial\bar{x}^k}\frac{\partial x^r}{\partial\bar{x}^i}+g_{sr}\left(\frac{\partial^2 x^s}{\partial\bar{x}^i\partial\bar{x}^j}\frac{\partial x^r}{\partial\bar{x}^k}+\frac{\partial^2 x^r}{\partial\bar{x}^i\partial\bar{x}^k}\frac{\partial x^s}{\partial\bar{x}^j}\right)$$

$$\bar{g}_{kij}=g_{trs}\frac{\partial x^t}{\partial\bar{x}^k}\frac{\partial x^r}{\partial\bar{x}^i}\frac{\partial x^s}{\partial\bar{x}^j}+g_{sr}\left(\frac{\partial^2 x^s}{\partial\bar{x}^j\partial\bar{x}^k}\frac{\partial x^r}{\partial\bar{x}^i}+\frac{\partial^2 x^r}{\partial\bar{x}^j\partial\bar{x}^i}\frac{\partial x^s}{\partial\bar{x}^k}\right)$$

give

$$\bar{\Gamma}_{ijk}=\Gamma_{rst}\frac{\partial x^r}{\partial\bar{x}^i}\frac{\partial x^s}{\partial\bar{x}^j}\frac{\partial x^t}{\partial\bar{x}^k}+g_{rs}\frac{\partial^2 x^r}{\partial\bar{x}^i\partial\bar{x}^j}\frac{\partial x^s}{\partial\bar{x}^k} \tag{6.3}$$

From the form of (*6.3*) it is clear that the set of Christoffel symbols is a third-order covariant affine tensor *but is not a general tensor*. Here again, conventional differentiation—this time, partial differentiation with respect to a coordinate—fails to produce more than an affine tensor (recall Problem 2.23).

6.3 CHRISTOFFEL SYMBOLS OF THE SECOND KIND

Definition and Basic Properties

The n^3 functions

$$\Gamma^i_{jk} = g^{ir}\Gamma_{jkr} \tag{6.4}$$

are the *Christoffel symbols of the second kind*. It should be noted that formula (*6.4*) is simply the result of raising the third subscript of the Christoffel symbol of the first kind, although here we are not dealing with tensors.

EXAMPLE 6.3 Calculate the Christoffel symbols of the second kind for the Euclidean metric in polar coordinates.

Since

$$G = \begin{bmatrix} 1 & 0 \\ 0 & (x^1)^2 \end{bmatrix}$$

we have:

$$\Gamma_{111} = \tfrac{1}{2}\, g_{111} = 0 \qquad \Gamma_{121} = \Gamma_{211} = \tfrac{1}{2}\,(-g_{121} + g_{211} + g_{112}) = 0$$

$$\bullet\ \Gamma_{221} = \tfrac{1}{2}\,(-g_{221} + g_{212} + g_{122}) = -x^1 \qquad \Gamma_{112} = \tfrac{1}{2}\,(-g_{112} + g_{121} + g_{211}) = 0$$

$$\bullet\ \Gamma_{122} = \Gamma_{212} = \tfrac{1}{2}\,(-g_{122} + g_{221} + g_{212}) = x^1 \qquad \Gamma_{222} = \tfrac{1}{2}\, g_{222} = 0$$

To continue,

$$G^{-1} = \begin{bmatrix} 1 & 0 \\ 0 & (x^1)^{-2} \end{bmatrix}$$

From $g_{12} = 0 = g_{21}$, it follows that $\Gamma^i_{jk} = g^{ir}\Gamma_{jkr} = g^{ii}\Gamma_{jki}$ (no sum). Therefore, when $i = 1$,

$$\bullet\ \Gamma^1_{22} = -x^1 \qquad \Gamma^1_{jk} = 0 \quad \text{otherwise}$$

and when $i = 2$,

$$\bullet\ \Gamma^2_{12} = \Gamma^2_{21} = 1/x^1 \qquad \Gamma^2_{jk} = 0 \quad \text{otherwise}$$

The basic properties of Γ_{ijk} carry over to Γ^i_{jk}:

(i) $\Gamma^i_{jk} = \Gamma^i_{kj}$ (symmetry in the lower indices)

(ii) all Γ^i_{jk} vanish if all g_{ij} are constant

Furthermore, by Problem 6.25, Lemma 6.2 holds for both first and second kinds of Christoffel symbols.

Transformation Law

Starting with

$$\bar{\Gamma}^i_{jk} = \bar{g}^{ir}\bar{\Gamma}_{jkr} = \left(g^{st}\,\frac{\partial \bar{x}^i}{\partial x^s}\,\frac{\partial \bar{x}^r}{\partial x^t}\right)\bar{\Gamma}_{jkr}$$

substitute for $\bar{\Gamma}_{jkr}$ from (6.3) to obtain

$$\bar{\Gamma}^i_{jk} = \left(g^{st}\frac{\partial \bar{x}^i}{\partial x^s}\frac{\partial \bar{x}^r}{\partial x^t}\right)\left(\Gamma_{uvw}\frac{\partial x^u}{\partial \bar{x}^j}\frac{\partial x^v}{\partial \bar{x}^k}\frac{\partial x^w}{\partial \bar{x}^r}\right) + \left(g^{st}\frac{\partial \bar{x}^i}{\partial x^s}\frac{\partial \bar{x}^r}{\partial x^t}\right)\left(g_{uv}\frac{\partial^2 x^u}{\partial \bar{x}^j \partial \bar{x}^k}\frac{\partial x^v}{\partial \bar{x}^r}\right)$$

$$= g^{st}\Gamma_{uvw}\delta^w_t \frac{\partial \bar{x}^i}{\partial x^s}\frac{\partial x^u}{\partial \bar{x}^j}\frac{\partial x^v}{\partial \bar{x}^k} + g^{st}g_{uv}\delta^v_t \frac{\partial \bar{x}^i}{\partial x^s}\frac{\partial^2 x^u}{\partial \bar{x}^j \partial \bar{x}^k}$$

$$= g^{st}\Gamma_{uvt}\frac{\partial \bar{x}^i}{\partial x^s}\frac{\partial x^u}{\partial \bar{x}^j}\frac{\partial x^v}{\partial \bar{x}^k} + g^{st}g_{ut}\frac{\partial \bar{x}^i}{\partial x^s}\frac{\partial^2 x^u}{\partial \bar{x}^j \partial \bar{x}^k}$$

Since $g^{st}\Gamma_{uvt} = \Gamma^s_{uv}$ and $g^{st}g_{ut} = \delta^s_u$, after changing indices this becomes

$$\bar{\Gamma}^i_{jk} = \Gamma^r_{st}\frac{\partial \bar{x}^i}{\partial x^r}\frac{\partial x^s}{\partial \bar{x}^j}\frac{\partial x^t}{\partial \bar{x}^k} + \frac{\partial^2 x^r}{\partial \bar{x}^j \partial \bar{x}^k}\frac{\partial \bar{x}^i}{\partial x^r} \tag{6.5}$$

The transformation law (6.5) shows that, like (Γ_{ijk}), (Γ^i_{jk}) is merely an affine tensor.

An Important Formula

$$\frac{\partial^2 x^r}{\partial \bar{x}^i \partial \bar{x}^j} = \bar{\Gamma}^s_{ij}\frac{\partial x^r}{\partial \bar{x}^s} - \Gamma^r_{st}\frac{\partial x^s}{\partial \bar{x}^i}\frac{\partial x^t}{\partial \bar{x}^j} \tag{6.6}$$

See Problem 6.24. Needless to say, (6.6) holds when barred and unbarred coordinates are interchanged.

6.4 COVARIANT DIFFERENTIATION

Of a Vector

Partial differentiation of the transformation law

$$\bar{T}_i = T_r \frac{\partial x^r}{\partial \bar{x}^i}$$

of a covariant vector $\mathbf{T} = (T_i)$ yields

$$\frac{\partial \bar{T}_i}{\partial \bar{x}^k} = \frac{\partial T_r}{\partial \bar{x}^k}\frac{\partial x^r}{\partial \bar{x}^i} + T_r \frac{\partial^2 x^r}{\partial \bar{x}^k \partial \bar{x}^i}$$

Using the chain rule on the first term on the right, and formula (6.6) on the second, results in the equations

$$\frac{\partial \bar{T}_i}{\partial \bar{x}^k} = \frac{\partial T_r}{\partial x^s}\frac{\partial x^r}{\partial \bar{x}^i}\frac{\partial x^s}{\partial \bar{x}^k} + T_r\left(\bar{\Gamma}^s_{ik}\frac{\partial x^r}{\partial \bar{x}^s} - \Gamma^r_{st}\frac{\partial x^s}{\partial \bar{x}^i}\frac{\partial x^t}{\partial \bar{x}^k}\right)$$

$$= \frac{\partial T_r}{\partial x^s}\frac{\partial x^r}{\partial \bar{x}^i}\frac{\partial x^s}{\partial \bar{x}^k} + \bar{\Gamma}^t_{ik}\bar{T}_t - \Gamma^t_{rs}T_t\frac{\partial x^r}{\partial \bar{x}^i}\frac{\partial x^s}{\partial \bar{x}^k}$$

which rearrange to

$$\frac{\partial \bar{T}_i}{\partial \bar{x}^k} - \bar{\Gamma}^t_{ik}\bar{T}_t = \left(\frac{\partial T_r}{\partial x^s} - \Gamma^t_{rs}T_t\right)\frac{\partial x^r}{\partial \bar{x}^i}\frac{\partial x^s}{\partial \bar{x}^k}$$

which is the defining law of a covariant tensor of order two. In other words, when the components of $\partial \mathbf{T}/\partial x^k$ are corrected by subtracting certain linear combinations of the components of $\mathbf{T}$ itself, the result is a *tensor* (and not just an affine tensor).

Definition 1: In any coordinate system (x^i), the *covariant derivative with respect to* x^k of a covariant vector $\mathbf{T} = (T_i)$ is the tensor

$$\mathbf{T}_{,k} = (T_{i,k}) \equiv \left(\frac{\partial T_i}{\partial x^k} - \Gamma^t_{ik}T_t\right)$$

Remark 1: The two covariant indices are notated i and $,k$ to emphasize that the second index arose from an *operation with respect to the* k*th coordinate.*

Remark 2: From Lemma 6.2, the covariant derivative and the partial derivative coincide when the g_{ij} are constants (as in a rectangular coordinate system).

A similar manipulation (Problem 6.7) of the contravariant vector law leads to

Definition 2: In any coordinate system (x^i), the *covariant derivative with respect to* x^k of a contravariant vector $\mathbf{T} = (T^i)$ is the tensor

$$\mathbf{T}_{,k} = (T^i_{,k}) \equiv \left(\frac{\partial T^i}{\partial x^k} + \Gamma^i_{tk} T^t \right)$$

Of Any Tensor

In the general definition, each covariant index (subscript) gives rise to a linear "correction term" of the form given in Definition 1, and each contravariant index (superscript) gives rise to a term of the form given in Definition 2.

Definition 3: In any coordinate system (x^i), the *covariant derivative with respect to* x^k of a tensor $\mathbf{T} = (T^{i_1 i_2 \cdots i_p}_{j_1 j_2 \cdots j_q})$ is the tensor $\mathbf{T}_{,k} = (T^{i_1 i_2 \cdots i_p}_{j_1 j_2 \cdots j_q, k})$, where

$$\begin{aligned} T^{i_1 i_2 \cdots i_p}_{j_1 j_2 \cdots j_q, k} = {} & \frac{\partial T^{i_1 i_2 \cdots i_p}_{j_1 j_2 \cdots j_q}}{\partial x^k} + \Gamma^{i_1}_{tk} T^{t i_2 \cdots i_p}_{j_1 j_2 \cdots j_q} + \Gamma^{i_2}_{tk} T^{i_1 t \cdots i_p}_{j_1 j_2 \cdots j_q} + \cdots + \Gamma^{i_p}_{tk} T^{i_1 i_2 \cdots t}_{j_1 j_2 \cdots j_q} \\ & - \Gamma^t_{j_1 k} T^{i_1 i_2 \cdots i_p}_{t j_2 \cdots j_q} - \Gamma^t_{j_2 k} T^{i_1 i_2 \cdots i_p}_{j_1 t \cdots j_q} - \cdots - \Gamma^t_{j_q k} T^{i_1 i_2 \cdots i_p}_{j_1 j_2 \cdots t} \end{aligned} \qquad (6.7)$$

That $\mathbf{T}_{,k}$ actually is a tensor must, of course, be proved. This can be accomplished basically as in Problem 6.8, by use of Theorem 4.2 and an induction on the number of indices.

Theorem 6.3: The covariant derivative of an arbitrary tensor is a tensor of which the covariant order exceeds that of the original tensor by exactly one.

6.5 ABSOLUTE DIFFERENTIATION ALONG A CURVE

Because $(T^i_{,j})$ is a tensor, the inner product of $(T^i_{,j})$ with another tensor is also a tensor. Suppose that the other tensor is (dx^i/dt), the tangent vector of the curve $\mathscr{C} : x^i = x^i(t)$. Then the inner product

$$\left(T^i_{,r} \frac{dx^r}{dt} \right)$$

is a tensor of the same type and order as the original tensor (T^i). This tensor is known as the *absolute derivative of* (T^i) *along* $\mathscr{C}$, with components written as

$$\left(\frac{\delta T^i}{\delta t} \right) \equiv \left(\frac{dT^i}{dt} + \Gamma^i_{rs} T^r \frac{dx^s}{dt} \right) \qquad \text{where} \qquad T^i = T^i(\mathbf{x}(t)) \qquad (6.8)$$

(see Problem 6.12). It is clear that, again, in coordinate systems in which the g_{ij} are constant, absolute differentiation reduces to ordinary differentiation.

The definition (6.8) is not an arbitrary one; in Problem 6.18 is proved

Theorem 6.4 (*Uniqueness of the Absolute Derivative*): The only tensor derivable from a given tensor (T^i) that coincides with the ordinary derivative (dT^i/dt) along some curve in a rectangular coordinate system is the absolute derivative of (T^i) along that curve.

Remark 3: Theorem 6.4 concerns tensors with a given form *in rectangular coordinates*. Thus it presumes the Euclidean metric (see Section 3.1).

Acceleration in General Coordinates

In rectangular coordinates, the *acceleration vector* of a particle is the time derivative of its velocity vector, or the second time derivative of its position function $\mathbf{x} = (x^i(t))$:

$$\mathbf{a} = (a^i) \equiv \left(\frac{d}{dt}\frac{dx^i}{dt}\right) = \left(\frac{d^2x^i}{dt^2}\right)$$

The (Euclidean) length of this vector at time t is the instantaneous *acceleration* of the particle:

$$a = \sqrt{\delta_{ij}a^i a^j}$$

Since derivatives are taken along the particle's trajectory, the natural generalization of $\frac{d}{dt}\left(\frac{dx^i}{dt}\right)$ is

$$\frac{\delta}{\delta t}\left(\frac{dx^i}{dt}\right) = \frac{d^2x^i}{dt^2} + \Gamma^i_{rs}\frac{dx^r}{dt}\frac{dx^s}{dt}$$

Hence, in general coordinates, we take as the acceleration vector and the acceleration

$$\mathbf{a} = (a^i) \equiv \left(\frac{d^2x^i}{dt^2} + \Gamma^i_{rs}\frac{dx^r}{dt}\frac{dx^s}{dt}\right) \tag{6.9}$$

$$a = \sqrt{|g_{ij}a^i a^j|} \tag{6.10}$$

Note that positive-definiteness of the metric is not assumed in *(6.10)*.

Curvature in General Coordinates

In Euclidean geometry an important role is played by the *curvature* of a curve $\mathscr{C} : x^i = x^i(t)$, commonly defined as the norm of the second derivative of $(x^i(s))$:

$$\kappa(s) = \sqrt{\delta_{ij}\frac{d^2x^i}{ds^2}\frac{d^2x^j}{ds^2}}$$

where $ds/dt = \sqrt{\delta_{ij}(dx^i/dt)(dx^j/dt)}$ gives the arc-length parameter. The obvious way to extend this concept as an invariant is again to use absolute differentiation. Writing

$$(b^i) \equiv \left(\frac{\delta}{\delta s}\frac{dx^i}{ds}\right) = \left(\frac{d^2x^i}{ds^2} + \Gamma^i_{pq}\frac{dx^p}{ds}\frac{dx^q}{ds}\right) \tag{6.11}$$

where the arc-length parameter $s = s(t)$ is given by *(5.6)*, we have:

$$\kappa(s) = \sqrt{|g_{ij}b^i b^j|} \tag{6.12}$$

Geodesics

An important application of *(6.12)* in curvilinear coordinates is the following. Suppose that we seek those curves for which $\kappa = 0$ (that is, the "straight" lines or *geodesics*). For positive definite metrics, this condition is equivalent to requiring that

$$b^i = \frac{d^2x^i}{ds^2} + \Gamma^i_{pq}\frac{dx^p}{ds}\frac{dx^q}{ds} = 0 \qquad (i = 1, 2, \ldots, n) \tag{6.13}$$

The solution of this system of second-order differential equations will define the geodesics $x^i = x^i(s)$.

EXAMPLE 6.4 In affine coordinates, where all g_{ij} are constant and all Christoffel symbols vanish, integration of (*6.13*) is immediate:

$$x^i = \alpha^i s + \beta^i \qquad (i = 1, 2, \ldots, n)$$

where, s being arc length, $g_{ij}\alpha^i\alpha^j = 1$. Thus, from each point $\mathbf{x} = \boldsymbol{\beta}$ of space there emanates a geodesic ray in every direction (unit vector) $\boldsymbol{\alpha}$.

6.6 RULES FOR TENSOR DIFFERENTIATION

Confidence in the preceding differentiation formulas should be considerably improved when it is learned (see Problem 6.15) that the same basic rules for differentiation from calculus carry over to covariant and absolute differentiation of tensors. For arbitrary tensors **T** and **S**, we have:

Rules for Covariant Differentiation

sum	$(\mathbf{T} + \mathbf{S})_{,k} = \mathbf{T}_{,k} + \mathbf{S}_{,k}$
outer product	$[\mathbf{TS}]_{,k} = [\mathbf{T}_{,k}\mathbf{S}] + [\mathbf{TS}_{,k}]$
inner product	$(\mathbf{TS})_{,k} = \mathbf{T}_{,k}\mathbf{S} + \mathbf{TS}_{,k}$

Since the absolute derivative along a curve is the inner product of the covariant derivative and the tangent vector, the above rules for differentiation repeat:

Rules for Absolute Differentiation

sum	$\dfrac{\delta}{\delta t}(\mathbf{T} + \mathbf{S}) = \dfrac{\delta \mathbf{T}}{\delta t} + \dfrac{\delta \mathbf{S}}{\delta t}$
outer product	$\dfrac{\delta}{\delta t}[\mathbf{TS}] = \left[\dfrac{\delta \mathbf{T}}{\delta t}\mathbf{S}\right] + \left[\mathbf{T}\dfrac{\delta \mathbf{S}}{\delta t}\right]$
inner product	$\dfrac{\delta}{\delta t}(\mathbf{TS}) = \dfrac{\delta \mathbf{T}}{\delta t}\mathbf{S} + \mathbf{T}\dfrac{\delta \mathbf{S}}{\delta t}$

In addition, we note that $\dfrac{\delta}{\delta t}\mathbf{g} = 0$ where **g** is the metric tensor (see Problem 6.11).

Solved Problems

CHRISTOFFEL SYMBOLS OF THE FIRST KIND

6.1 Verify that $\Gamma_{ijk} = \Gamma_{jik}$.

By definition,

$$\Gamma_{ijk} = \tfrac{1}{2}(-g_{ijk} + g_{jki} + g_{kij}) \qquad \text{and} \qquad \Gamma_{jik} = \tfrac{1}{2}(-g_{jik} + g_{ikj} + g_{kji})$$

But $g_{ijk} = g_{jik}$, $g_{jki} = g_{kji}$, and $g_{kij} = g_{ikj}$, by symmetry of g_{ij}, and the result follows.

6.2 Show that if (g_{ij}) is a diagonal matrix, then for all fixed subscripts α and $\beta \neq \alpha$ in the range $1, 2, \ldots, n$,

(*a*) $\Gamma_{\alpha\alpha\alpha} = \tfrac{1}{2}\, g_{\alpha\alpha\alpha}$ (not summed on α)

(*b*) $-\Gamma_{\alpha\alpha\beta} = \Gamma_{\alpha\beta\alpha} = \Gamma_{\beta\alpha\alpha} = \tfrac{1}{2}\, g_{\alpha\alpha\beta}$ (not summed on α)

(*c*) All remaining Christoffel symbols Γ_{ijk} are zero.

(*a*) By definition, $\Gamma_{\alpha\alpha\alpha} = \frac{1}{2}(-g_{\alpha\alpha\alpha} + g_{\alpha\alpha\alpha} + g_{\alpha\alpha\alpha}) = \frac{1}{2} g_{\alpha\alpha\alpha}$.

(*b*) Since $\alpha \neq \beta$,

$$-\Gamma_{\alpha\alpha\beta} = -\frac{1}{2}(-g_{\alpha\alpha\beta} + g_{\alpha\beta\alpha} + g_{\beta\alpha\alpha}) = -\frac{1}{2}(-g_{\alpha\alpha\beta} + 0 + 0) = \frac{1}{2} g_{\alpha\alpha\beta}$$

$$\Gamma_{\alpha\beta\alpha} = \Gamma_{\beta\alpha\alpha} = \frac{1}{2}(-g_{\alpha\beta\alpha} + g_{\beta\alpha\alpha} + g_{\alpha\alpha\beta}) = \frac{1}{2}(-0 + 0 + g_{\alpha\alpha\beta}) = \frac{1}{2} g_{\alpha\alpha\beta}$$

(*c*) Let i, j, k be distinct subscripts. Then $g_{ij} = 0$ and $g_{ijk} = 0$, implying that $\Gamma_{ijk} = 0$.

6.3 Is it true that if all Γ_{ijk} vanish in some coordinate system, then the metric tensor has constant components in every coordinate system?

By Lemma 6.2, the conclusion would be valid if the Γ_{ijk} vanished in every coordinate system. But (Γ_{ijk}) is not a tensor, and the conclusion is false. For instance, all $\bar{\Gamma}_{ijk} = 0$ for the Euclidean metric in rectangular coordinates, but $g_{22} = (x^1)^2$ in spherical coordinates.

CHRISTOFFEL SYMBOLS OF THE SECOND KIND

6.4 If (g_{ij}) is a diagonal matrix, show that for all fixed indices (no summation) in the range $1, 2, \ldots, n$,

(*a*) $\Gamma^{\alpha}_{\alpha\beta} = \Gamma^{\alpha}_{\beta\alpha} = \dfrac{\partial}{\partial x^{\beta}}\left(\dfrac{1}{2}\ln|g_{\alpha\alpha}|\right)$

(*b*) $\Gamma^{\alpha}_{\beta\beta} = -\dfrac{1}{2g_{\alpha\alpha}}\dfrac{\partial}{\partial x^{\alpha}}(g_{\beta\beta}) \quad (\alpha \neq \beta)$

(*c*) All other Γ^{i}_{jk} vanish.

(*a*) Both (g_{ij}) and $(g_{ij})^{-1} = (g^{ij})$ are diagonal, with nonzero diagonal elements. Thus,

$$\Gamma^{\alpha}_{\alpha\beta} = g^{\alpha j}\Gamma_{\alpha\beta j} = g^{\alpha\alpha}\Gamma_{\alpha\beta\alpha} = \frac{1}{g_{\alpha\alpha}}\left(\frac{1}{2}\frac{\partial g_{\alpha\alpha}}{\partial x^{\beta}}\right) = \frac{\partial}{\partial x^{\beta}}\left(\frac{1}{2}\ln|g_{\alpha\alpha}|\right)$$

(*b*)

$$\Gamma^{\alpha}_{\beta\beta} = g^{\alpha\alpha}\Gamma_{\beta\beta\alpha} = \frac{1}{g_{\alpha\alpha}}\left(-\frac{1}{2} g_{\beta\beta\alpha}\right)$$

(*c*) When i, j, k are distinct, $\Gamma^{i}_{jk} = g^{ir}\Gamma_{jkr} = g^{ii}\Gamma_{jki} = 0$ (not summed on i).

6.5 Calculate the Christoffel symbols of the second kind for the Euclidean metric in spherical coordinates, using Problem 6.4.

We have $g_{11} = 1$, $g_{22} = (x^1)^2$, and $g_{33} = (x^1)^2 \sin^2 x^2$. Noting that g_{11} is a constant and that all $g_{\alpha\alpha}$ are independent of x^3, we obtain the following nonzero symbols from Problem 6.4(*a*):

$$\Gamma^{2}_{21} = \Gamma^{2}_{12} = \frac{\partial}{\partial x^1}\left(\frac{1}{2}\ln(x^1)^2\right) = \frac{1}{x^1}$$

$$\Gamma^{3}_{31} = \Gamma^{3}_{13} = \frac{\partial}{\partial x^1}\left(\frac{1}{2}\ln((x^1)^2\sin^2 x^2)\right) = \frac{1}{x^1}$$

$$\Gamma^{3}_{32} = \Gamma^{3}_{23} = \frac{\partial}{\partial x^2}\left(\frac{1}{2}\ln((x^1)^2\sin^2 x^2)\right) = \cot x^2$$

Similarly, from Problem 6.4(*b*),

$$\Gamma^{1}_{22} = -\frac{1}{2(1)}\frac{\partial}{\partial x^1}(x^1)^2 = -x^1$$

$$\Gamma^{1}_{33} = -\frac{1}{2(1)}\frac{\partial}{\partial x^1}((x^1)^2\sin^2 x^2) = -x^1\sin^2 x^2$$

$$\Gamma^{2}_{33} = -\frac{1}{2(x^1)^2}\frac{\partial}{\partial x^2}((x^1)^2\sin^2 x^2) = -\sin x^2\cos x^2$$

6.6 Use (*6.6*) to find the most general 3-dimensional transformation $x^i = x^i(\bar{\mathbf{x}})$ of coordinates such that (x^i) is rectangular and $(\bar{x}^i)$ is any other coordinate system for which the Christoffel symbols are

$$\bar{\Gamma}^1_{11} = 1 \qquad \bar{\Gamma}^2_{22} = 2 \qquad \bar{\Gamma}^3_{33} = 3 \qquad \text{all others} = 0$$

Since $\Gamma^r_{st} = 0$, (*6.6*) reduces to the system of linear partial differential equations with constant coefficients:

$$\frac{\partial^2 x^r}{\partial \bar{x}^i \partial \bar{x}^j} = \bar{\Gamma}^s_{ij} \frac{\partial x^r}{\partial \bar{x}^s} \tag{1}$$

It is simplest first to solve the intermediate, first-order system

$$\frac{\partial \bar{u}^r_j}{\partial \bar{x}^i} = \bar{\Gamma}^s_{ij} \bar{u}^r_s \qquad \left(\bar{u}^r_s = \frac{\partial x^r}{\partial \bar{x}^s}\right) \tag{2}$$

Since the systems (*2*) for $r = 1, 2, 3$ are the same, temporarily replace $\bar{u}^r_s$ by $\bar{u}_s$, and x^r by a single variable x; thus,

$$\frac{\partial \bar{u}_j}{\partial \bar{x}^i} = \bar{\Gamma}^s_{ij} \bar{u}_s \tag{3}$$

For $j = 1$, (*3*) becomes

$$\frac{\partial \bar{u}_1}{\partial \bar{x}^1} = \bar{\Gamma}^1_{11}\bar{u}_1 + \bar{\Gamma}^2_{11}\bar{u}_2 + \bar{\Gamma}^3_{11}\bar{u}_3 = \bar{u}_1$$

$$\frac{\partial \bar{u}_1}{\partial \bar{x}^2} = \bar{\Gamma}^1_{21}\bar{u}_1 + \bar{\Gamma}^2_{21}\bar{u}_2 + \bar{\Gamma}^3_{21}\bar{u}_3 = 0$$

$$\frac{\partial \bar{u}_1}{\partial \bar{x}^3} = \bar{\Gamma}^1_{31}\bar{u}_1 + \bar{\Gamma}^2_{31}\bar{u}_2 + \bar{\Gamma}^3_{31}\bar{u}_3 = 0$$

Hence $\bar{u}_1$ is a function of $\bar{x}^1$ alone, and the first differential equation integrates to give

$$\bar{u}_1 = b_1 \exp \bar{x}^1 \qquad (b_1 = \text{constant})$$

In the same way, we find for $j = 2$ and $j = 3$:

$$\bar{u}_2 = b_2 \exp 2\bar{x}^2 \qquad (b_2 = \text{constant})$$
$$\bar{u}_3 = b_3 \exp 3\bar{x}^3 \qquad (b_3 = \text{constant})$$

Now we return to the equations $\partial x / \partial \bar{x}^i = \bar{u}_i$ with the solutions just found for the $\bar{u}_i$.

$$\frac{\partial x}{\partial \bar{x}^1} = b_1 \exp \bar{x}^1 \qquad \frac{\partial x}{\partial \bar{x}^2} = b_2 \exp 2\bar{x}^2 \qquad \frac{\partial x}{\partial \bar{x}^3} = b_3 \exp 3\bar{x}^3 \tag{4}$$

Integration of the first equation (*4*) yields

$$x = b_1 \exp \bar{x}^1 + \varphi(\bar{x}^2, \bar{x}^3)$$

and then the second and third equations give:

$$\frac{\partial \varphi}{\partial \bar{x}^2} = b_2 \exp 2\bar{x}^2 \qquad \text{or} \qquad \varphi = a_2 \exp 2\bar{x}^2 + \psi(\bar{x}^3)$$

$$\frac{d\psi}{d\bar{x}^3} = b_3 \exp 3\bar{x}^3 \qquad \text{or} \qquad \psi = a_3 \exp 3\bar{x}^3 + a_4$$

This means that, with $a_1 = b_1$,

$$x = a_1 \exp \bar{x}^1 + a_2 \exp 2\bar{x}^2 + a_3 \exp 3\bar{x}^3 + a_4$$

so that the general solution of (*1*) is

$$x^r = a^r_1 \exp \bar{x}^1 + a^r_2 \exp 2\bar{x}^2 + a^r_3 \exp 3\bar{x}^3 + a^r_4 \tag{5}$$

for $r = 1, 2, 3$.

The constants a^r_4 in (*5*) are unimportant; they merely allow any point in $\mathbf{R}^3$ to serve as the origin of the rectangular system (x^r). The remaining constants may be chosen at will, subject to a single condition (see Problem 6.27).

COVARIANT DIFFERENTIATION

6.7 Establish the tensor character of $\mathbf{T}_{,k}$ (Definition 2), where $\mathbf{T}$ is a contravariant vector.

Beginning with the transformation law

$$\bar{T}^i = T^r \frac{\partial \bar{x}^i}{\partial x^r}$$

take the partial derivative with respect to $\bar{x}^k$ and use the chain rule:

$$\frac{\partial \bar{T}^i}{\partial \bar{x}^k} = \frac{\partial}{\partial x^s}\left(T^r \frac{\partial \bar{x}^i}{\partial x^r}\right)\frac{\partial x^s}{\partial \bar{x}^k} = \frac{\partial T^r}{\partial x^s}\frac{\partial \bar{x}^i}{\partial x^r}\frac{\partial x^s}{\partial \bar{x}^k} + T^r \frac{\partial^2 \bar{x}^i}{\partial x^s \partial x^r}\frac{\partial x^s}{\partial \bar{x}^k}$$

Now use (*6.6*), with barred and unbarred systems interchanged:

$$\frac{\partial \bar{T}^i}{\partial \bar{x}^k} = \frac{\partial T^r}{\partial x^s}\frac{\partial \bar{x}^i}{\partial x^r}\frac{\partial x^s}{\partial \bar{x}^k} + T^r\left(\Gamma^t_{sr}\frac{\partial \bar{x}^i}{\partial x^t} - \bar{\Gamma}^i_{uv}\frac{\partial \bar{x}^u}{\partial x^s}\frac{\partial \bar{x}^v}{\partial x^r}\right)\frac{\partial x^s}{\partial \bar{x}^k}$$

Since $(\partial \bar{x}^u/\partial x^s)(\partial x^s/\partial \bar{x}^k) = \delta^u_k$ and $T^r(\partial \bar{x}^v/\partial x^r) = \bar{T}^v$, this becomes

$$\frac{\partial \bar{T}^i}{\partial \bar{x}^k} = \frac{\partial T^r}{\partial x^s}\frac{\partial \bar{x}^i}{\partial x^r}\frac{\partial x^s}{\partial \bar{x}^k} + \Gamma^t_{sr}T^r\frac{\partial \bar{x}^i}{\partial x^t}\frac{\partial x^s}{\partial \bar{x}^k} - \bar{\Gamma}^i_{kv}\bar{T}^v$$

or (by factoring and using the symmetry of the Christoffel symbols)

$$\frac{\partial \bar{T}^i}{\partial \bar{x}^k} + \bar{\Gamma}^i_{tk}\bar{T}^t = \left(\frac{\partial T^r}{\partial x^s} + \Gamma^r_{ts}T^t\right)\frac{\partial \bar{x}^i}{\partial x^r}\frac{\partial x^s}{\partial \bar{x}^k} \qquad \text{or} \qquad \bar{T}^i_{,k} = T^r_{,s}\frac{\partial \bar{x}^i}{\partial x^r}\frac{\partial x^s}{\partial \bar{x}^k}$$

6.8 Show that $(T^i_{j,k})$, as defined by (*6.7*), is a tensor, using the previously proven facts that $T^i_{,k}$ and $T_{i,k}$ are tensorial for all tensors (T^i) and (T_i).

Let (V_i) be any vector and set $U_j = T^r_j V_r$. The covariant derivative of the tensor (U_j) is the tensor $(U_{j,k})$, where

$$U_{j,k} = \frac{\partial U_j}{\partial x^k} - \Gamma^r_{jk}U_r = \frac{\partial}{\partial x^k}(T^r_j V_r) - \Gamma^r_{jk}(T^s_r V_s) = \frac{\partial T^s_j}{\partial x^k}V_s + T^r_j\frac{\partial V_r}{\partial x^k} - \Gamma^r_{jk}T^s_r V_s$$

But

$$V_{r,k} = \frac{\partial V_r}{\partial x^k} - \Gamma^s_{rk}V_s \qquad \text{or} \qquad \frac{\partial V_r}{\partial x^k} = V_{r,k} + \Gamma^s_{rk}V_s$$

When the above expression for $\partial V_r/\partial x^k$ is substituted into the preceding equation and the terms rearranged, the result is:

$$\left(\frac{\partial T^s_j}{\partial x^k} + \Gamma^s_{rk}T^r_j - \Gamma^r_{jk}T^s_r\right)V_s = U_{j,k} - T^r_j V_{r,k}$$

i.e., $$T^s_{j,k}V_s = \text{tensor component}$$

It follows at once from the Quotient Theorem (Theorem 4.2) that $(T^i_{j,k})$ is a tensor.

6.9 Extend the notion of covariant differentiation so that it will apply to invariants.

First note that the partial derivative of an invariant is a tensor:

$$\frac{\partial \bar{E}}{\partial \bar{x}^i} = \frac{\partial E}{\partial \bar{x}^i} = \frac{\partial E}{\partial x^r}\frac{\partial x^r}{\partial \bar{x}^i}$$

Now, under any reasonable definition, $(E_{,i})$ must (1) be a tensor; (2) coincide with $(\partial E/\partial x^i)$ in rectangular coordinates. The obvious choice is therefore

$$(E_{,i}) \equiv \left(\frac{\partial E}{\partial x^i}\right)$$

6.10 Write the formula for the covariant derivative indicated by $T^{ij}_{k,l}$.

$$T^{ij}_{k,l} = \frac{\partial T^{ij}_k}{\partial x^l} + \Gamma^i_{rl} T^{rj}_k + \Gamma^j_{rl} T^{ir}_k - \Gamma^r_{kl} T^{ij}_r$$

6.11 Prove that the metric tensor behaves like a constant under covariant differentiation; i.e., $g_{ij,k} = 0$ for all i, j, k. Hence it follows that the absolute derivative of g_{ij} along any curve is zero.

By definition, since (g_{ij}) is covariant of order 2,

$$g_{ij,k} = \frac{\partial g_{ij}}{\partial x^k} - \Gamma^r_{ik} g_{rj} - \Gamma^r_{jk} g_{ir} = g_{ijk} - \Gamma_{ikj} - \Gamma_{jki} = 0$$

by (*6.2*). (In a similar manner, it follows that $g^{ij}_{,k} = 0$; see Problem 6.34.)

Because of the above property of the metric tensor and its inverse, *the operation of covariant differentiation commutes with those of raising and lowering indices*. For example,

$$T^i_{j,k} = (g^{ir} T_{rj})_{,k} = g^{ir} T_{rj,k}$$

ABSOLUTE DIFFERENTIATION

6.12 Prove that (*6.8*) is the result of forming the inner product of the covariant derivative $(T^i_{,j})$ with the tangent vector (dx^i/dt) of the curve.

$$T^i_{,j} \frac{dx^j}{dt} = \left(\frac{\partial T^i}{\partial x^j} + \Gamma^i_{rj} T^r \right) \frac{dx^j}{dt} = \frac{\partial T^i}{\partial x^j} \frac{dx^j}{dt} + \Gamma^i_{rj} T^r \frac{dx^j}{dt} = \frac{dT^i}{dt} + \Gamma^i_{rj} T^r \frac{dx^j}{dt}$$

6.13 A particle is in motion along the circular arc given parametrically in spherical coordinates by $x^1 = b,\ x^2 = \pi/4,\ x^3 = \omega t\quad (t = \text{time})$. Find its acceleration using the formula (*6.10*) and compare with the result $a = r\omega^2$ from elementary mechanics.

From Problem 6.5, we have along the circle

$$\Gamma^1_{22} = -x^1 = -b \qquad \Gamma^1_{33} = -x^1 \sin^2 x^2 = -b \sin^2 \frac{\pi}{4} = -\frac{b}{2}$$

$$\Gamma^2_{12} = \Gamma^2_{21} = \frac{1}{x^1} = \frac{1}{b} \qquad \Gamma^2_{33} = -\sin x^2 \cos x^2 = -\sin \frac{\pi}{4} \cos \frac{\pi}{4} = -\frac{1}{2}$$

$$\Gamma^3_{13} = \Gamma^3_{31} = \frac{1}{x^1} = \frac{1}{b} \qquad \Gamma^3_{23} = \Gamma^3_{32} = \cot x^2 = \cot \frac{\pi}{4} = 1$$

with all other symbols vanishing. The components of acceleration are, from (*6.9*),

$$a^1 = \frac{d^2x^1}{dt^2} + \Gamma^1_{rs} \frac{dx^r}{dt} \frac{dx^s}{dt} = 0 + \Gamma^1_{22} \left(\frac{dx^2}{dt} \right)^2 + \Gamma^1_{33} \left(\frac{dx^3}{dt} \right)^2 = 0 + \left(-\frac{b}{2} \right) (\omega)^2 = -\frac{b\omega^2}{2}$$

$$a^2 = \frac{d^2x^2}{dt^2} + \Gamma^2_{rs} \frac{dx^r}{dt} \frac{dx^s}{dt} = 0 + 2\Gamma^2_{12} \frac{dx^1}{dt} \frac{dx^2}{dt} + \Gamma^2_{33} \left(\frac{dx^3}{dt} \right)^2 = 0 + \left(-\frac{1}{2} \right) (\omega)^2 = -\frac{\omega^2}{2}$$

$$a^3 = \frac{d^2x^3}{dt^2} + \Gamma^3_{rs} \frac{dx^r}{dt} \frac{dx^s}{dt} = 0 + 2\Gamma^3_{13} \frac{dx^1}{dt} \frac{dx^3}{dt} + 2\Gamma^3_{23} \frac{dx^2}{dt} \frac{dx^3}{dt} = 0$$

Together with the metric components along the circle,

$$g_{11} = 1 \qquad g_{22} = (x^1)^2 = b^2 \qquad g_{33} = (x^1)^2 \sin^2 x^2 = \frac{b^2}{2}$$

the acceleration components give, via (*6.10*),

$$a = \sqrt{g_{ij} a^i a^j} = \sqrt{(1)(-b\omega^2/2)^2 + (b^2)(-\omega^2/2)^2 + 0} = b\omega^2/\sqrt{2}$$

Upon introducing the radius of the circle, using (*3.4*) with $\bar{x}^1 = x = r$ and $x^3 = 0$,

$$r = b \sin \frac{\pi}{4} = \frac{b}{\sqrt{2}}$$

we obtain $a = r\omega^2$.

6.14 Verify that $x^1 = a \sec x^2$ is a geodesic for the Euclidean metric in polar coordinates. [In rectangular coordinates (x, y), the curve is $x = a$, a vertical line.]

First choose a parameterization for the curve; say,

$$\begin{aligned} x^1 &= a \sec t \\ x^2 &= t \end{aligned} \qquad (-\pi/2 < t < \pi/2) \tag{1}$$

Parameter t is related to the arc-length parameter s via

$$\frac{ds}{dt} = \sqrt{g_{ij}\frac{dx^i}{dt}\frac{dx^j}{dt}} = \sqrt{\left(\frac{dx^1}{dt}\right)^2 + (a \sec t)^2\left(\frac{dx^2}{dt}\right)^2} = \sqrt{a^2 \sec^2 t \tan^2 t + a^2 \sec^2 t}$$
$$= (a \sec t)\sqrt{1 + \tan^2 t} = a \sec^2 t$$

or

$$\frac{dt}{ds} = \frac{\cos^2 t}{a}$$

so that for any function $x(t)$

$$\frac{dx}{ds} = \frac{dx}{dt}\frac{dt}{ds} = \frac{\cos^2 t}{a}\frac{dx}{dt}$$
$$\frac{d^2x}{ds^2} = \frac{d}{dt}\left(\frac{\cos^2 t}{a}\frac{dx}{dt}\right)\frac{dt}{ds} = \frac{\cos^4 t}{a^2}\frac{d^2x}{dt^2} - \frac{2 \sin t \cos^3 t}{a^2}\frac{dx}{dt}$$

Now, taking the nonzero Christoffel symbols from Example 6.3, we can rewrite the geodesic equations (*6.13*) in terms of the independent variable t:

$$0 = \frac{d^2x^1}{ds^2} + \Gamma^1_{22}\left(\frac{dx^2}{ds}\right)^2 = \frac{\cos^4 t}{a^2}\frac{d^2x^1}{dt^2} - \frac{2 \sin t \cos^3 t}{a^2}\frac{dx^1}{dt} + (-x^1)\frac{\cos^4 t}{a^2}\left(\frac{dx^2}{dt}\right)^2$$

or

$$\frac{d^2x^1}{dt^2} - (2 \tan t)\frac{dx^1}{dt} - x^1\left(\frac{dx^2}{dt}\right)^2 = 0 \tag{2}$$

and

$$0 = \frac{d^2x^2}{ds^2} + 2\Gamma^2_{12}\frac{dx^1}{ds}\frac{dx^2}{ds} = \frac{\cos^4 t}{a^2}\frac{d^2x^2}{dt^2} - \frac{2 \sin t \cos^3 t}{a^2}\frac{dx^2}{dt} + 2\left(\frac{1}{x^1}\right)\left(\frac{\cos^2 t}{a}\right)^2\frac{dx^1}{dt}\frac{dx^2}{dt}$$

or

$$\frac{d^2x^2}{dt^2} - (2 \tan t)\frac{dx^2}{dt} + \left(\frac{2}{x^1}\right)\frac{dx^1}{dt}\frac{dx^2}{dt} = 0 \tag{3}$$

All that remains is to verify that the functions (*1*) satisfy the system (*2*)–(*3*). Substituting in (*2*):

$$a(\sec t + 2 \sec t \tan^2 t) - (2 \tan t)(a \sec t \tan t) - (a \sec t)(1) = 0$$

Substituting in (3):

$$0 - (2 \tan t)(1) + \left(\frac{2}{a \sec t}\right)(a \sec t \tan t)(1) = 0 \qquad \text{QED}$$

DIFFERENTIATION RULES

6.15 Prove the rules for covariant differentiation stated in Section 6.6.

(*a*) The ***sum rule*** obviously holds, as (*6.7*) is linear in the tensor components.

(*b*) Let $\mathbf{T} = (T^i_j)$ and $\mathbf{S} = (S^i_j)$ be two mixed tensors of order 2, with outer product $\mathbf{U} = (T^p_r S^q_s)$. Then,

$$T^p_{r,k}S^q_s + T^p_r S^q_{s,k}$$
$$= \left(\frac{\partial T^p_r}{\partial x^k} + \Gamma^p_{tk}T^t_r - \Gamma^t_{rk}T^p_t\right)S^q_s + T^p_r\left(\frac{\partial S^q_s}{\partial x^k} + \Gamma^q_{tk}S^t_s - \Gamma^t_{sk}S^q_t\right)$$

$$= \underbrace{\left(\frac{\partial T^p_r}{\partial x^k} S^q_s + T^p_r \frac{\partial S^q_s}{\partial x^k}\right)}_{\partial U^{pq}_{rs}/\partial x^k} + \Gamma^p_{tk} U^{tq}_{rs} + \Gamma^q_{tk} U^{pt}_{rs} - \Gamma^t_{rk} U^{pq}_{ts} - \Gamma^t_{sk} U^{pq}_{rt}$$

$$\equiv U^{pq}_{rs,k}$$

and this proof of the ***outer-product rule*** extends to arbitrary **T** and **S**.

(*c*) The ***inner-product rule*** follows from the outer-product rule and the following useful result: Contraction of indices and covariant differentiation commute. To prove this last, let $\mathbf{R} = (R^{ij}_k)$. Then,

$$R^{ij}_{k,l}\delta^k_j = \left(\frac{\partial R^{ij}_k}{\partial x^l} + \Gamma^i_{tl} R^{tj}_k + \Gamma^j_{tl} R^{it}_k - \Gamma^t_{kl} R^{ij}_t\right)\delta^k_j$$

$$= \frac{\partial R^{ik}_k}{\partial x^l} + \Gamma^i_{tl} R^{tk}_k + 0 = (R^{ik}_k)_{,l} \qquad \text{QED}$$

6.16 Instead of Problem 6.15, why not: "Each rule is a tensor equation that is valid in rectangular coordinates, where covariant differentiation reduces to partial differentiation. Therefore, each rule holds in every coordinate system."?

If the space metric is non-Euclidean, there is no way to transform to a rectangular coordinate system (in which the rules would indeed hold).

6.17 Infer the outer-product rule for absolute differentation from the corresponding rule for covariant differentiation.

Let $\mathbf{x} = \mathbf{x}(t)$ be any curve, and $\mathbf{T}(\mathbf{x}(t))$ and $\mathbf{S}(\mathbf{x}(t))$ two tensors defined on the curve. Then,

$$\frac{\delta}{\delta t}[\mathbf{TS}] = [\mathbf{TS}]_{,k}\frac{dx^k}{dt} = ([\mathbf{T}_{,k}\mathbf{S}] + [\mathbf{TS}_{,k}])\frac{dx^k}{dt} = \left[\mathbf{T}_{,k}\frac{dx^k}{dt}\mathbf{S}\right] + \left[\mathbf{T}\,\mathbf{S}_{,k}\frac{dx^k}{dt}\right] = \left[\frac{\delta\mathbf{T}}{\delta t}\mathbf{S}\right] + \left[\mathbf{T}\frac{\delta\mathbf{S}}{\delta t}\right]$$

UNIQUENESS OF THE ABSOLUTE DERIVATIVE

6.18 Prove Theorem 6.4.

Denote by $\Delta\mathbf{T}/\Delta t$ any tensor that satisfies the hypothesis of the theorem. The tensor equation

$$\frac{\Delta\mathbf{T}}{\Delta t} = \frac{\delta\mathbf{T}}{\delta t}$$

is valid in rectangular coordinates (x^i), since, in (x^i), both sides coincide with $d\mathbf{T}/dt$. But then (Section 4.3) the equation holds in every coordinate system; i.e.,

$$\frac{\Delta\mathbf{T}}{\Delta t} \equiv \frac{\delta\mathbf{T}}{\delta t}$$

Supplementary Problems

6.19 Find the general solution of the linear system

$$\frac{\partial^2 \bar{x}^i}{\partial x^j \partial x^k} = a^i_{jk} = \text{const.}$$

with a^i_{jk} symmetric in the two lower subscripts. [*Hint*: Set $y^i_k = \partial\bar{x}^i/\partial x^k - a^i_{rk}x^r$.]

6.20 A two-dimensional coordinate system (x^i) is connected to a rectangular coordinate system $(\bar{x}^i)$ through

$$\bar{x}^1 = 2(x^1)^2 + x^2 \qquad \bar{x}^2 = -x^1 + 3x^2$$

(*a*) Exhibit the metric tensor in (x^i).

(*b*) Calculate the Christoffel symbols of the first kind for (x^i) directly from the definition (*6.1*).

6.21 (*a*) Derive the formula

$$\Gamma_{ijk} = \frac{\partial^2 \bar{x}^r}{\partial x^i \partial x^j} \frac{\partial \bar{x}^r}{\partial x^k}$$

when $(\bar{x}^i)$ is rectangular and (x^i) is any other coordinate system. [*Hint*: Interchange barred and unbarred coordinate systems in (*6.3*), and use $g_{ij} = \delta_{ij}$]. (*b*) Derive the analogous formula

$$\Gamma^i_{jk} = \frac{\partial^2 \bar{x}^r}{\partial x^j \partial x^k} \frac{\partial x^i}{\partial \bar{x}^r}$$

when $(\bar{x}^i)$ is such that all $\bar{g}_{ij}$ are constant. [*Hint*: Interchange barred and unbarred coordinate systems in (*6.5*).]

6.22 Let the coordinate system (x^i) be connected to a system of rectangular coordinates $(\bar{x}^i)$ via

$$\bar{x}^1 = \exp(x^1 + x^2) \qquad \bar{x}^2 = \exp(x^1 - x^2)$$

Use Problem 6.21(*b*) to compute the nonzero Christoffel symbols of the second kind for (x^i).

6.23 If

$$\begin{aligned}
\bar{x}^1 &= -\exp d_1x^1 + \exp d_2x^2 + \exp d_3x^3 \\
\bar{x}^2 &= 2\exp d_1x^1 - \exp d_2x^2 + \exp d_3x^3 \\
\bar{x}^3 &= \exp d_1x^1 - 2\exp d_2x^2 + 3\exp d_3x^3
\end{aligned}$$

and if all $\bar{\Gamma}^i_{jk} = 0$, find the Γ^i_{jk}.

6.24 Derive (*6.6*) by solving (*6.5*) for the second derivative and then changing indices.

6.25 Prove that all Γ^i_{jk} vanish only if all g_{ij} are constant.

6.26 Calculate the nonzero Christoffel symbols of both kinds for the Euclidean metric in cylindrical coordinates, (*5.3*).

6.27 Express the condition that the transformation (*5*) of Problem 6.6 be bijective (Section 2.6).

6.28 Show that if Γ_{ijk} are constant, then g_{ij} are linear in the variables (x^i); but that this is not necessarily true if Γ^i_{jk} are constant. (For a counterexample, use the metric $g_{11} = \exp 2x^1$, $g_{12} = g_{21} = 0$, $g_{22} = 1$.)

6.29 What is the most general two-dimensional transformation $\bar{x}^i = \bar{x}^i(\mathbf{x})$ of coordinates such that $(\bar{x}^i)$ are rectangular and the Christoffel symbols Γ^i_{jk} in (x^i) are those for the metric of polar coordinates (Example 6.3)?

6.30 Is the covariant derivative of a tensor with constant components equal to zero as in ordinary differentiation? Explain your answer.

6.31 If T^i_{jrs} are tensor components, write out the components of the covariant derivative, $T^i_{jrs,k}$.

6.32 Show that $\delta^i_{j,k} = 0$ for all i, j, k.

6.33 For any tensor $\mathbf{T}$, verify that $(\mathbf{g} * \mathbf{T})_{,k} = \mathbf{g} * \mathbf{T}_{,k}$, where $*$ denotes either an outer or inner product.

6.34 Use Problem 6.32 and $g_{jr}g^{ri} = \delta^i_j$ to show that the covariant derivative of $\mathbf{g}^{-1}$ is zero.

6.35 Use the recursive method of Problem 6.8 to verify that $(T_{ij,k})$ is a tensor.

6.36 Using tensor methods in polar coordinates, find the curvature of the circle

$$x^1 = b \qquad x^2 = t$$

6.37 If the metric for (x^i) is

$$G = \begin{bmatrix} (x^1)^2 & 0 \\ 0 & 1 \end{bmatrix}$$

(*a*) write the differential equations of the geodesics in terms of the dependent variables $u = (x^1)^2$ and $v = x^2$; (*b*) integrate these equations and eliminate the arc-length parameter from the solution.

6.38 Find the geodesics on the surface of a sphere of radius a by (*a*) writing the geodesic equations for the spherical coordinates x^2 and x^3 (the x^1-equation is trivial for $x^1 = a =$ const. and may be ignored); (*b*) exhibiting a particular solution of these two equations; and (*c*) generalizing on (*b*). Use Problem 6.5 for the Christoffel symbols.

Chapter 7

Riemannian Geometry of Curves

7.1 INTRODUCTION

At this point, some new terminology is introduced, which commemorates the general formulation of n-dimensional geometry by Bernhard Riemann (1826–1866).

Definition 1: A *Riemannian space* is the space $\mathbf{R}^n$ coordinatized by (x^i), together with a *fundamental form* or *Riemannian metric*, $g_{ij}\,dx^i\,dx^j$, where $\mathbf{g} = (g_{ij})$ obeys conditions A–D of Section 5.3.

Thus, in our preliminary treatment of angles, tangents, normals, and geodesic curves, in Chapters 5 and 6, we already entered Riemannian geometry—though largely restricted to familiar 3-dimensional coordinate systems and a positive definite (Euclidean) metric. The present chapter focuses on the theory of curves in a Riemannian space with an indefinite metric. It also takes up geodesics from a different viewpoint.

7.2 LENGTH AND ANGLE UNDER AN INDEFINITE METRIC

Formulas (*5.10*) and (*5.11*) must be generalized to allow for changes in sign of the fundamental form.

Definition 2: The *norm* of an arbitrary (contravariant or covariant) vector $\mathbf{V}$ is

$$\|\mathbf{V}\| \equiv \sqrt{\varepsilon \mathbf{V}^2} = \sqrt{\varepsilon V_i V^i} \qquad (\varepsilon = \varepsilon(\mathbf{V}))$$

where $\varepsilon(\ )$ is the indicator function (Section 5.3).

Under this definition, $\|\mathbf{V}\| \geqq 0$, but it is possible that $\|\mathbf{V}\| = 0$ for $\mathbf{V} \neq \mathbf{0}$; such a vector is called a *null vector*. Moreover, the triangle inequality is not necessarily obeyed by this norm (see Problem 7.8).

If $\mathbf{V}(t)$ is the tangent field to the curve $x^i = x^i(t)$ $\quad (a \leqq t \leqq b)$, then the length formula (*5.1a*) may be written as

$$L = \int_a^b \sqrt{\varepsilon g_{ij} \frac{dx^i}{dt} \frac{dx^j}{dt}}\, dt = \int_a^b \|\mathbf{V}(t)\|\, dt \tag{7.1}$$

The angle between non-null contravariant vectors is still defined by (*5.11*), provided the new norm is understood:

$$\cos\theta = \frac{\mathbf{UV}}{\|\mathbf{U}\|\,\|V\|} = \frac{g_{ij}U^iV^j}{\sqrt{\varepsilon_1 g_{pq} U^p U^q}\,\sqrt{\varepsilon_2 g_{rs} V^r V^s}} \tag{7.2}$$

where $\varepsilon_1 = \varepsilon(\mathbf{U})$ and $\varepsilon_2 = \varepsilon(\mathbf{V})$. Because of the indefiniteness of the metric, we must distinguish two possibilities in the application of (*7.2*).

Case 1: $|\mathbf{UV}| \leqq \|\mathbf{U}\|\,\|\mathbf{V}\|$ (the Cauchy–Schwarz inequality holds for $\mathbf{U}$ and $\mathbf{V}$). Then θ is uniquely determined as a real number in the interval $[0, \pi]$.

Case 2: $|\mathbf{UV}| > \|\mathbf{U}\|\,\|\mathbf{V}\|$ (the Cauchy–Schwarz does not hold). Then (*7.2*) takes the form

$$\cos\theta = k \qquad (|k| > 1)$$

which has an infinite number of solutions for θ, all of them complex. By convention, we always choose the solution

$$\theta = \begin{cases} i \ln (k + \sqrt{k^2 - 1}) & k > 1 \\ \pi + i \ln (-k + \sqrt{k^2 - 1}) & k < -1 \end{cases}$$

that exhibits the proper limiting behavior as $k \to 1^+$ or $k \to -1^-$.

EXAMPLE 7.1 At the points of intersection, find the angles between the curves (i.e., between their tangents)

$$\mathscr{C}_1 \; : \; (x_1^i) = (t, 0, 0, t^2) \qquad \mathscr{C}_2 \; : \; (x_2^i) = (u, 0, 0, 2 - u^2)$$

(t, u real), if the Riemannian metric is

$$\varepsilon \, ds^2 = (dx^1)^2 + (dx^2)^2 + (dx^3)^2 - (dx^4)^2$$

[This is the metric of Special Relativity, with $x^4 \equiv$ (speed of light) $\times$ (time).]

The curves meet in the two points $P(1, 0, 0, 1)$ and $Q(-1, 0, 0, 1)$. At P, (where $t = u = 1$) the two tangent vectors are

$$\mathbf{U}_P = (dx_1^i/dt)_P = (1, 0, 0, 2t)_P = (1, 0, 0, 2)$$
$$\mathbf{V}_P = (dx_2^i/du)_P = (1, 0, 0, -2u)_P = (1, 0, 0, -2)$$

so that (*7.2*) gives

$$\cos \theta_P = \frac{1(1)(1) + 1(0)(0) + 1(0)(0) - 1(2)(-2)}{\sqrt{\varepsilon_1[1(1)^2 + 1(0)^2 + 1(0)^2 - 1(2)^2]} \, \sqrt{\varepsilon_2[1(1)^2 + 1(0)^2 + 1(0)^2 - 1(-2)^2]}}$$
$$= \frac{5}{\sqrt{+3}\,\sqrt{+3}} = \frac{5}{3}$$

and $\theta_P = i \ln [(5/3) + \sqrt{(5/3)^2 - 1}] = i \ln 3$.

Similarly we calculate (for $t = u = -1$)

$$\mathbf{U}_Q = (1, 0, 0, -2) = \mathbf{V}_P \qquad \mathbf{V}_Q = (1, 0, 0, 2) = \mathbf{U}_P$$

so that $\theta_Q = \theta_P$.

7.3 NULL CURVES

If $\mathbf{g}$ is not required to be positive definite, a curve can have zero length.

EXAMPLE 7.2 In $\mathbf{R}^4$, under the metric of Example 7.1, consider the curve

$$x^1 = 3 \cos t \qquad x^2 = 3 \sin t \qquad x^3 = 4t \qquad x^4 = 5t$$

for $0 \leqq t \leqq 1$. Along the curve,

$$\left(\frac{dx^i}{dt}\right) = (-3 \sin t, 3 \cos t, 4, 5)$$

$$\varepsilon \left(\frac{ds}{dt}\right)^2 = g_{ij} \frac{dx^i}{dt} \frac{dx^j}{dt} = (-3 \sin t)^2 + (3 \cos t)^2 + (4)^2 - (5)^2 = 0$$

and so the arc length is

$$L = \int_0^1 0 \, dt = 0$$

A curve is *null* if it or any of its subarcs has zero length. Here, a subarc is understood to be nontrivial; that is, it consists of more than one point and corresponds to an interval $c \leqq t \leqq d$, where $c < d$. A curve is *null at a point* if for some value of the parameter t the tangent vector is a null vector; i.e.,

$$g_{ij} \frac{dx^i}{dt} \frac{dx^j}{dt} = 0$$

The set of t-values at which the curve is null is known as the *null set* of the curve.

Under the above definitions, a curve can be null without having zero length (if there is a subarc with zero length); but a curve having zero length is necessarily null at every point, and hence a null curve. Example 7.3 exhibits a null curve having positive length.

EXAMPLE 7.3 Under the Riemannian metric

$$G = \begin{bmatrix} (x^1)^2 & -1 \\ -1 & 0 \end{bmatrix}$$

the curve $(x^1, x^2) = (t, |t^3|/6)$ possesses a null subarc that renders the length of the curve much smaller than might be expected. In fact, because $dx^1/dt = 1$ and $dx^2/dt = \delta\, t^2/2$, where $\delta = \pm 1$ and is positive only if $t \geqq 0$,

$$\varepsilon g_{ij} \frac{dx^i}{dt}\frac{dx^j}{dt} = \varepsilon\left[(x^1)^2\left(\frac{dx^1}{dt}\right)^2 - 2\frac{dx^1}{dt}\frac{dx^2}{dt}\right] = \varepsilon[t^2(1) - 2(1)(\delta\, t^2/2)] = \varepsilon(t^2 - \delta\, t^2)$$

Since the quantity following the indicator is nonnegative, $\varepsilon = +1$ everywhere. But note that $t^2 - \delta\, t^2 = 0$ if $t \geqq 0$. Hence,

$$L = \int_{-1}^{999} \sqrt{t^2 - \delta\, t^2}\, dt = \int_{-1}^{0} \sqrt{2t^2}\, dt + \int_0^{999} 0\, dt = \sqrt{2}\int_{-1}^{0} (-t)\, dt$$
$$= -\sqrt{2}\, t^2/2\Big|_{-1}^{0} = \sqrt{2}/2 \approx 0.707$$

The interpretation in rectangular coordinates (x^1, x^2) is curious: As a particle travels less than a millimeter along the curve, its "shadow" on the x^1-axis travels a meter!

Nonexistence of an Arc-Length Parameter

For a positive definite metric, the arc-length parameter s is well-defined by *(5.6)* as a strictly increasing function of the curve parameter t. (Then it is also the case that t is a strictly increasing function of s.) This fact allowed us freely to convert between the two parameterizations in the Solved Problems to Chapter 6. However, it is clear that on a null curve, which possesses at least one interval $t_1 < t < t_2$ of null points, it is impossible to define arc length s.

Indeed, even isolated points of nullity pose analytical problems. For if $s'(t_0) = 0$, then the chain rule,

$$\frac{dx^i}{ds} = \frac{dx^i}{dt}\frac{1}{s'(t)} \tag{7.3}$$

breaks down at s_0, the image of t_0. When necessary, we get around the difficulty by restricting attention to curves that are *regular*.

Definition 3: A curve is regular if it has no null points (i.e., $ds/dt > 0$).

It will be further assumed that all curves are of sufficiently high differentiability class to permit the theory considered; usually, this will require the assumption that curves are of class C^2.

7.4 REGULAR CURVES: UNIT TANGENT VECTOR

Let a regular curve $\mathscr{C}$: $x^i = x^i(s)$ be given in terms of the arc-length parameter; the tangent field is $\mathbf{T} \equiv (dx^i/ds)$. By definition of arc length,

$$s = \int_0^s \|\mathbf{T}(u)\|\, du$$

and differentiation gives $1 = \|\mathbf{T}(s)\|$, showing that $\mathbf{T}$ has unit length at each point of $\mathscr{C}$.

When it is inconvenient or impossible to convert to the arc-length parameter, we can, by *(7.3)*, obtain $\mathbf{T}$ by normalizing the tangent vector $\mathbf{U} = (dx^i/dt)$:

$$\mathbf{T} = \frac{1}{\|\mathbf{U}\|}\mathbf{U} = \frac{1}{s'(t)}\mathbf{U} \tag{7.4}$$

In Problem 7.20 is proved the useful

Theorem 7.1: The absolute derivative $\delta\mathbf{T}/\delta s$ of the unit tangent vector $\mathbf{T}$ is orthogonal to $\mathbf{T}$.

7.5 REGULAR CURVES: UNIT PRINCIPAL NORMAL AND CURVATURE

Also associated with a regular curve $\mathscr{C}$ is a vector orthogonal to the tangent vector. It may be introduced in two ways: (1) as the normalized $\delta\mathbf{T}/\delta s$, if it exists; (2) as any differentiable unit vector orthogonal to $\mathbf{T}$ and proportional to $\delta\mathbf{T}/\delta s$ when $\|\delta\mathbf{T}/\delta s\| \neq 0$. The latter definition is global in nature, and it applies to a larger class of curves than does the former.

Analytical (Local) Approach

At any point of $\mathscr{C}$ at which $\|\delta\mathbf{T}/\delta s\| \neq 0$, define the *unit principal normal* as the vector

$$\mathbf{N}_0 \equiv \frac{\delta\mathbf{T}}{\delta s}\bigg/\left\|\frac{\delta\mathbf{T}}{\delta s}\right\| \tag{7.5}$$

The *absolute curvature* is the scale factor in (*7.5*):

$$\kappa_0 \equiv \left\|\frac{\delta\mathbf{T}}{\delta s}\right\| = \sqrt{\varepsilon g_{ij}\frac{\delta T^i}{\delta s}\frac{\delta T^j}{\delta s}} \tag{7.6}$$

This notion of curvature was informally defined in (*6.12*).

Calling this quantity "curvature" is suggestive of the fact that in rectangular coordinates $\|\delta\mathbf{T}/\delta s\| = \|d\mathbf{T}/ds\|$ measures the rate of change of the tangent vector with respect to distance, or how sharply $\mathscr{C}$ "bends" at each point. Substitution of (*7.6*) into (*7.5*) yields one of the *Frenet equations*:

$$\frac{\delta\mathbf{T}}{\delta s} = \kappa_0\mathbf{N}_0 \qquad (\kappa_0 \neq 0) \tag{7.7}$$

While this approach is simple and concise, it does not apply to many curves we want to consider; for instance, a geodesic—as defined by (*6.13*)—will not possess a local normal $\mathbf{N}_0$ at any point. Even if there is only one point of zero curvature and the metric is Euclidean, $\mathbf{N}_0$ can have an essential point of discontinuity there.

EXAMPLE 7.4 The simple cubic $y = x^3$ has an inflection point at the origin, or $s = 0$ (by arrangement). As shown in Fig. 7-1,

$$\lim_{s\to -0}\mathbf{N}_0 = (0, -1) \qquad \lim_{s\to +0}\mathbf{N}_0 = (0, 1)$$

To verify this analytically, make the parameterization $x = t$, $y = t^3$, and calculate $\mathbf{N}_0$ as a function of t ($s'(t) = \sqrt{1 + 9t^4}$).

$$\mathbf{U} = (x'(t), y'(t)) = (1, 3t^2)$$

$$\mathbf{T} = \frac{1}{s'(t)}\mathbf{U} = \frac{1}{\sqrt{1+9t^4}}(1, 3t^2)$$

$$\frac{d\mathbf{T}}{ds} = \frac{1}{s'(t)}\frac{d\mathbf{T}}{dt} = \frac{6t}{(1+9t^4)^2}(-3t^2, 1)$$

$$\kappa_0 = \left\|\frac{d\mathbf{T}}{ds}\right\| = \frac{6|t|}{(1+9t^4)^{3/2}}$$

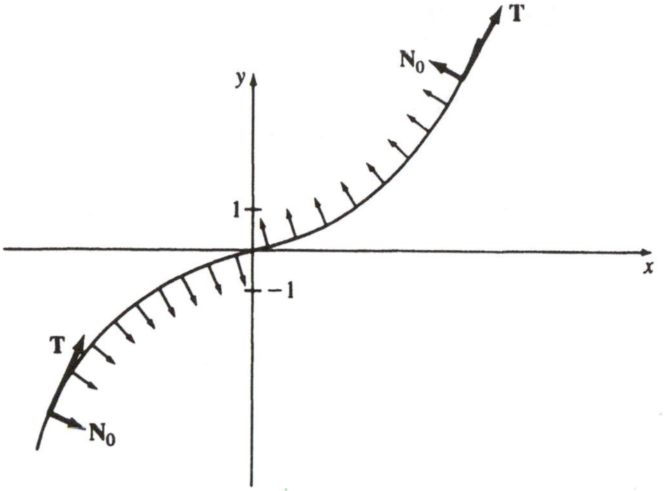

Fig. 7-1

$$\mathbf{N}_0 = \frac{1}{\kappa_0}\frac{d\mathbf{T}}{ds} = \frac{t/|t|}{\sqrt{1+9t^4}}(-3t^2, 1) \qquad (t \neq 0)$$

The scalar factor $t/|t|$ accounts for the discontinuity in $\mathbf{N}_0$ at $t = 0$ $(s = 0)$.

Geometric (Global) Approach

A *unit principal normal* to a regular curve $\mathscr{C}$ is any contravariant vector $\mathbf{N} = (N^i(s))$ such that, along $\mathscr{C}$,

A. N^i is continuously differentiable (class C^1) for each i;

B. $\|\mathbf{N}\| = 1$;

C. $\mathbf{N}$ is orthogonal to the unit tangent vector $\mathbf{T}$, and is a scalar multiple of $\delta\mathbf{T}/\delta s$ wherever $\|\delta\mathbf{T}/\delta s\| \neq 0$.

The *curvature* under this development (motivated by (7.7) using $\mathbf{N}(\kappa\mathbf{N}) = \kappa\mathbf{N}^2 = \kappa$) is defined as

$$\kappa \equiv \varepsilon\mathbf{N}\frac{\delta\mathbf{T}}{\delta s} = \varepsilon g_{ij}N^i\frac{\delta T^j}{\delta s} \qquad (\varepsilon = \varepsilon(\mathbf{N})) \tag{7.8}$$

If the metric is positive definite, the Frenet equation

$$\frac{\delta\mathbf{T}}{\delta s} = \kappa\mathbf{N} \tag{7.9}$$

holds unrestrictedly along a regular curve (see Problem 7.13).

EXAMPLE 7.5 For the curve of Example 7.4, conditions A, B, and C allow precisely two possibilities for $\mathbf{N}$:

$$\mathbf{N} = \frac{+1}{\sqrt{1+9t^4}}(-3t^2, 1) \qquad \text{or} \qquad \mathbf{N} = \frac{-1}{\sqrt{1+9t^4}}(-3t^2, 1)$$

for $-\infty < t < \infty$. Geometrically, these amount to reversing the normal arrows in either the left half or the right half of Fig. 7-1. The corresponding formulas for curvature are $(\varepsilon \equiv 1)$

$$\kappa = \frac{6t}{(1+9t^4)^{3/2}} \qquad \text{or} \qquad \kappa = \frac{-6t}{(1+9t^4)^{3/2}}$$

On curves having everywhere a non-null $\delta\mathbf{T}/\delta s$, either $\mathbf{N} \equiv \mathbf{N}_0$ (with $\kappa = \kappa_0$) or $\mathbf{N} \equiv -\mathbf{N}_0$ (with $\kappa = -\kappa_0$). Thus, the global concept applies to all curves covered by the local concept and, in addition, to all regular planar curves (see Problem 7.14) and all analytic curves (curves for which the x^i are representable as convergent Taylor series in s).

7.6 GEODESICS AS SHORTEST ARCS

When the metric is positive definite, a geodesic may be defined by the zero-curvature conditions (*6.13*), or, equivalently, by the condition that for any two of its points sufficiently close together, its length between the two points is least among all curves joining those points.

The minimum-length development employs a variational argument. We need to assume that all curves under consideration are class C^2 (that is, the parametric functions which represent them have continuous second-order derivatives). Let $x^i = x^i(t)$ represent a shortest curve (geodesic) passing through $A = (x^i(a))$ and $B = (x^i(b))$, where $b - a$ is as small as necessary. Embed the geodesic in a one-parameter family of C^2 curves passing through A and B:

$$x^i = X^i(t, u) \equiv x^i(t) + (t-a)(b-t)u\phi^i(t)$$

where the multipliers $\phi^i(t)$ are arbitrary twice-differentiable functions. The length of a curve in this family is given by

$$L(u) = \int_a^b \sqrt{\varepsilon g_{ij} \frac{\partial X^i}{\partial t} \frac{\partial X^j}{\partial t}}\, dt \equiv \int_a^b \sqrt{w(t, u)}\, dt$$

with $\varepsilon = 1$ for a positive-definite metric. Since $X^i(t, 0) = x^i(t)$ $(i = 1, 2, \ldots, n)$, the function $L(u)$ must have a local minimum at $u = 0$. Standard calculus techniques yield the following expression of the necessary condition $L'(0) = 0$:

$$\int_a^b \left[w^{-1/2} \frac{\partial g_{ij}}{\partial x^k} \frac{dx^i}{dt} \frac{dx^j}{dt} - \frac{d}{dt}\left(2w^{-1/2} g_{ik} \frac{dx^i}{dt}\right)\right](t-a)(b-t)\phi^k(t)\, dt = 0 \qquad (7.10)$$

in which

$$w \equiv w(t, 0) = g_{ij} \frac{dx^i}{dt} \frac{dx^j}{dt} \qquad (7.11)$$

Since $(t-a)(b-t) > 0$ on (a, b) and $\phi^k(t)$ may be chosen arbitrarily, the bracketed expression in (*7.10*) must vanish identically over (a, b), for $k = 1, 2, \ldots, n$; this leads to (Problem 7.21)

$$\frac{d^2x^i}{dt^2} + \Gamma^i_{jk} \frac{dx^j}{dt} \frac{dx^k}{dt} = \frac{1}{2w} \frac{dw}{dt} \frac{dx^i}{dt} \qquad (i = 1, 2, \ldots, n) \qquad (7.12)$$

System (*7.12*), with w defined by (*7.11*), are the differential equations for the geodesics of Riemannian space, in terms of the arbitrary curve parameter t. Assuming that these geodesics will be regular curves, we may choose $t = s =$ arc length. Then

$$w = \left(\frac{ds}{dt}\right)^2 = \left(\frac{ds}{ds}\right)^2 = (1)^2 = 1 \qquad \text{and} \qquad \frac{dw}{ds} = 0$$

so that (*7.12*) becomes

$$\frac{d^2x^i}{ds^2} + \Gamma^i_{jk} \frac{dx^j}{ds} \frac{dx^k}{ds} = 0 \qquad (i = 1, 2, \ldots, n) \qquad (7.13)$$

which is precisely (*6.13*).

It must be emphasized that $L'(0) = 0$ is only a necessary condition for minimum length, so that the geodesics are found *among* the solutions of (*7.12*) or (*7.13*).

Null Geodesics

Consider the case of indefinite metrics and class $\mathscr{C}^2$ curves which may have one or more null points. Since, at a null point, $w = 0$ in (*7.11*) the variational theory breaks down, because $L(u)$ fails to be differentiable at such a point. Analogous to the zero-curvature approach, we consider the more general condition for geodesics

$$\frac{d^2x^i}{dt^2} + \Gamma^i_{jk}\frac{dx^j}{dt}\frac{dx^k}{dt} \equiv \frac{\delta U^i}{\delta t} = 0 \qquad (i = 1, 2, \ldots, n) \tag{7.14}$$

where $U = U^i = (dx^i/dt)$ is the tangent vector field. By properties of absolute differentiation,

$$\frac{dw}{dt} = \frac{d}{dt}(\varepsilon g_{ij}U^iU^j) = \frac{\delta}{\delta t}(\varepsilon g_{ij}U^iU^j) = 2\varepsilon g_{ij}U^i\frac{\delta U^j}{\delta t} = 0$$

along a solution curve to (*7.14*); so $w = \text{const.}$ along the curve. Since the curve has at least one point of nullity, $w = 0$ at all points, whence the curve is a null curve—called a null geodesic. In summary, the following system of $n + 1$ ordinary differential equations in the n unknown functions $x^i(t)$ will determine the null geodesics:

$$\frac{d^2x^i}{dt^2} + \Gamma^i_{rs}\frac{dx^r}{dt}\frac{dx^s}{dt} = 0 \qquad (i = 1, 2, \ldots, n)$$

$$g_{rs}\frac{dx^r}{dt}\frac{dx^s}{dt} = 0 \tag{7.15}$$

EXAMPLE 7.6 If the g_{ij} are constants, (*7.15*) has the expected general solution

$$x^i = x^i_0 + \alpha^i t \qquad \text{with} \qquad g_{ij}\alpha^i\alpha^j = 0$$

Imagining the x^i to be rectangular coordinates, we interpret the null geodesics as a bundle of straight lines issuing from the arbitrary point $\mathbf{x}_0$; each line is in the direction of some null vector $\boldsymbol{\alpha}$. By elimination of the α^i the equation of the bundle is found to be

$$g_{ij}(x^i - x^i_0)(x^j - x^j_0) = 0$$

In particular, for the space of Special Relativity ($g_{11} = g_{22} = g_{33} = -g_{44} = 1$, $g_{ij} = 0$ for $i \neq j$), the null geodesics compose the 45° cone

$$(x^1 - x^1_0)^2 + (x^2 - x^2_0)^2 + (x^3 - x^3_0)^2 = (x^4 - x^4_0)^2$$

—see Fig. 7-2.

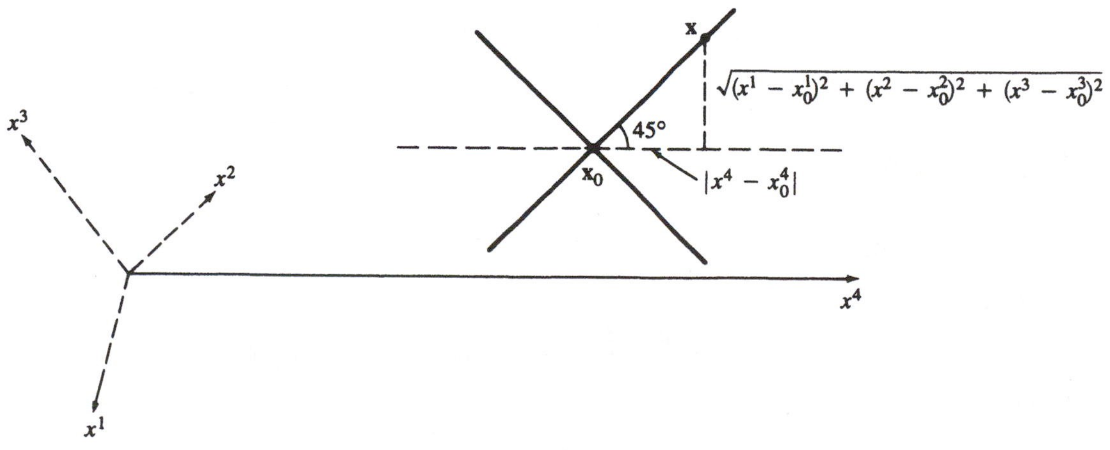

Fig. 7-2

Solved Problems

LENGTH IN RIEMANNIAN SPACE

7.1 Determine the indicator of the tangent vector **U** to the curve

$$x^1 = t^3 \qquad x^2 = t^2 \qquad x^3 = t$$

$(-\infty < t < \infty)$ if the fundamental form is

(*a*) $(dx^1)^2 + (x^2)^2(dx^2)^2 + (x^1)^2(dx^3)^2 - 6\,dx^1\,dx^3 + 2x^1x^2\,dx^2\,dx^3$

(*b*) $(dx^1)^2 + 2(dx^2)^2 + 3(dx^3)^2$

(*a*) $(3t^2)^2 + t^4(2t)^2 + t^6(1)^2 - 6(3t^2)(1) + 2(t^3)(t^2)(2t)(1) = 9t^6 + 9t^4 - 18t^2 = 9t^2(t^2+2)(t^2-1)$

Since $t^2 + 2$ is always positive,

$$\varepsilon(\mathbf{U}) = \begin{cases} +1 & t \geqq 1 \\ -1 & 0 < t < 1 \\ +1 & t = 0 \\ -1 & -1 < t < 0 \\ +1 & t \leqq -1 \end{cases}$$

(*b*) $\varepsilon(\mathbf{U}) \equiv +1$, because the form is positive definite.

7.2 Show that the following matrix defines a Riemannian metric on $\mathbf{R}^2$:

$$G = \begin{bmatrix} x^2 & -x^1 \\ -x^1 & x^2 \end{bmatrix} \qquad (x^1 > 0,\ -x^1 < x^2 < x^1)$$

We must show that conditions A–D of Section 5.3 are satisfied.

A. Since each g_{ij} is linear in the x^i, it is differentiable to any order.

B. By observation, the matrix is symmetric.

C. $|g_{ij}| = (x^2)^2 - (x^1)^2 < 0$ over the given domain.

D. Extend the matrix to a tensor **g** by using the tensor transformation laws to define the $\bar{g}_{ij}$ in terms of the g_{ij}. This will then make the quadratic form $g_{ij}\,dx^i\,dx^j$, hence the distance formula, an invariant.

7.3 Find the null set of the curve $\mathscr{C}$: $x^2 = (x^1)^2$ $(x^1 > 0)$ under the metric of Problem 7.2.

Let $\mathscr{C}$ be parameterized by $x^1 = t$, $x^2 = t^2$ $(t > 0)$. Then, along $\mathscr{C}$,

$$g_{ij}\frac{dx^i}{dt}\frac{dx^j}{dt} = [1 \quad 2t]\begin{bmatrix} t^2 & -t \\ -t & t^2 \end{bmatrix}\begin{bmatrix} 1 \\ 2t \end{bmatrix} = [1 \quad 2t]\begin{bmatrix} -t^2 \\ -t+2t^3 \end{bmatrix} = t^2(4t^2-3)$$

which, for positive t, vanishes only at $t = \sqrt{3}/2$.

7.4 Find the arc length of the curve $\mathscr{C}$ in Problem 7.3 from $x^1 = 0$ to $x^1 = 1$.

Again using $t = x^1$, observe that

$$g_{ij}\frac{dx^i}{dt}\frac{dx^j}{dt} < 0 \qquad \text{for} \qquad 0 < t < \sqrt{3}/2$$

Hence,

$$L = \int_0^1 \sqrt{\varepsilon t^2(4t^2-3)}\,dt = \int_0^{\sqrt{3}/2} t\sqrt{-(4t^2-3)}\,dt + \int_{\sqrt{3}/2}^1 t\sqrt{4t^2-3}\,dt$$

$$= -\frac{1}{12}(3-4t^2)^{3/2}\Big|_0^{\sqrt{3}/2} + \frac{1}{12}(4t^2-3)^{3/2}\Big|_{\sqrt{3}/2}^1 = \frac{3\sqrt{3}+1}{12} \approx 0.516$$

7.5 Write $g \equiv \det G$ for the determinant of a Riemannian metric. Prove that $|g|$ is a differentiable function of the coordinates.

Applying the chain rule to $|g| = \sqrt{g^2}$, we have

$$\frac{\partial |g|}{\partial x^i} = \frac{g}{|g|}\frac{\partial g}{\partial x^i} \qquad (1)$$

Since $\partial g/\partial x^i$ exists (Property A) and $|g| \neq 0$ (Property C), the right-hand side of (*1*) is well-defined.

7.6 Show that under the metric $\varepsilon\, ds^2 = (dx^1)^2 - (dx^2)^2 - (dx^3)^2 - (dx^4)^2$ (another version of the metric for Special Relativity), the curve

$$x^1 = A\sinh t \quad x^2 = A\cosh t \quad x^3 = Bt \quad x^4 = Ct \qquad (0 \leqq t \leqq 1)$$

with $A^2 = B^2 + C^2$, is null at each of its points.

$$\begin{aligned} g_{ij}\frac{dx^i}{dt}\frac{dx^j}{dt} &= \left(\frac{dx^1}{dt}\right)^2 - \left(\frac{dx^2}{dt}\right)^2 - \left(\frac{dx^3}{dt}\right)^2 - \left(\frac{dx^4}{dt}\right)^2 \\ &= (A\cosh t)^2 - (A\sinh t)^2 - B^2 - C^2 \\ &= A^2(\cosh^2 t - \sinh^2 t) - B^2 - C^2 = A^2 - B^2 - C^2 \equiv 0 \end{aligned}$$

7.7 At the point of intersection (0, 0), find the angle between the curves

$$\mathscr{C}_1 : \begin{cases} x^1 = 2t - 2 \\ x^2 = t^2 - 1 \end{cases} \qquad \mathscr{C}_2 : \begin{cases} x^1 = u^4 - 1 \\ x^2 = 25u^2 + 50u - 75 \end{cases}$$

if the Riemannian metric is given by $\varepsilon\, ds^2 = (dx^1)^2 - 2\, dx^1\, dx^2$.

At $t = 1$, $\mathbf{T} \equiv (dx^i/dt) = (2, 2)$; at $u = 1$, $\mathbf{U} \equiv (dx^i/du) = (4, 100)$. Hence, using matrices,

$$\mathbf{TU} = [2 \quad 2]\begin{bmatrix} 1 & -1 \\ -1 & 0 \end{bmatrix}\begin{bmatrix} 4 \\ 100 \end{bmatrix} = -200$$

$$\|\mathbf{T}\|^2 = \varepsilon_1[2 \quad 2]\begin{bmatrix} 1 & -1 \\ -1 & 0 \end{bmatrix}\begin{bmatrix} 2 \\ 2 \end{bmatrix} = (\varepsilon_1)(-4) = 4$$

$$\|\mathbf{U}\|^2 = \varepsilon_2[4 \quad 100]\begin{bmatrix} 1 & -1 \\ -1 & 0 \end{bmatrix}\begin{bmatrix} 4 \\ 100 \end{bmatrix} = (\varepsilon_2)(-784) = 784$$

and

$$\cos\theta = \frac{-200}{\sqrt{4}\sqrt{784}} = -\frac{25}{7}$$

This is Case 2 of Section 7.2; we have

$$\theta = \pi + i\ln\left(\frac{25}{7} + \sqrt{\left(\frac{25}{7}\right)^2 - 1}\right) = \pi + i\ln 7$$

7.8 Verify that the vectors of Problem 7.7 do not obey the triangle inequality.

As calculated, $\|\mathbf{T}\| + \|\mathbf{U}\| = 2 + 28 = 30$. But

$$\|\mathbf{T} + \mathbf{U}\|^2 = \varepsilon_3[6 \quad 102]\begin{bmatrix} 1 & -1 \\ -1 & 0 \end{bmatrix}\begin{bmatrix} 6 \\ 102 \end{bmatrix} = \varepsilon_3(-1188) = 1188$$

whence $\|\mathbf{T} + \mathbf{U}\| \approx 34.46 > \|\mathbf{T}\| + \|\mathbf{U}\|$.

ARC-LENGTH PARAMETER, UNIT TANGENT VECTOR

7.9 Let $\mathscr{C} : x^i = x^i(t)$ be any non-null curve. (*a*) Prove that arc length along $\mathscr{C}$ is defined as a strictly increasing function of t. (*b*) Exhibit the arc-length parameterization of $\mathscr{C}$.

(*a*) For $t_1 < t_2$, the Mean-Value Theorem of calculus gives

$$s(t_2) - s(t_1) = (t_2 - t_1)s'(\tau) \qquad (t_1 < \tau < t_2)$$

The right-hand side is nonnegative, so that $s(t_1) \leqq s(t_2)$. But, in view of the identity

$$s(t_2) - s(t_1) = [s(t_2) - s(t_3)] + [s(t_3) - s(t_1)]$$

where t_3 is any point in (t_1, t_2), the equality $s(t_1) = c = s(t_2)$ would imply $s(t_3) = c$; i.e., $s(t)$ would be constant on $[t_1, t_2]$, making $s'(t) \equiv 0$ on (t_1, t_2) and thus making $\mathscr{C}$ a null curve. We conclude that

$$s(t_1) < s(t_2) \qquad \text{whenever} \qquad t_1 < t_2$$

(*b*) The strictly increasing function $s(t)$ will possess a strictly increasing inverse; denote it as $t = \theta(s)$. Then $\mathscr{C}$ admits the parameterization $x^i = x^i(\theta(s))$.

7.10 **(*a*) In rectangular coordinates (x^1, x^2) but adopting the metric of Problem 7.7, find the null points of the parabola $\mathscr{C}\ :\ x^1 = t, x^2 = t^2 \quad (0 \leqq t \leqq \frac{1}{2})$. (*b*) Show that the arc-length parameterization of $\mathscr{C}$ is differentiable to all orders except at the null points. (*c*) Find the length of $\mathscr{C}$.**

(*a*)
$$\varepsilon\left(\frac{ds}{dt}\right)^2 = \left(\frac{dx^1}{dt}\right)^2 - 2\frac{dx^1}{dt}\frac{dx^2}{dt} = 1 - 4t$$

so there is only one null point, at $t = 1/4$.

(*b*)
$$s = \int_0^t \sqrt{\varepsilon(1-4u)}\, du$$

Thus, for $0 \leqq t \leqq 1/4$,

$$s = \int_0^t \sqrt{1-4u}\, du = \frac{1}{6}[1 - (1-4t)^{3/2}]$$

and, for $1/4 \leqq t \leqq 1/2$,

$$s = \int_0^{1/4} \sqrt{1-4u}\, du + \int_{1/4}^t \sqrt{4u-1}\, du = \frac{1}{6}[1 + (4t-1)^{3/2}]$$

Inversion of these formulas gives

$$t = \theta(s) \equiv \begin{cases} \frac{1}{4}[1 - (1-6s)^{2/3}] & 0 \leqq s \leqq 1/6 \\ \frac{1}{4}[1 + (6s-1)^{2/3}] & 1/6 \leqq s \leqq 1/3 \end{cases} \tag{1}$$

It is evident that $\theta(s)$ is infinitely differentiable except at the null point $s = 1/6$ (the image of $t = 1/4$); the same will be true of the functions $x^1 = \theta(s)$, $x^2 = \theta^2(s)$.

(*c*) Set $t = 1/2$ in the applicable expression for s:

$$s = \frac{1}{6}[1 + (2-1)^{3/2}] = \frac{1}{3}$$

7.11 **Find the arc length of the same curve $\mathscr{C}$ as in Problem 7.10, but with the normal Euclidean metric, $ds^2 = (dx^1)^2 + (dx^2)^2$.**

Now
$$\frac{ds}{dt} = \sqrt{\left(\frac{dx^1}{dt}\right)^2 + \left(\frac{dx^2}{dt}\right)^2} = \sqrt{4t^2+1}$$

so that

$$L = \int_0^{1/2} \sqrt{4t^2+1}\, dt = \left[\frac{t}{2}\sqrt{4t^2+1} + \frac{1}{4}\ln(2t + \sqrt{4t^2+1})\right]_0^{1/2} = \frac{\sqrt{2} + \ln(1+\sqrt{2})}{4} \approx 0.574$$

as compared to $L \approx 0.333$ in Problem 7.10.

7.12 Using the arc-length parameterization found for the curve $\mathscr{C}$ in Problem 7.10(*b*), compute the components $T^i(s)$ of the tangent vector and verify that this vector has unit length for all $s \neq 1/6$.

We have $(T^i) = (\theta', 2\theta\theta')$, where $\theta = \theta(s)$ is the function defined by (*1*) in Problem 7.10(*b*). Hence,

$$\|\mathbf{T}\|^2 = \varepsilon(\theta'^2 - 4\theta\theta'^2) = \varepsilon(1 - 4\theta)\theta'^2$$

But, by (*1*) of Problem 7.10(*b*),

$$1 - 4\theta = \begin{cases} (1-6s)^{2/3} & 0 \leqq s \leqq 1/6 \\ -(6s-1)^{2/3} & 1/6 \leqq s \leqq 1/3 \end{cases} \qquad \theta' = \begin{cases} (1-6s)^{-1/3} & 0 < s < 1/6 \\ (6s-1)^{-1/3} & 1/6 < s < 1/3 \end{cases}$$

Therefore, $\|\mathbf{T}\|^2 = (\varepsilon)(\pm 1) = +1$, or $\|\mathbf{T}\| \equiv 1 \quad (s \neq 1/6)$.

UNIT PRINCIPAL NORMAL, CURVATURE

7.13 Prove that the Frenet equation (*7.9*) holds at each point of a regular curve when the metric is positive definite.

At a point where $\|\delta\mathbf{T}/\delta s\| \neq 0$, we have (from property C of **N**),

$$\mathbf{N} = \lambda \frac{\delta \mathbf{T}}{\delta s} \tag{1}$$

from some real λ. Take the inner product with the vector **N** in (*1*); with $\varepsilon = \varepsilon(\mathbf{N})$,

$$\varepsilon\mathbf{N}^2 = \varepsilon\lambda\mathbf{N}\frac{\delta\mathbf{T}}{\delta s} = \lambda\kappa \qquad \text{or} \qquad 1 = \lambda\kappa \tag{2}$$

Then $\lambda = 1/\kappa$, and substitution into (*1*) yields (*7.9*).

At a point where $\|\delta\mathbf{T}/\delta s\| = 0$, both $\delta\mathbf{T}/\delta s = \mathbf{0}$ (because the metric is positive definite) and $\kappa = 0$ (by (*7.8*)); the Frenet equation then holds trivially.

7.14 For any regular two-dimensional curve $\mathscr{C}$: $x^i = x^i(s)$, define the contravariant vector

$$\mathbf{N} = (N^i) \equiv (-T_2/\sqrt{|g|},\ T_1/\sqrt{|g|}) \tag{7.16}$$

where $\mathbf{T} = (T^i)$ is the unit tangent vector along $\mathscr{C}$ and $g = \det(g_{ij})$. Show that **N** is a global unit normal for $\mathscr{C}$.

We must show that the three properties of Section 7.5 are possessed by the given vector (except possibly at null points).

A. Since $\mathscr{C}$ is regular, the T^i, and with them the $T_i = g_{ij}T^j$, are in C^1. The same is true of $|g|$ (Problem 7.5), which function is strictly positive. Therefore, the N^i are also in C^1.

B. By (*2.11*), $g^{11} = g_{22}/g$, $g^{12} = g^{21} = -g_{12}/g$, and $g^{22} = g_{11}/g$. Hence,

$$\begin{aligned} \|\mathbf{N}\|^2 &= |g_{11}(T_2^2/|g|) + 2g_{12}(-T_1T_2/|g|) + g_{22}(T_1^2/|g|)| \\ &= \frac{1}{|g|}\,|gg^{22}T_2^2 + 2gg^{12}T_1T_2 + gg^{11}T_1^2| \\ &= \frac{|g|}{|g|}\,|g^{ij}T_iT_j| = |T^jT_j| = \|\mathbf{T}\|^2 \end{aligned}$$

and so $\|\mathbf{N}\| = \|\mathbf{T}\| = 1$.

C. **N** is orthogonal to **T**:

$$N^iT_i = -\frac{T_2}{\sqrt{|g|}}\,T_1 + \frac{T_1}{\sqrt{|g|}}\,T_2 = 0$$

Furthermore, when $\|\delta\mathbf{T}/\delta s\| \neq 0$, then $\mathbf{N}_0$ is defined and is also a vector orthogonal to **T** (by Theorem 7.1). In two dimensions this implies $\mathbf{N} = \pm\mathbf{N}_0 = \lambda(\delta\mathbf{T}/\delta s)$.

7.15 For the curve and metric of Problem 7.10, determine the local normal $\mathbf{N}_0$ and, using Problem 7.14, a global normal $\mathbf{N}$. Verify that the two stand in the proper relationship.

We have $g_{11}=1$, $g_{12}=g_{21}=-1$, $g_{22}=0$ (all constants), and $\mathbf{T}=(\theta', 2\theta\theta')$; therefore, for $s \neq 1/6$,

$$\frac{\delta \mathbf{T}}{\delta s}=\frac{d\mathbf{T}}{ds}=(\theta'', 2\theta'^2+2\theta\theta'')$$

$$=\begin{cases}(2(1-6s)^{-4/3}, (1-6s)^{-2/3}+(1-6s)^{-4/3}) & 0<s<1/6\\ (-2(6s-1)^{-4/3}, (6s-1)^{-2/3}-(6s-1)^{-4/3}) & 1/6<s<1/3\end{cases}$$

and
$$\left\|\frac{\delta \mathbf{T}}{\delta s}\right\|=\sqrt{\varepsilon g_{ij}\frac{dT^i}{ds}\frac{dT^j}{ds}}=\begin{cases}2(1-6s)^{-1}\\ 2(6s-1)^{-1}\end{cases}$$

Thus,
$$\mathbf{N}_0=\frac{\delta \mathbf{T}}{\delta s}\Big/\left\|\frac{\delta \mathbf{T}}{\delta s}\right\|=\begin{cases}((1-6s)^{-1/3}, \frac{1}{2}(1-6s)^{1/3}+\frac{1}{2}(1-6s)^{-1/3})\\ (-(6s-1)^{-1/3}, \frac{1}{2}(6s-1)^{1/3}-\frac{1}{2}(6s-1)^{-1/3})\end{cases}$$

With $g=-1$, Problem 7.14 gives ($s\neq 1/6$):

$$T_1=g_{1j}T^j=T^1-T^2=\theta'(1-2\theta)=\begin{cases}\frac{1}{2}(1-6s)^{-1/3}+\frac{1}{2}(1-6s)^{1/3}\\ \frac{1}{2}(6s-1)^{-1/3}-\frac{1}{2}(6s-1)^{1/3}\end{cases}$$

$$T_2=g_{2j}T^j=-T^1=-\theta'=\begin{cases}-(1-6s)^{-1/3}\\ -(6s-1)^{-1/3}\end{cases}$$

$$\mathbf{N}=(-T_2, T_1)=\begin{cases}((1-6s)^{-1/3}, \frac{1}{2}(1-6s)^{-1/3}+\frac{1}{2}(1-6s)^{1/3})\\ ((6s-1)^{-1/3}, \frac{1}{2}(6s-1)^{-1/3}-\frac{1}{2}(6s-1)^{1/3})\end{cases}$$

It is seen that, as expected, $\mathbf{N}=+\mathbf{N}_0$ for $s<1/6$ and $\mathbf{N}=-\mathbf{N}_0$ for $s>1/6$. neither $\mathbf{N}_0$ nor $\mathbf{N}$ is defined at the null point $s=1/6$. For comparison, recall the situation in Examples 7.4 and 7.5: there the discontinuity in $\mathbf{N}_0$ occurred at a regular point (the cubic has no null points under the Euclidean metric), and $\mathbf{N}$ (either choice) was defined everywhere.

7.16 Under the metric of Special Relativity (Example 7.1), a regular curve $\mathscr{C}$ is given by

$$x^1=s^2 \qquad x^2=\frac{3s}{5} \qquad x^3=\frac{4s}{5} \qquad x^4=s^2$$

for $0\leqq s\leqq 1$. (*a*) Verify that s is arc length for $\mathscr{C}$, and show that the absolute derivative, $\delta\mathbf{T}/\delta s$, of $\mathbf{T}$ is a null vector at every point of the curve (hence, a local principal normal $\mathbf{N}_0$ is nowhere defined on $\mathscr{C}$). Construct a global principal normal for $\mathscr{C}$ in such a manner that the corresponding curvature function is nonzero. Is more than one curvature function possible?

(*a*) We have $(T^i)=(2s, 3/5, 4/5, 2s)$ and

$$|g_{ij}T^iT^j|=|4s^2+(9/25)+(16/25)-4s^2|=1$$

hence, s is an arc-length parameter. Also, since the g_{ij} are constant, all Christoffel symbols vanish and

$$\frac{\delta \mathbf{T}}{\delta s}=\frac{d\mathbf{T}}{ds}=(2,0,0,2) \qquad \left\|\frac{\delta T}{\delta s}\right\|=\sqrt{|2^2+0^2+0^2-2^2|}=0$$

for all s.

(*b*) Any differentiable unit vector orthogonal to $\mathbf{T}$ will do for $\mathbf{N}$, which then determines the curvature through (*7.8*). In the orthonormality conditions

$$2sN^1+\frac{3}{5}N^2+\frac{4}{5}N^3-2sN^4=0$$

$$(N^1)^2+(N^2)^2+(N^3)^2-(N^4)^2=\pm 1$$

we may successively set $N^1=N^4=0$, $N^3=N^4=0$, and $N^2=N^4=0$ to obtain three candidate normals:

$$\mathbf{N}_1 = \left(0, -\frac{4}{5}, \frac{3}{5}, 0\right) \qquad \mathbf{N}_2 = \frac{1}{\sqrt{(9/25) + 4s^2}} \left(-\frac{3}{5}, 2s, 0, 0\right)$$

$$\mathbf{N}_3 = \frac{1}{\sqrt{(16/25) + 4s^2}} \left(-\frac{4}{5}, 0, 2s, 0\right)$$

The constant $\mathbf{N}_1$ yields $\kappa_1 \equiv 0$; but $\mathbf{N}_2$ and $\mathbf{N}_3$ yield the distinct curvature functions

$$\kappa_2 = \frac{-1}{\sqrt{\frac{1}{4} + \frac{25}{9}s^2}} \qquad \kappa_3 = \frac{-1}{\sqrt{\frac{1}{4} + \frac{25}{16}s^2}}$$

Note that the Frenet equation is invalid for all these normals.

7.17 Refer to Problems 7.10 and 7.15. Calculate the curvature functions κ_0 and κ, and discuss the variability of κ_0 over the parabolic arc $0 \leqq s \leqq 1/3$.

Our previous results show that both curvatures are defined everywhere except $s = 1/6$, with $\kappa = \kappa_0$ on $0 \leqq s < 1/6$ and $\kappa = -\kappa_0$ on $1/6 < s \leqq 1/3$ (cf. Problem 7.13). By Problem 7.15,

$$\kappa_0 = \left\| \frac{\delta \mathbf{T}}{\delta s} \right\| = \frac{2}{|1 - 6s|} \qquad (s \neq 1/6)$$

whence

$$\kappa = \frac{2}{1 - 6s} \qquad (s \neq 1/6)$$

It is seen that κ_0 has the same value, 2, at $s = 0$ (the vertex, or point of greatest Euclidean curvature) and the undistinguished point $s = 1/3$. Moreover, near the ordinary (from the Euclidean viewpoint) point $s = 1/6$, the absolute curvature becomes arbitrarily large.

7.18 (*a*) For any regular two-dimensional curve, derive the formula for absolute curvature

$$\kappa_0 = \sqrt{|g|} \left| T^1 \frac{\delta T^2}{\delta s} - T^2 \frac{\delta T^1}{\delta s} \right| \tag{7.17}$$

(*b*) Use (*7.17*) to check Problem 7.17.

(*a*) By (*7.8*) and the remarks made following Example 7.5,

$$\kappa_0 = |\kappa| = \left| N_j \frac{\delta T^j}{\delta s} \right| = \left| N_1 \frac{\delta T^1}{\delta s} + N_2 \frac{\delta T^2}{\delta s} \right| \tag{1}$$

Choosing the global normal (N^i) established in Problem 7.14, we have:

$$N_1 = g_{11}N^1 + g_{12}N^2 = (gg^{22})\left(\frac{-T_2}{\sqrt{|g|}}\right) + (-gg^{21})\left(\frac{T_1}{\sqrt{|g|}}\right) = -\frac{g}{\sqrt{|g|}}(g^{21}T_1 + g^{22}T_2) = -\sqrt{|g|}\, T^2$$

$$N_2 = g_{21}N^1 + g_{22}N^2 = (-gg^{12})\left(\frac{-T_2}{\sqrt{|g|}}\right) + (gg^{11})\left(\frac{T_1}{\sqrt{|g|}}\right)$$

$$= \frac{g}{\sqrt{|g|}}(g^{11}T_1 + g^{12}T_2) = \sqrt{|g|}\, T^1$$

Substitution of these components in (*1*) yields (*7.17*).

(*b*) For the metric of Problem 7.10,

$$G = \begin{bmatrix} 1 & -1 \\ -1 & 0 \end{bmatrix}$$

$g = \det G = -1$ and absolute derivatives reduce to ordinary derivatives. Thus we can rewrite (*7.17*) in terms of the curve parameter t, as follows:

$$\kappa_0 = \left| T^1 \frac{dT^2}{dt}\frac{dt}{ds} - T^2 \frac{dT^1}{dt}\frac{dt}{ds} \right| = \frac{1}{s'(t)} \left| T^1 \frac{dT^2}{dt} - T^2 \frac{dT^1}{dt} \right| = \frac{(T^1)^2}{s'(t)} \left| \frac{d}{dt}\left(\frac{T^2}{T^1}\right) \right|$$

Substitution of $s'(t) = \sqrt{|1-4t|}$ $(t \neq 1/4)$ and the components of the *unit* tangent vector,

$$T^1 = \frac{1}{s'(t)} \frac{dx^1}{dt} \equiv \frac{1}{s'(t)} \qquad T^2 = \frac{1}{s'(t)} \frac{dx^2}{dt} = \frac{2t}{s'(t)}$$

gives:

$$\kappa_0 = \frac{2}{(s'(t))^3} = \frac{2}{|1-4t|^{3/2}} \qquad (t \neq 1/4)$$

From Problem 7.12,

$$|1-4t| = |1-4\theta(s)| = |1-6s|^{2/3}$$

yielding exact agreement with Problem 7.17.

7.19 Compute the absolute curvature of the logarithmic curve $\mathscr{C}$: $x^1 = t$, $x^2 = a \ln t$, for $\frac{1}{2} \leqq t < a$, if the Riemannian metric is

$$\varepsilon\, ds^2 = (dx^1)^2 - (dx^2)^2$$

As the g_{ij} are constants (with $g = -1$), we can proceed as in Problem 7.18(*b*). This time, the most convenient version of (*7.17*) is

$$\kappa_0 = \sqrt{|g|} \frac{(T^2)^2}{s'(t)} \left| \frac{d}{dt} \left(\frac{T^1}{T^2} \right) \right|$$

Substituting

$$s'(t) = \sqrt{\left| \left(\frac{dx^1}{dt} \right)^2 - \left(\frac{dx^2}{dt} \right)^2 \right|} = \sqrt{\left| 1 - \frac{a^2}{t^2} \right|} = \frac{1}{t} \sqrt{a^2 - t^2} \, (\neq 0)$$

$$T^1 = \frac{1}{s'(t)} \frac{dx^1}{dt} = \frac{t}{\sqrt{a^2 - t^2}}$$

$$T^2 = \frac{1}{s'(t)} \frac{dx^2}{dt} = \frac{a}{\sqrt{a^2 - t^2}}$$

we find: $\kappa_0 = at(a^2 - t^2)^{-3/2}$.

7.20 Prove Theorem 7.1.

Along a regular curve we have

$$\|\mathbf{T}\|^2 = \varepsilon \mathbf{T}\mathbf{T} = 1 \qquad \text{or} \qquad \mathbf{T}\mathbf{T} = \varepsilon$$

where the indicator ε is constant, $|\varepsilon| = 1$, on the curve. By the inner-product rule for absolute differentiation, and the fact that the absolute derivative of an invariant is the ordinary derivative,

$$\frac{\delta \mathbf{T}}{\delta s} \mathbf{T} + \mathbf{T} \frac{\delta \mathbf{T}}{\delta s} \equiv 2\mathbf{T} \frac{\delta \mathbf{T}}{\delta s} = \frac{d}{ds}(\varepsilon) = 0 \qquad \text{or} \qquad \mathbf{T} \frac{\delta \mathbf{T}}{\delta s} = 0$$

GEODESICS

7.21 Establish (*7.12*).

Start with the conditions

$$w^{-1/2} \frac{\partial g_{ij}}{\partial x^k} \frac{dx^i}{dt} \frac{dx^j}{dt} = \frac{d}{dt} \left(2w^{-1/2} g_{ik} \frac{dx^i}{dt} \right) \tag{1}$$

By use of the product and chain rules, the expression on the right may be written

$$-w^{-3/2}\frac{dw}{dt}\left(g_{ik}\frac{dx^i}{dt}\right)+2w^{-1/2}\left(\frac{\partial g_{ik}}{\partial x^j}\frac{dx^j}{dt}\right)\frac{dx^i}{dt}+2w^{-1/2}g_{ik}\frac{d^2x^i}{dt^2}$$

Put this back in (1), multiply both sides by $w^{1/2}$, and go over to the notation $g_{ijk}\equiv \partial g_{ij}/\partial x^k$:

$$g_{ijk}\frac{dx^i}{dt}\frac{dx^j}{dt}=-w^{-1}g_{ik}\frac{dw}{dt}\frac{dx^i}{dt}+2g_{ikj}\frac{dx^j}{dt}\frac{dx^i}{dt}+2g_{ik}\frac{d^2x^i}{dt^2}$$

which rearranges to

$$2g_{ik}\frac{d^2x^i}{dt^2}-g_{ijk}\frac{dx^i}{dt}\frac{dx^j}{dt}+2g_{ikj}\frac{dx^j}{dt}\frac{dx^i}{dt}=\frac{1}{w}g_{ik}\frac{dw}{dt}\frac{dx^i}{dt}$$

Making use of the symmetry of g_{ij}, the third term on the left may be split into two similar terms, yielding

$$2g_{ik}\frac{d^2x^i}{dt^2}-g_{ijk}\frac{dx^i}{dt}\frac{dx^j}{dt}+g_{jki}\frac{dx^i}{dt}\frac{dx^j}{dt}+g_{kij}\frac{dx^i}{dt}\frac{dx^j}{dt}=\frac{1}{w}g_{ik}\frac{dw}{dt}\frac{dx^i}{dt}$$

Divide by 2, multiply by g^{pk}, and sum on k:

$$\delta^p_i\frac{d^2x^i}{dt^2}-g^{pk}\Gamma_{ijk}\frac{dx^i}{dt}\frac{dx^j}{dt}=\frac{1}{2w}\delta^p_i\frac{dw}{dt}\frac{dx^i}{dt}\qquad\text{or}\qquad\frac{d^2x^p}{dt^2}+\Gamma^p_{jk}\frac{dx^j}{dt}\frac{dx^k}{dt}=\frac{1}{2w}\frac{dw}{dt}\frac{dx^p}{dt}$$

which is (7.12).

7.22 In a Riemannian 2-space with fundamental form $(dx^1)^2-(x^2)^{-2}(dx^2)^2$, determine (a) the regular geodesics, (b) the null geodesics.

Here $g_{11}=1$, $g_{12}=g_{21}=0$, $g_{22}=-(x^2)^{-2}$; Problem 6.4 gives

$$\Gamma^2_{22}=\frac{d}{dx^2}\left[\frac{1}{2}\ln(x^2)^{-2}\right]=-\frac{1}{x^2}$$

as the only nonvanishing Christoffel symbol.

(a) The system (7.13) becomes

$$\frac{d^2x^1}{ds^2}=0\qquad\frac{d^2x^2}{ds^2}-\frac{1}{x^2}\left(\frac{dx^2}{ds}\right)^2=0$$

The first equation integrates to $x^1=as+x^1_0$. In the second, let $u\equiv dx^2/ds$:

$$u\frac{du}{dx^2}-\frac{1}{x^2}u^2=0\qquad\text{or}\qquad\frac{du}{u}=\frac{dx^2}{x^2}\qquad\text{or}\qquad u=cx^2$$

from which

$$\frac{dx^2}{ds}=cx^2\qquad\text{or}\qquad\frac{dx^2}{x^2}=c\,ds\qquad\text{or}\qquad x^2=x^2_0e^{cs}$$

As our notation indicates, an arbitrary point (x^1_0, x^2_0) is the origin $(s=0)$ of a family of geodesics that seems to depend on two parameters, a and c. However, s must represent arc length, so that

$$\pm1=\left(\frac{dx^1}{ds}\right)^2-(x^2)^{-2}\left(\frac{dx^2}{ds}\right)^2=a^2-c^2$$

Hence, either $a^2=c^2+1$ (the fundamental form is positive) or $c^2=a^2+1$ (the fundamental form is negative). Both cases may be accounted for by a single parameter, λ, if s is eliminated between the parametric equations for x^1 and x^2:

$$\textit{regular geodesics}\qquad x^2=x^2_0\exp[\lambda(x^1-x^1_0)]\quad(|\lambda|\neq1)$$

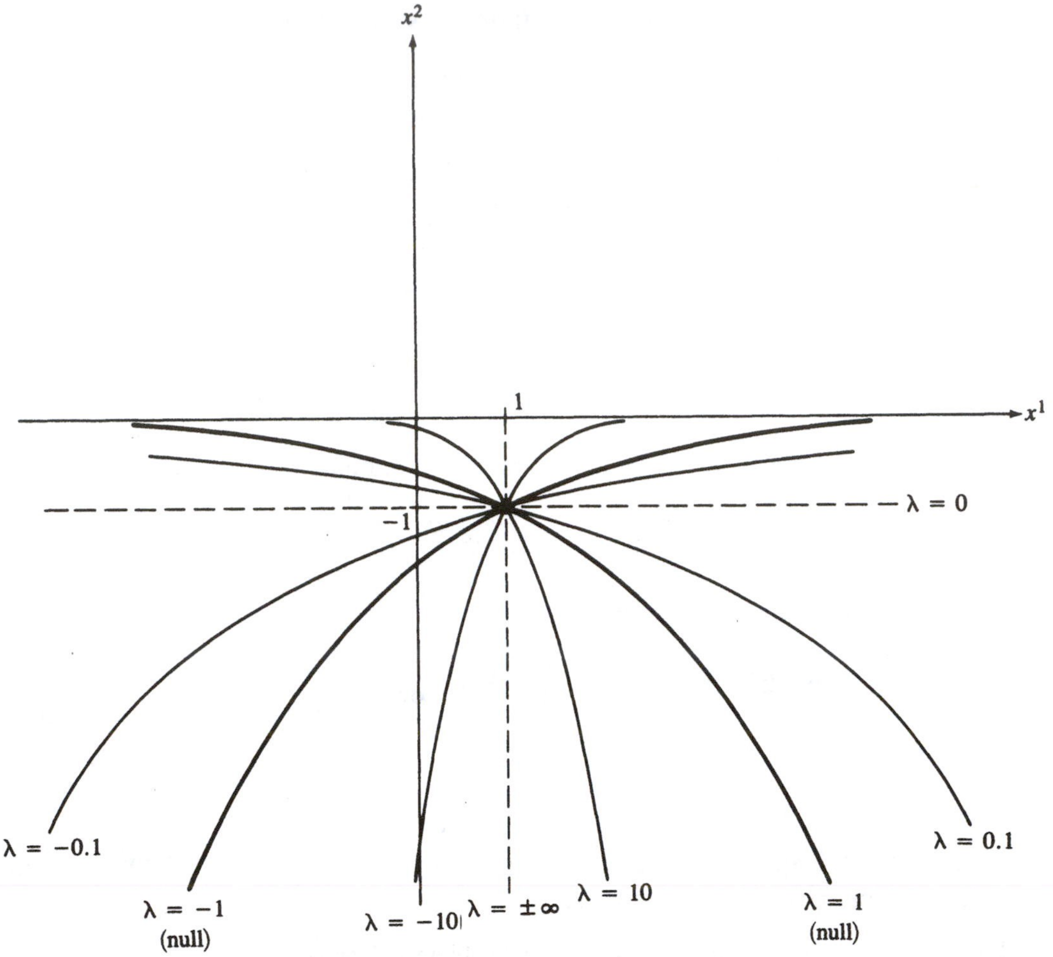

Fig. 7-3

(*b*) System (*7.14*), in t, becomes

$$\frac{d^2x^1}{dt^2}=0 \qquad \frac{d^2x^2}{dt^2}-\frac{1}{x^2}\left(\frac{dx^2}{dt}\right)^2=0$$

$$\left(\frac{dx^1}{dt}\right)^2-(x^2)^{-2}\left(\frac{dx^2}{dt}\right)^2=0$$

It is clear that the solution may be found by formally replacing s by t in part (*a*) and setting $a^2=c^2$. Thus, the null geodesics through (x_0^1, x_0^2) are given by

$$\textbf{\textit{null geodesics}} \qquad x^2=x_0^2\exp\left[+(x^1-x_0^1)\right] \qquad \text{and} \qquad x^2=x_0^2\exp\left[-(x^1-x_0^1)\right]$$

Note that the null geodesics correspond to the exceptional values $\lambda=\pm1$ in part (*a*). Figure 7-3 is a sketch of the geodesics through the point $(1,-1)$ in cartesian coordinates.

7.23 Without converting to arc length, verify that in spherical coordinates, under the Euclidean metric

$$ds^2=(dx^1)^2+(x^1\,dx^2)^2+(x^1\sin x^2\,dx^3)^2$$

any curve of the form $\mathscr{C}\ :\ x^1=a\sec t,\ x^2=t+b,\ x^3=c$ (a, b, c constant) is a geodesic. (It should be apparent that $\mathscr{C}$ is a straight line.)

The equations (*7.12*) must be verified. The Christoffel symbols Γ^i_{jk} for spherical coordinates are (Problem 6.5):

$$\boldsymbol{i=1} \qquad \Gamma^1_{22}=-x^1, \quad \Gamma^1_{33}=-x^1\sin^2 x^2$$

$$i = 2 \qquad \Gamma^2_{12} = \Gamma^2_{21} = \frac{1}{x^1}, \quad \Gamma^2_{33} = -\sin x^2 \cos x^2$$

$$i = 3 \qquad \Gamma^3_{13} = \Gamma^3_{31} = \frac{1}{x^1}, \quad \Gamma^3_{23} = \Gamma^3_{32} = \cot x^2$$

The derivatives of the $x^i(t)$ are:

$$\frac{dx^1}{dt} = a \sec t \tan t, \qquad \frac{d^2x^1}{dt^2} = (a \sec t)(\tan^2 t + \sec^2 t)$$

$$\frac{dx^2}{dt} = 1, \quad \frac{d^2x^2}{dt^2} = 0 \qquad \text{and} \qquad \frac{dx^3}{dt} = \frac{d^2x^3}{dt^2} = 0$$

With $\varepsilon \equiv 1$, (*7.11*) gives

$$w = g_{ij} \frac{dx^i}{dt}\frac{dx^j}{dt} = \left(\frac{dx^1}{dt}\right)^2 + (x^1)^2\left(\frac{dx^2}{dt}\right)^2 + (x^1 \sin x^2)^2\left(\frac{dx^3}{dt}\right)^2$$
$$= (a \sec t \tan t)^2 + (a \sec t)^2(1)^2 + 0 = a^2 \sec^4 t$$

and
$$\frac{1}{2w}\frac{dw}{dt} = \frac{(4a^2 \sec^3 t)(\sec t \tan t)}{2a^2 \sec^4 t} = 2 \tan t$$

For convenience in the verification of (*7.12*), let LS denote the left side, and RS the right side, of the equation in question. We obtain:

$$i = 1 \qquad \text{LS} = \frac{d^2x^1}{dt^2} + \Gamma^1_{22}\left(\frac{dx^2}{dt}\right)^2 + \Gamma^1_{33}\left(\frac{dx^3}{dt}\right)^2$$
$$= (a \sec t)(\tan^2 t + \sec^2 t) - (a \sec t)(1)^2 + 0 = 2a \sec t \tan^2 t$$

$$\text{RS} = (2 \tan t)\frac{dx^1}{dt} = (2 \tan t)(a \sec t \tan t) = 2a \sec t \tan^2 t = \text{LS}$$

$$i = 2 \qquad \text{LS} = \frac{d^2x^2}{dt^2} + 2\Gamma^2_{12}\frac{dx^1}{dt}\frac{dx^2}{dt} + \Gamma^2_{33}\left(\frac{dx^3}{dt}\right)^2 = 0 + \frac{2}{a \sec t}(a \sec t \tan t)(1) + 0 = 2 \tan t$$

$$\text{RS} = (2 \tan t)\frac{dx^2}{dt} = 2 \tan t = \text{LS}$$

$$i = 3 \qquad \text{LS} = \frac{d^2x^3}{dt^2} + 2\Gamma^3_{13}\frac{dx^1}{dt}\frac{dx^3}{dt} + 2\Gamma^3_{23}\frac{dx^2}{dt}\frac{dx^3}{dt} = 0$$

$$\text{RS} = (2 \tan t)\frac{dx^3}{dt} = 0 = \text{LS}$$

Supplementary Problems

7.24 Determine the fundamental indicator $\varepsilon(\mathbf{U})$ if $(U^i) = (2t, -2t, 1)$ at the point $(x^i) = (t^2, -t^2, t)$. The Riemannian metric is given by

$$(g_{ij}) = \begin{bmatrix} 2x^1 & x^3 & 0 \\ x^3 & 2x^2 & 0 \\ 0 & 0 & 1 \end{bmatrix} \qquad (4x^1x^2 \neq (x^3)^2)$$

7.25 Find the null points of the curve $\mathscr{C} : x^1 = t, \; x^2 = t^4$ (t real), if the metric is

$$\varepsilon \, ds^2 = 8(x^1 \, dx^1)^2 - 2 \, dx^1 \, dx^2$$

7.26 Find the arc length of the curve in Problem 7.25 if $0 \leqq t \leqq 2$.

7.27 Find the null points of the curve $\mathscr{C}$: $x^1 = t^3 + 1$, $x^2 = t^2$, $x^3 = t$, if the metric is

$$\varepsilon\, ds^2 = (dx^1)^2 - (dx^2)^2 - (x^3\, dx^3)^2$$

7.28 Find the arc length of the curve in Problem 7.27 if $\frac{1}{2} \leqq t \leqq 1$.

7.29 Find the angle between the curves

$$\mathscr{C}_1 : \begin{cases} x^1 = 5t \\ x^2 = 2 \\ x^3 = 3t \end{cases} \qquad \mathscr{C}_2 : \begin{cases} x^1 = u \\ x^2 = 2 \\ x^3 = 3u^2/25 \end{cases}$$

at each of the points of intersection, if the fundamental form is $(dx^1)^2 - (dx^2)^2 - (dx^3)^2$.

7.30 If $\varepsilon\, ds^2 = (dx^1)^2 - (dx^2)^2$, (*a*) find the length L of the curve $\mathscr{C}$: $x^1 = 12t^2$, $x^2 = 8t^3$, for $0 \leqq t \leqq 2$. (*b*) Find an arc-length parameterization, $x^i = x^i(s)$, for $\mathscr{C}$, with $s = 0$ corresponding to $t = 1$. (*c*) Show that the $x^i(s)$ are differentiable to all orders except at points of nullity.

7.31 Find the arc length of the curve of Problem 7.30, but with the Euclidean metric.

7.32 Compute $\mathbf{T} = (dx^i/ds)$ from the arc-length parameterization found in Problem 7.30 and verify that $\mathbf{T}$ has unit length at all points except $s = 0$.

7.33 Calculate the components N^i of the unit principal normal of the curve of Problem 7.30, using (*7.16*) (Problem 7.14).

7.34 Calculate both the curvature κ and the absolute curvature κ_0 for the curve of Problem 7.30. Discuss the numerical behavior of κ_0 along the curve.

7.35 Use the formula of Problem 7.18(*b*) to confirm the value of κ_0 found in Problem 7.34.

7.36 Compute κ_0 under the Euclidean metric for the curve of Problem 7.30; compare with the result obtained in Problem 7.34. For convenience, let $t = 0$ correspond to $s = 8$.

7.37 Without calculating an arc-length parameter, find the vectors $\mathbf{T}$ and $\mathbf{N}$, and the curvature κ, for the "parabola" $x^1 = t$, $x^2 = t^2$ $(0 \leqq t < \frac{1}{4})$ under the Riemannian metric

$$\varepsilon\, ds^2 = (dx^1)^2 - 2\, dx^1\, dx^2$$

7.38 Show that the first-quadrant portion $(x^i > 0)$ of the hypocycloid $\mathscr{H}$ of four cusps

$$(x^1)^{2/3} + (x^2)^{2/3} = a^{2/3} \qquad (a > 0)$$

may be parameterized as $x^1 = a \cos^3 t$, $x^2 = a \sin^3 t$, with $0 \leqq t \leqq \pi/2$. Find the arc length under the two metrics

$$(a)\quad \varepsilon\, ds^2 = (dx^1)^2 - (dx^2)^2 \qquad (b)\quad ds^2 = (dx^1)^2 + (dx^2)^2$$

(*c*) Without computing an arc-length parameter, find $\mathbf{T}$ and κ_0 for $\mathscr{H}$ under both metrics.

7.39 (*a*) Determine the Christoffel symbols of the second kind for the Riemannian metric $\varepsilon\, ds^2 = x^1(dx^1)^2 + x^2(dx^2)^2$. (*b*) Without converting to an arc-length parameter, verify that all curves $x^1 = t^2$, $x^2 = (at^3 + b)^{2/3}$, where a and b are arbitrary constants, are geodesics.

Chapter 8

Riemannian Curvature

8.1 THE RIEMANN TENSOR

The Riemann tensor emerges from an analysis of a simple question. Starting with a covariant vector (V_i) and taking the covariant derivative with respect to x^j and then with respect to x^k produces the third-order tensor

$$((V_i)_{,j})_{,k} \equiv (V_{i,jk})$$

Does the order of differentiation matter, or does $V_{i,jk} = V_{i,kj}$ hold in general?

Standard hypotheses concerning differentiability suffice to guarantee that the partial derivative of order two is order-independent,

$$\frac{\partial^2 V_i}{\partial x^j \partial x^k} = \frac{\partial^2 V_i}{\partial x^k \partial x^j}$$

but due to the presence of Christoffel symbols, such hypotheses do not extend to covariant differentiation. The following formula is established in Problem 8.1:

$$V_{j,kl} - V_{j,lk} = R^i_{jkl} V_i \tag{8.1}$$

where

$$R^i_{jkl} \equiv \frac{\partial \Gamma^i_{jl}}{\partial x^k} - \frac{\partial \Gamma^i_{jk}}{\partial x^l} + \Gamma^r_{jl}\Gamma^i_{rk} - \Gamma^r_{jk}\Gamma^i_{rl} \tag{8.2}$$

The Quotient Theorem (covariant form) immediately implies

Theorem 8.1: The n^4 components defined by (*8.2*) are those of a fourth-order tensor, contravariant of order one, covariant of order three.

(R^i_{jkl}) is called the *Riemann* (or *Riemann–Christoffel*) *tensor of the second kind*; lowering the contravariant index produces

$$R_{ijkl} \equiv g_{ir} R^r_{jkl} \tag{8.3}$$

the *Riemann tensor of the first kind.*

In answer to our original question, we may now say that covariant differentiation is order-dependent unless the metric is such as to make the Riemann tensor (either kind) vanish.

8.2 PROPERTIES OF THE RIEMANN TENSOR

Two Important Formulas

The Riemann tensor of the first kind can be introduced independently via the following formula (see Problem 8.4):

$$R_{ijkl} = \frac{\partial \Gamma_{jli}}{\partial x^k} - \frac{\partial \Gamma_{jki}}{\partial x^l} + \Gamma_{ilr}\Gamma^r_{jk} - \Gamma_{ikr}\Gamma^r_{jl} \tag{8.4}$$

From (*8.4*) there follows

$$R_{ijkl} = \frac{1}{2}\left(\frac{\partial^2 g_{il}}{\partial x^j \partial x^k} + \frac{\partial^2 g_{jk}}{\partial x^i \partial x^l} - \frac{\partial^2 g_{ik}}{\partial x^j \partial x^l} - \frac{\partial^2 g_{jl}}{\partial x^i \partial x^k}\right) + \Gamma_{ilr}\Gamma^r_{jk} - \Gamma_{ikr}\Gamma^r_{jl} \tag{8.5}$$

EXAMPLE 8.1 Calculate the components R_{ijkl} of the Riemann tensor for the metric of Problem 7.22,

$$\varepsilon\, ds^2 = (dx^1)^2 - (x^2)^{-2}(dx^2)^2$$

The nonvanishing Christoffel symbols are $\Gamma^2_{22} = -(x^2)^{-1}$ and $\Gamma_{222} = g_{22}\Gamma^2_{22} = (x^2)^{-3}$. The partial-derivative terms in (*8.4*) vanish unless all indices are 2; but then the two terms cancel. Likewise the Christoffel-symbol terms either vanish or cancel. We conclude that all sixteen components $R_{ijkl} = 0$.

Symmetry Properties

Interchange of k and l in (*8.2*) shows that $R^i_{jkl} = -R^i_{jlk}$, whence $R_{ijkl} = -R_{ijlk}$. This and two other symmetry properties are easily established at this point; Bianchi's (first) identity will be demonstrated in Chapter 9.

$$\begin{array}{rl} \textit{first skew symmetry} & R_{ijkl} = -R_{jikl} \\ \textit{second skew symmetry} & R_{ijkl} = -R_{ijlk} \\ \textit{block symmetry} & R_{ijkl} = R_{klij} \\ \textit{Bianchi's identity} & R_{ijkl} + R_{iklj} + R_{iljk} = 0 \end{array} \tag{8.6}$$

Number of Independent Components

We shall count the separate types of potentially nonzero components, using the above symmetry properties. The first two properties imply that R_{aacd} and R_{abcc} (not summed on a or c) are zero. In the following list, we agree not to sum on repeated indices.

(A) Type R_{abab}, $a < b$: $n_A = {}_nC_2 = n(n-1)/2$

(B) Type R_{abac}, $b < c$: $n_B = 3 \cdot {}_nC_3 = n(n-1)(n-2)/2$

(C) Type R_{abcd} or R_{acbd}, $a < b < c < d$ (for type R_{adbc}, use Bianchi's identity): $n_C = 2 \cdot {}_nC_4 = n(n-1)(n-2)(n-3)/12$

In (A) the count is of combinations of n numbers two at a time (for a and b). In (B) one partitions the index strings into the three groups (with ${}_nC_3$ in each group) for which

$$a < b < c \qquad b < a < c \qquad b < c < a$$

Either subtype of (C) has as many members as there are combinations of n numbers four at a time (for a, b, c, and d).

Summing n_A, n_B, and n_C, we prove

Theorem 8.2: There are a total of $n^2(n^2-1)/12$ components of the Riemann tensor (R_{ijkl}) that are not identically zero and that are independent from the rest.

Corollary 8.3: In two-dimensional Riemannian space, the only components of the Riemann tensor not identically zero are $R_{1212} = R_{2121} = -R_{1221} = -R_{2112}$.

EXAMPLE 8.2 For the metric of spherical coordinates,

$$ds^2 = (dx^1)^2 + (x^1\,dx^2)^2 + (x^1 \sin x^2\,dx^3)^2$$

list and calculate the nonzero components R_{ijkl}, if any.

By Theorem 8.2 with $n = 3$, there are six potentially nonzero components:

(A) R_{1212}, R_{1313}, R_{2323}

(B) R_{1213}, $R_{1232}(= R_{2123})$, $R_{1323}(= R_{3132})$

Because $R_{ijkl} = g_{ii}R^i_{jkl}$ (diagonal metric tensor; no summation), we may instead compute the mixed components. From Problem 6.5,

$$\begin{array}{ll} i = 1 & \Gamma^1_{22} = -x^1, \quad \Gamma^1_{33} = -x^1 \sin^2 x^2 \\ i = 2 & \Gamma^2_{12} = \Gamma^2_{21} = \Gamma^2_{13} = \Gamma^2_{31} = \dfrac{1}{x^1}, \quad \Gamma^2_{33} = -\sin x^2 \cos x^2 \\ i = 3 & \Gamma^3_{13} = \Gamma^3_{31} = \dfrac{1}{x^1}, \quad \Gamma^3_{23} = \Gamma^3_{32} = \cot x^2 \end{array}$$

and (*8.2*) gives:

$$R^1_{212} = \frac{\partial \Gamma^1_{22}}{\partial x^1} - \frac{\partial \Gamma^1_{21}}{\partial x^2} + \Gamma^r_{22}\Gamma^1_{r1} - \Gamma^r_{21}\Gamma^1_{r2} = -1 - \Gamma^1_{22}\Gamma^1_{11} - \Gamma^2_{21}\Gamma^1_{22} = 0$$

$$R^1_{313} = \frac{\partial \Gamma^1_{33}}{\partial x^1} - \frac{\partial \Gamma^1_{31}}{\partial x^3} + \Gamma^r_{33}\Gamma^1_{r1} - \Gamma^r_{31}\Gamma^1_{r3} = -\sin^2 x^2 + \Gamma^1_{33}\Gamma^1_{11} + \Gamma^3_{31}\Gamma^1_{33} = 0$$

$$R^2_{323} = \frac{\partial \Gamma^2_{33}}{\partial x^2} - \frac{\partial \Gamma^2_{32}}{\partial x^3} + \Gamma^r_{33}\Gamma^2_{r2} - \Gamma^r_{32}\Gamma^2_{r3} = -\cos 2x^2 + \Gamma^1_{33}\Gamma^2_{12} - \Gamma^3_{32}\Gamma^2_{33} = -\cos 2x^2 - \sin^2 x^2 + \cos^2 x^2 = 0$$

$$R^1_{213} = \frac{\partial \Gamma^1_{23}}{\partial x^1} - \frac{\partial \Gamma^1_{21}}{\partial x^3} + \Gamma^r_{23}\Gamma^1_{r1} - \Gamma^r_{21}\Gamma^1_{r3} = \Gamma^3_{23}\Gamma^1_{31} - \Gamma^2_{21}\Gamma^1_{23} = 0$$

$$R^1_{232} = \frac{\partial \Gamma^1_{22}}{\partial x^3} - \frac{\partial \Gamma^1_{23}}{\partial x^2} + \Gamma^r_{22}\Gamma^1_{r3} - \Gamma^r_{23}\Gamma^1_{r2} = \Gamma^1_{22}\Gamma^1_{13} - \Gamma^3_{23}\Gamma^1_{32} = 0$$

$$R^1_{323} = \frac{\partial \Gamma^1_{33}}{\partial x^2} - \frac{\partial \Gamma^1_{32}}{\partial x^3} + \Gamma^r_{33}\Gamma^1_{r2} - \Gamma^r_{32}\Gamma^1_{r3} = -2x^1 \sin x^2 \cos x^2 + \Gamma^2_{33}\Gamma^1_{22} - \Gamma^3_{32}\Gamma^1_{33}$$

$$= -2x^1 \sin x^2 \cos x^2 + x^1 \sin x^2 \cos x^2 + (\cot x^2)(x^1 \sin^2 x^2) = 0$$

Therefore, $R_{ijkl} = 0$ for all i, j, k, l.

8.3 RIEMANNIAN CURVATURE

The *Riemannian* (or *sectional*) *curvature* relative to a given metric (g_{ij}) is defined for each pair of (contravariant) vectors $\mathbf{U} = (U^i)$, $\mathbf{V} = (V^i)$ as

$$\mathrm{K} = \mathrm{K}(\mathbf{x}; \mathbf{U}, \mathbf{V}) = \frac{R_{ijkl}U^iV^jU^kV^l}{G_{pqrs}U^pV^qU^rV^s} \qquad (G_{pqrs} \equiv g_{pr}g_{qs} - g_{ps}g_{qr}) \tag{8.7}$$

This sort of curvature depends not only on position, but also on a pair of directions selected at each point (the vectors **U** and **V**). By contrast, the curvature κ of a curve depends only on the points along the curve. Although it would seem desirable for K to depend only on the points of space, to demand this would impose severe and unrealistic restrictions, as will become apparent in Chapter 9.

EXAMPLE 8.3 The numerator of (*8.7*) is an invariant, because (R_{ijkl}) is a tensor. As for the denominator, the identity

$$G_{pqrs}V^p_{(1)}V^q_{(2)}V^r_{(3)}V^s_{(4)} = (\mathbf{V}_{(1)}\mathbf{V}_{(3)})(\mathbf{V}_{(2)}\mathbf{V}_{(4)}) - (\mathbf{V}_{(1)}\mathbf{V}_{(4)})(\mathbf{V}_{(2)}\mathbf{V}_{(3)}) \tag{8.8}$$

implies that the denominator is an invariant and proves (Lemma 4.1) that (G_{ijkl}) is also a tensor. It follows that $\mathrm{K}(\mathbf{x}; \mathbf{U}, \mathbf{V})$ *is an invariant*, and thus it serves to generalize the *Gaussian curvature* of a surface to higher dimensions.

Helpful in the calculation of K is the fact that the G_{ijkl} possess exactly the same symmetries as the R_{ijkl} (see Problem 8.19). Moreover, if $g = (g_{ij})$ is diagonal, all the nonzero G_{ijkl} will be derivable from the type-A terms

$$G_{abab} = g_{aa}g_{bb} \qquad (a < b;\ \text{no summation})$$

EXAMPLE 8.4 Evaluate the Riemannian curvature at any point (x^i) of Riemannian 3-space in the directions (*a*) $\mathbf{U} = (1, 0, 0)$ and $\mathbf{V} = (0, 1, 1)$, and (*b*) $\mathbf{U} = (0, 1, 0)$ and $\mathbf{V} = (1, 1, 0)$, if the metric is given by

$$g_{11} = 1 \qquad g_{22} = 2x^1 \qquad g_{33} = 2x^2 \qquad g_{ij} = 0 \quad \text{if} \quad i \neq j$$

From Problem 6.4, the nonzero Christoffel symbols are:

$$\Gamma^1_{22} = -1 \qquad \Gamma^2_{12} = \Gamma^2_{21} = \frac{1}{2x^1} \qquad \Gamma^2_{33} = -\frac{1}{2x^1} \qquad \Gamma^3_{23} = \Gamma^3_{32} = \frac{1}{2x^2}$$

Since $n = 3$, only six (by Theorem 8.2) components of the Riemann tensor need be considered: R_{1212}, R_{1313}, R_{2323}, R_{1213}, R_{2123} and R_{3132}. The metric being diagonal, we compute:

$$R^1_{212} = \frac{\partial \Gamma^1_{22}}{\partial x^1} - \frac{\partial \Gamma^1_{21}}{\partial x^2} + \Gamma^r_{22}\Gamma^1_{r1} - \Gamma^r_{21}\Gamma^1_{r2} = 0 - 0 + 0 - \Gamma^2_{21}\Gamma^1_{22} = \frac{1}{2x^1}$$

$$R^1_{313} = 0 \qquad R^2_{323} = \frac{1}{4x^1x^2} \qquad R^1_{213} = 0 \qquad R^2_{123} = 0 \qquad R^3_{132} = \frac{1}{4x^1x^2}$$

which yield the three terms

(A) $R_{1212} = g_{11}R^1_{212} = 1/2x^1, \;\; R_{2323} = g_{22}R^2_{323} = 1/2x^2$

(B) $R_{3132} = g_{33}R^3_{132} = 1/2x^1$

Theorem 8.2 also applies to the G_{ijkl}; but we may take the shortcut indicated in Example 8.3:

(A) $G_{1212} = g_{11}g_{22} = 2x^1, \;\; G_{1313} = g_{11}g_{33} = 2x^2, \;\; G_{2323} = g_{22}g_{33} = 4x^1x^2$

Let us now give an expanded form of (8.7) in the case that type-C terms are absent. It is convenient to define the n^4 functions

$$W_{ijkl} \equiv \begin{vmatrix} U^i & U^j \\ V^i & V^j \end{vmatrix} \begin{vmatrix} U^k & U^l \\ V^k & V^l \end{vmatrix} \tag{8.9}$$

of two vectors $\mathbf{U} = (U^i)$ and $\mathbf{V} = (V^i)$. Observe that the W_{ijkl} possess all the symmetries of the R_{ijkl} (or the G_{ijkl}). Looking at the numerator of (8.7), we see that a given type-A coefficient from the basic set generates, via its skew symmetries, the 4 terms

$$R_{abab}(U^aV^bU^aV^b - U^bV^aU^aV^b - U^aV^bU^bV^a + U^bV^aU^bV^a) = R_{abab}W_{abab}$$

and these precisely exhaust the $2 \times 2 = 4$ terms in which the coefficient R_{ijkl} involves the same distinct integers, a and b, in the first pair of indices as in the second pair. A given type-B coefficient from the basic set will generate, via its skew symmetries and block symmetry, the 8 terms

$$R_{abac}(W_{abac} + W_{acab}) = 2R_{abac}W_{abac}$$

and these exactly correspond to the $2^2 \times 2 = 8$ ways of writing R_{ijkl} such that the first and second index-pairs contain the common integer a but are otherwise composed of distinct integers b and c. Analyzing the denominator of (8.7) in the same fashion, we obtain as the desired formula:

$$\mathrm{K} = \frac{\sum\limits_{\text{type A}} R_{abab}W_{abab} \;+\; 2 \sum\limits_{\text{type B}} R_{abac}W_{abac}}{\sum\limits_{\text{type A}} G_{abab}W_{abab} \;+\; 2 \sum\limits_{\text{type B}} G_{abac}W_{abac}} \tag{8.10}$$

It is understood that the summation convention does not operate in (8.10); the indicated summations are over all the nonzero, independent R_{ijkl} (G_{ijkl}), according to type. Now to the problem at hand.

(*a*) For the data, (8.10) becomes

$$\mathrm{K} = \frac{R_{1212}W_{1212} + R_{2323}W_{2323} + 2R_{3132}W_{3132}}{G_{1212}W_{1212} + G_{1313}W_{1313} + G_{2323}W_{2323}}$$

For

$$\begin{bmatrix} \mathbf{U} \\ \mathbf{V} \end{bmatrix} = \begin{bmatrix} 1 & 0 & 0 \\ 0 & 1 & 1 \end{bmatrix}$$

we have

$$W_{1212} = \begin{vmatrix} 1 & 0 \\ 0 & 1 \end{vmatrix}^2 = 1 \qquad W_{2323} = \begin{vmatrix} 0 & 0 \\ 1 & 1 \end{vmatrix}^2 = 0$$

$$W_{3132} = \begin{vmatrix} 0 & 1 \\ 1 & 0 \end{vmatrix} \begin{vmatrix} 0 & 0 \\ 1 & 1 \end{vmatrix} = 0 \qquad W_{1313} = \begin{vmatrix} 1 & 0 \\ 0 & 1 \end{vmatrix}^2 = 1$$

and so

$$\mathrm{K} = \frac{(1/2x^1)(1) + (1/2x^2)(0) + 2(1/2x^1)(0)}{(2x^1)(1) + (2x^2)(1) + (4x^1x^2)(0)} = \frac{1}{4x^1(x^1 + x^2)}$$

(b) For

$$\begin{bmatrix} \mathbf{U} \\ \mathbf{V} \end{bmatrix} = \begin{bmatrix} 0 & 1 & 0 \\ 1 & 1 & 0 \end{bmatrix}$$

we have

$$W_{1212} = \begin{vmatrix} 0 & 1 \\ 1 & 1 \end{vmatrix}^2 = 1 \qquad W_{2323} = \begin{vmatrix} 1 & 0 \\ 1 & 0 \end{vmatrix}^2 = 0$$

$$W_{3132} = \begin{vmatrix} 0 & 0 \\ 0 & 1 \end{vmatrix} \begin{vmatrix} 0 & 1 \\ 0 & 1 \end{vmatrix} = 0 \qquad W_{1313} = \begin{vmatrix} 0 & 0 \\ 1 & 0 \end{vmatrix}^2 = 0$$

whence

$$\mathrm{K} = \frac{(1/2x^1)(1) + 0 + 0}{(2x^1)(1) + 0 + 0} = \frac{1}{4(x^1)^2}$$

Observations on the Curvature Formula

I. If $n = 2$, (*8.7*) reduces to

$$\mathrm{K} = \frac{R_{1212}}{g_{11}g_{22} - g_{12}^2} \equiv \frac{R_{1212}}{g} \tag{8.11}$$

(see Problem 8.7). Thus, at a given point in Riemannian 2-space, the curvature is determined by the g_{ij} and their derivatives, and is independent of the directions **U** and **V**.

II. The extension of (*8.10*) to include type-C terms is as follows:

$$\mathrm{K} = \frac{\sum\limits_{\text{type A}} R_{abab}W_{abab} + 2\sum\limits_{\text{type B}} R_{abac}W_{abac} + 2\sum\limits_{\text{type C}} R_{abcd}(W_{abcd} - W_{adbc}) + 2\sum\limits_{\text{type C}} R_{acbd}(W_{acbd} - W_{adcb})}{\sum\limits_{\text{type A}} G_{abab}W_{abab} + 2\sum\limits_{\text{type B}} G_{abac}W_{abac} + 2\sum\limits_{\text{type C}} G_{abcd}(W_{abcd} - W_{adbc}) + 2\sum\limits_{\text{type C}} G_{acbd}(W_{acbd} - W_{adcb})} \tag{8.12}$$

(see Problem 8.9).

III. If linearly independent **U** and **V** are replaced by independent linear combinations of themselves, the curvature is unaffected; i.e.,

$$\mathrm{K}(\mathbf{x}; \lambda\mathbf{U} + \nu\mathbf{V}, \mu\mathbf{U} + \omega\mathbf{V}) = \mathrm{K}(\mathbf{x}; \mathbf{U}, \mathbf{V}) \tag{8.13}$$

Therefore, at a given point **x**, the curvature will have a value, not for each pair of vectors **U** and **V**, but for each 2-flat passing through **x**.

Isotropic Points

If the Riemannian curvature at **x** does not change with the orientation of a 2-flat through **x**, then **x** is called *isotropic*. From (*8.11*), we have

Theorem 8.4: All points of a two-dimensional Riemannian space are isotropic.

It is not immediately clear whether any metric (g_{ij}) could lead to isotropic points in $\mathbf{R}^n$, $n \geqq 3$. But such is the case. Indeed, as is shown in Problem 8.12, $\mathbf{R}^3$ under a hyperbolic metric is isotropic *at any point*.

8.4 THE RICCI TENSOR

A brief look will be given a tensor that is of importance in Relativity. The *Ricci tensor of the first kind* is defined as a contraction of the Riemann tensor of the second kind:

$$R_{ij} \equiv R^k_{ijk} = \frac{\partial \Gamma^k_{ik}}{\partial x^j} - \frac{\partial \Gamma^k_{ij}}{\partial x^k} + \Gamma^r_{ik}\Gamma^k_{rj} - \Gamma^r_{ij}\Gamma^k_{rk} \tag{8.14}$$

Raising an index yields the *Ricci tensor of the second kind*:

$$R^i_j \equiv g^{ik}R_{kj} \tag{8.15}$$

By use of the following simple consequence of Laplace's expansion (*2.5*):

Lemma 8.5: Let $A = [a_{ij}(\mathbf{x})]_{nn}$ be a nonsingular matrix of multivariate functions, with inverse $B = [b_{ij}(\mathbf{x})]_{nn}$. Then by the chain rule,

$$\frac{\partial}{\partial x^i}(\ln|\det A|) \equiv \frac{1}{\det A}\frac{\partial}{\partial x^i}(\det A) = \frac{1}{\det A}\frac{\partial}{\partial a_{rs}}(\det A)\frac{\partial a_{rs}}{\partial x^i} = \frac{A_{rs}}{\det A}\frac{\partial a_{rs}}{\partial x^i} = b_{sr}\frac{\partial a_{rs}}{\partial x^1}$$

the definition (*8.14*) may be put into a form (Problem 8.14) that makes evident the symmetry of the R_{ij}.

$$R_{ij} = \frac{\partial^2}{\partial x^i \partial x^j}(\ln\sqrt{|g|}) - \frac{1}{\sqrt{|g|}}\frac{\partial}{\partial x^r}(\sqrt{|g|}\,\Gamma^r_{ij}) + \Gamma^r_{is}\Gamma^s_{rj} \qquad (8.16)$$

Here, as always, $g \equiv \det G$.

Theorem 8.6: The Ricci tensor is symmetric.

After raising a subscript to define the Ricci tensor of the second kind, $R^i_j = g^{is}R_{sj}$, and then contracting on the remaining pair of indices, the important invariant $R \equiv R^i_i$ results, called the *Ricci* (or *scalar*) *curvature*. By (*8.16*),

$$R = g^{ij}\left[\frac{\partial^2}{\partial x^i \partial x^j}(\ln\sqrt{|g|}) - \frac{1}{\sqrt{|g|}}\frac{\partial}{\partial x^r}(\sqrt{|g|}\,\Gamma^r_{ij}) + \Gamma^r_{is}\Gamma^s_{rj}\right] \qquad (8.17)$$

Solved Problems

THE RIEMANN TENSOR

8.1 Prove (*8.1*).

By definition of the covariant derivative,

$$V_{i,jk} = (V_{i,j})_{,k} = \frac{\partial}{\partial x^k}(V_{i,j}) - \Gamma^r_{ik}(V_{r,j}) - \Gamma^r_{jk}(V_{i,r}) \qquad (1)$$

Substitute

$$V_{i,j} = \frac{\partial V_i}{\partial x^j} - \Gamma^s_{ij}V_s$$

in (*1*), carry out the differentiation, and remove parentheses:

$$V_{i,jk} = \frac{\partial^2 V_i}{\partial x^k \partial x^j} - \frac{\partial \Gamma^s_{ij}}{\partial x^k}V_s - \Gamma^s_{ij}\frac{\partial V_s}{\partial x^k} - \Gamma^r_{ik}\frac{\partial V_r}{\partial x^j} + \Gamma^r_{ik}\Gamma^s_{rj}V_s - \Gamma^r_{jk}\frac{\partial V_i}{\partial x^r} + \Gamma^r_{jk}\Gamma^s_{ir}V_s \qquad (2)$$

Interchanging j and k yields

$$V_{i,kj} = \frac{\partial^2 V_i}{\partial x^j \partial x^k} - \frac{\partial \Gamma^s_{ik}}{\partial x^j}V_s - \Gamma^s_{ik}\frac{\partial V_s}{\partial x^j} - \Gamma^r_{ij}\frac{\partial V_r}{\partial x^k} + \Gamma^r_{ij}\Gamma^s_{rk}V_s - \Gamma^r_{kj}\frac{\partial V_i}{\partial x^r} + \Gamma^r_{kj}\Gamma^s_{ir}V_s \qquad (3)$$

Subtracting (*3*) from (*2*), one sees that the first, third, fourth, sixth, and seventh terms on the right of (*2*) cancel with the first, fourth, third, sixth, and seventh terms on the right of (*3*), leaving

$$V_{i,jk} - V_{i,kj} = -\frac{\partial \Gamma^s_{ij}}{\partial x^k}V_s + \Gamma^r_{ik}\Gamma^s_{rj}V_s + \frac{\partial \Gamma^s_{ik}}{\partial x^j}V_s - \Gamma^r_{ij}\Gamma^s_{rk}V_s$$

$$= \left(\frac{\partial \Gamma^s_{ik}}{\partial x^j} - \frac{\partial \Gamma^s_{ij}}{\partial x^k} + \Gamma^r_{ik}\Gamma^s_{rj} - \Gamma^r_{ij}\Gamma^s_{rk}\right)V_s = R^s_{ijk}V_s$$

8.2 Show that at any point where the Christoffel symbols vanish,

$$R^i_{jkl} + R^i_{klj} + R^i_{ljk} = 0$$

In this case the expression for R^i_{jkl} reduces to just $\partial\Gamma^i_{jl}/\partial x^k - \partial\Gamma^i_{jk}/\partial x^l$. Therefore,

$$R^i_{jkl} + R^i_{klj} + R^i_{ljk} = \frac{\partial\Gamma^i_{jl}}{\partial x^k} - \frac{\partial\Gamma^i_{jk}}{\partial x^l} + \frac{\partial\Gamma^i_{kj}}{\partial x^l} - \frac{\partial\Gamma^i_{kl}}{\partial x^j} + \frac{\partial\Gamma^i_{lk}}{\partial x^j} - \frac{\partial\Gamma^i_{lj}}{\partial x^k}$$

As all the terms cancel, the desired relationship is proved.

8.3 Prove that for an arbitrary second-order covariant tensor (T_{ij})

$$T_{ij,kl} - T_{ij,lk} = R^s_{ikl}T_{sj} + R^s_{jkl}T_{is}$$

(The general formula,

$$T_{i_1i_2\ldots i_p,kl} - T_{i_1i_2\ldots i_p,lk} = \sum_{q=1}^{p} R^s_{i_qkl}T_{i_1\ldots i_{q-1}si_{q+1}\ldots i_p} \tag{8.18}$$

which is credited to Ricci, is similarly established.)

A direct approach would be quite tedious; instead, first establish that

$$V^i_{,jk} - V^i_{,kj} = -R^i_{sjk}V^s \tag{1}$$

for any contravariant vector (V^i) (see Problem 8.16). Now observe that (V^qT_{iq}) is a covariant vector, to which (*8.1*) applies. Thus,

$$(V^qT_{iq})_{,kl} - (V^qT_{iq})_{,lk} = R^s_{ikl}V^qT_{sq} \tag{2}$$

By the inner-product rule for covariant differentiation,

$$\begin{aligned}(V^qT_{iq})_{,k} &= V^q_{,k}T_{iq} + V^qT_{iq,k}\\ (V^qT_{iq})_{,kl} &= V^q_{,kl}T_{iq} + V^q_{,k}T_{iq,l} + V^q_{,l}T_{iq,k} + V^qT_{iq,kl}\end{aligned} \tag{3}$$

Interchange k and l:

$$(V^qT_{iq})_{,lk} = V^q_{,lk}T_{iq} + V^q_{,l}T_{iq,k} + V^q_{,k}T_{iq,l} + V^qT_{iq,lk} \tag{4}$$

Subtraction of (*4*) from (*3*) will cancel the middle two terms on the right-hand sides, leaving

$$R^s_{ikl}V^qT_{sq} = (V^q_{,kl} - V^q_{,lk})T_{iq} + (T_{iq,kl} - T_{iq,lk})V^q \tag{5}$$

Now use (*1*) in the right member of (*5*):

$$R^s_{ikl}V^qT_{sq} = -R^s_{qkl}V^qT_{is} + (T_{iq,kl} - T_{iq,lk})V^q$$

which may be rearranged into

$$[(T_{iq,kl} - T_{iq,lk}) - (R^s_{ikl}T_{sq} + R^s_{qkl}T_{is})]V^q = 0$$

But (V^i) is arbitrary, so the bracketed expression must vanish. QED

PROPERTIES OF THE RIEMANN TENSOR

8.4 Establish (*8.4*).

By definition,

$$\begin{aligned}R_{ijkl} = g_{is}R^s_{jkl} &= g_{is}\frac{\partial\Gamma^s_{jl}}{\partial x^k} - g_{is}\frac{\partial\Gamma^s_{jk}}{\partial x^l} + g_{is}\Gamma^r_{jl}\Gamma^s_{rk} - g_{is}\Gamma^r_{jk}\Gamma^s_{rl}\\ &= \frac{\partial(g_{is}\Gamma^s_{jl})}{\partial x^k} - \frac{\partial g_{is}}{\partial x^k}\Gamma^s_{jl} - \frac{\partial(g_{is}\Gamma^s_{jk})}{\partial x^l} + \frac{\partial g_{is}}{\partial x^l}\Gamma^s_{jk} + \Gamma^r_{jl}\Gamma_{rki} - \Gamma^r_{jk}\Gamma_{rli}\end{aligned}$$

$$= \frac{\partial \Gamma_{jli}}{\partial x^k} - \frac{\partial \Gamma_{jki}}{\partial x^l} + \Gamma^r_{jk}\left(\frac{\partial g_{ir}}{\partial x^l} - \Gamma_{rli}\right) - \Gamma^r_{jl}\left(\frac{\partial g_{ir}}{\partial x^k} - \Gamma_{rki}\right)$$

Recall from (*6.2*) that for arbitrary index l,

$$\frac{\partial g_{ir}}{\partial x^l} - \Gamma_{lri} = \Gamma_{ilr}$$

By substitution,

$$R_{ijkl} = \frac{\partial \Gamma_{jli}}{\partial x^k} - \frac{\partial \Gamma_{jki}}{\partial x^l} + \Gamma_{ilr}\Gamma^r_{jk} - \Gamma_{ikr}\Gamma^r_{jl}$$

8.5 Establish the first skew-symmetry property, $R_{ijkl} = -R_{jikl}$.

To save writing, let

$$G^{ij}_{kl} \equiv \frac{1}{2}\left(\frac{\partial^2 g_{ij}}{\partial x^k \partial x^l} + \frac{\partial^2 g_{kl}}{\partial x^i \partial x^j}\right) \qquad \text{and} \qquad H^{ij}_{kl} \equiv \Gamma_{ijr}\Gamma^r_{kl}$$

Note the obvious symmetry properties

$$G^{ij}_{kl} = G^{ji}_{kl} = G^{ij}_{lk} \qquad \text{and} \qquad H^{ij}_{kl} = H^{ji}_{kl} = H^{ij}_{lk}$$

Also, it is clear that $G^{ij}_{kl} = G^{kl}_{ij}$; furthermore,

$$H^{ij}_{kl} = (g_{rs}\Gamma^s_{ij})\Gamma^r_{kl} = \Gamma^s_{ij}(g_{sr}\Gamma^r_{kl}) = \Gamma^s_{ij}\Gamma_{kls} = H^{kl}_{ij}$$

Now, by (*8.5*),

$$R_{ijkl} = G^{il}_{jk} - G^{ik}_{jl} + H^{il}_{jk} - H^{ik}_{jl}$$

and

$$R_{jikl} = G^{jl}_{ik} - G^{jk}_{il} + H^{jl}_{ik} - H^{jk}_{il} = G^{ik}_{jl} - G^{il}_{jk} + H^{ik}_{jl} - H^{il}_{jk} = -R_{ijkl}$$

8.6 List the independent, potentially nonzero components of R_{ijkl} for $n = 5$ and verify the formula of Theorem 8.2 in this case.

Type A: R_{1212}, R_{1313}, R_{1414}, R_{1515}
R_{2323}, R_{2424}, R_{2525}
R_{3434}, R_{3535}
R_{4545}

Type B: R_{1213}, R_{1214}, R_{1215}, R_{1314}, R_{1315}, R_{1415}
R_{2123}, R_{2124}, R_{2125}, R_{2324}, R_{2325}, R_{2425}
R_{3132}, R_{3134}, R_{3135}, R_{3234}, R_{3235}, R_{3435}
R_{4142}, R_{4143}, R_{4145}, R_{4243}, R_{4245}, R_{4345}
R_{5152}, R_{5153}, R_{5154}, R_{5253}, R_{5254}, R_{5354}

Type C: R_{1234}, R_{1235}, R_{1245}, R_{1345}, R_{2345}
R_{1324}, R_{1325}, R_{1425}, R_{1435}, R_{2435}

There are 10 components of types A and C each, and 30 of type B; or 50 altogether. From the formula,

$$\frac{n^2(n^2-1)}{12} = \frac{5^2(5^2-1)}{12} = \frac{(25)(24)}{12} = 50$$

RIEMANNIAN CURVATURE

8.7 Prove (*8.11*).

By Corollary 8.3 and the corresponding result for the G_{ijkl},

$$\mathrm{K} = \frac{R_{ijkl}U^iV^jU^kV^l}{G_{pqrs}U^pV^qU^rV^s} = \frac{R_{1212}[(U^1)^2(V^2)^2 - 2U^1V^2U^2V^1 + (U^2)^2(V^1)^2]}{G_{1212}[(U^1)^2(V^2)^2 - 2U^1V^2U^2V^1 + (U^2)^2(V^1)^2]} = \frac{R_{1212}}{G_{1212}} = \frac{R_{1212}}{g_{11}g_{22} - g_{12}^2}$$

8.8 Calculate K for the Riemannian metric $\varepsilon\, ds^2 = (x^1)^{-2}(dx^1)^2 - (x^1)^{-2}(dx^2)^2$, using the result of Problem 8.7.

We have only to calculate $R_{1212} = g_{11}R^1_{212}$. The nonvanishing Christoffel symbols are, by Problem 6.4,

$$\Gamma^1_{11} = -\frac{1}{x^1} \qquad \Gamma^1_{22} = -\frac{1}{x^1} \qquad \Gamma^2_{12} = \Gamma^2_{21} = -\frac{1}{x^1}$$

Consequently,

$$R^1_{212} = \frac{\partial \Gamma^1_{22}}{\partial x^1} - \frac{\partial \Gamma^1_{21}}{\partial x^2} + \Gamma^r_{22}\Gamma^1_{r1} - \Gamma^r_{21}\Gamma^1_{r2} = \frac{1}{(x^1)^2} - 0 + \Gamma^1_{22}\Gamma^1_{11} - \Gamma^2_{21}\Gamma^1_{22}$$

$$= \frac{1}{(x^1)^2} - \frac{1}{x^1}\left(-\frac{1}{x^1}\right) - \left(-\frac{1}{x^1}\right)\left(-\frac{1}{x^1}\right) = \frac{1}{(x^1)^2}$$

and

$$\mathrm{K} = \frac{g_{11}R^1_{212}}{g_{11}g_{22}} = \frac{R^1_{212}}{g_{22}} = \frac{(x^1)^{-2}}{-(x^1)^{-2}} = -1$$

8.9 Derive the form (*8.12*) of the curvature equation.

We need only establish the summations over the type-C terms in the numerator; the rest of the work was done in Example 8.4.

First of all, let us verify that all R_{ijkl} with $ijkl$ a permutation of $abcd$, where $a<b<c<d$ are distinct integers, are generated by the skew and block symmetries of the three components R_{abcd}, R_{acbd}, and R_{adbc}. Examination of Table 8-1, which uses an obvious notation for the symmetry operators, shows that all $4! = 24$ permutations are accounted for. Consequently, the type-C part of the numerator of (*8.12*) is [cf. the equation preceding (*8.10*)]

$$2\sum R_{abcd}W_{abcd} + 2\sum R_{acbd}W_{acbd} + 2\sum R_{adbc}W_{adbc} \tag{1}$$

Table 8-1

Symmetry Operator	Subscript Chain		
	abcd	***acbd***	***adbc***
I	*abcd*	*acbd*	*adbc*
S_1	*bacd*	*cabd*	*dabc*
S_2	*abdc*	*acdb*	*adcb*
$S_1S_2 = S_2S_1$	*badc*	*cadb*	*dacb*
B	*cdab*	*bdac*	*bcad*
$BS_1 = S_2B$	*cdba*	*bdca*	*bcda*
$BS_2 = S_1B$	*dcab*	*dbac*	*cbad*
$BS_1S_2 = S_1S_2B$	*dcba*	*dbca*	*cbda*

The first summation is over all $a<b<c<d$ that yield a nonzero R_{abcd} in the basic set; similarly for the second summation. The third summation does not involve the basic set, but the symmetries of R_{ijkl} (shared by W_{ijkl}) allow its absorption in the first two summations. Thus, by Bianchi's identity,

$$2R_{adbc}W_{adbc} = 2(-R_{abcd} - R_{acdb})W_{adbc} = -2R_{abcd}W_{adbc} - 2R_{acbd}W_{adcb} \tag{2}$$

and substitution of (*2*) in (*1*) produces the expression given in (*8.12*).

8.10 Prove (*8.13*).

We have

$$W_{ijkl}(\lambda\mathbf{U}+\nu\mathbf{V},\mu\mathbf{U}+\omega\mathbf{V})=\begin{vmatrix}\lambda U^i+\nu V^i & \lambda U^j+\nu V^j\\ \mu U^i+\omega V^i & \mu U^j+\omega V^j\end{vmatrix}\begin{vmatrix}\lambda U^k+\nu V^k & \lambda U^l+\nu V^l\\ \mu U^k+\omega V^k & \mu U^l+\omega V^l\end{vmatrix}$$

$$=\begin{vmatrix}\lambda & \nu\\ \mu & \omega\end{vmatrix}^2\begin{vmatrix}U^i & U^j\\ V^i & V^j\end{vmatrix}\begin{vmatrix}U^k & U^l\\ V^k & V^l\end{vmatrix}=(\lambda\omega-\nu\mu)^2W_{ijkl}(\mathbf{U},\mathbf{V})$$

so that the quantity $(\lambda\omega-\nu\mu)^2$ factors out of all terms in (*8.12*) for $\mathrm{K}(\mathbf{x};\lambda\mathbf{U}+\nu\mathbf{V},\mu\mathbf{U}+\omega\mathbf{V})$, leaving $\mathrm{K}(\mathbf{x};\mathbf{U},\mathbf{V})$.

8.11 Find the isotropic points in the Riemannian space $\mathbf{R}^3$ with metric

$$g_{11}=1\qquad g_{22}=g_{33}=(x^1)^2+1\qquad g_{ij}=0\quad(i\neq j)$$

and calculate the curvature K at those points.

Follow Example 8.4. By Problem 6.4, the nonzero Christoffel symbols are

$$\Gamma^1_{22}=-x^1\qquad \Gamma^1_{33}=-x^1\qquad \Gamma^2_{12}=\Gamma^2_{21}=\frac{x^1}{(x^1)^2+1}\qquad \Gamma^3_{13}=\Gamma^3_{31}=\frac{x^1}{(x^1)^2+1}$$

Then:

$$R^1_{212}=\frac{\partial\Gamma^1_{22}}{\partial x^1}+\Gamma^1_{22}\Gamma^1_{11}-\Gamma^2_{21}\Gamma^1_{22}=-1-\frac{x^1}{(x^1)^2+1}(-x^1)=-\frac{1}{(x^1)^2+1}$$

$$R^1_{313}=\frac{\partial\Gamma^1_{33}}{\partial x^1}+\Gamma^1_{33}\Gamma^1_{11}-\Gamma^3_{31}\Gamma^1_{33}=-1-\frac{x^1}{(x^1)^2+1}(-x^1)=-\frac{1}{(x^1)^2+1}$$

$$R^2_{323}=\Gamma^1_{33}\Gamma^2_{12}=-x^1\cdot\frac{x^1}{(x^1)^2+1}=-\frac{(x^1)^2}{(x^1)^2+1}$$

$$R^1_{213}=R^2_{123}=R^3_{132}=0$$

which give

(A) $R_{1212}=g_{11}R^1_{212}=-[(x^1)^2+1]^{-1}$, $R_{1313}=g_{11}R^1_{313}=-[(x^1)^2+1]^{-1}$, $R_{2323}=g_{22}R^2_{323}=-(x^1)^2$

Types B and C vanish identically.

The corresponding terms for the denominator of (*8.10*) are

(A) $G_{1212}=g_{11}g_{22}=(x^1)^2+1,\ \ G_{1313}=g_{11}g_{33}=(x^1)^2+1,\ \ G_{2323}=g_{22}g_{33}=[(x^1)^2+1]^2$

Thus

$$\mathrm{K}=\frac{-[(x^1)^2+1]^{-1}W_{1212}-[(x^1)^2+1]^{-1}W_{1313}-(x^1)^2W_{2323}}{[(x^1)^2+1]W_{1212}+[(x^1)^2+1]W_{1313}+[(x^1)^2+1]^2W_{2323}}$$

$$=-[(x^1)^2+1]^{-2}\,\frac{W_{1212}+W_{1313}+(x^1)^2[(x^1)^2+1]W_{2323}}{W_{1212}+W_{1313}+[(x^1)^2+1]W_{2323}}$$

If K is to be independent of the W_{ijkl} (which vary with the direction of the 2-flat), then $(x^1)^2=1$, or $x^1=\pm1$. Therefore, the isotropic points compose two surfaces, on which the curvature has the value $\mathrm{K}=-[1+1]^{-2}\cdot 1=-1/4$.

8.12 Show that every point of $\mathbf{R}^3$ is isotropic for the metric

$$ds^2=(x^1)^{-2}(dx^1)^2+(x^1)^{-2}(dx^2)^2+(x^1)^{-2}(dx^3)^2$$

Problem 6.4 gives as the nonvanishing Christoffel symbols:

$$\Gamma^1_{11}=-\frac{1}{x^1}\qquad \Gamma^1_{22}=\frac{1}{x^1}\qquad \Gamma^1_{33}=\frac{1}{x^1}$$

$$\Gamma^2_{12}=\Gamma^2_{21}=-\frac{1}{x^1}\qquad \Gamma^3_{13}=\Gamma^3_{31}=-\frac{1}{x^1}$$

As in earlier problems, we proceed to calculate a basic set of R_{ijkl}, via $R_{ijkl} = g_{ii}R^i_{jkl}$ (no sum).

$$R^1_{212} = \frac{\partial \Gamma^1_{22}}{\partial x^1} - \frac{\partial \Gamma^1_{21}}{\partial x^2} + \Gamma^r_{22}\Gamma^1_{r1} - \Gamma^r_{21}\Gamma^1_{r2} = -\frac{1}{(x^1)^2} - 0 + \Gamma^1_{22}\Gamma^1_{11} - \Gamma^2_{21}\Gamma^1_{22}$$

$$= -\frac{1}{(x^1)^2} + \frac{1}{x^1}\left(-\frac{1}{x^1}\right) - \left(-\frac{1}{x^1}\right)\frac{1}{x^1} = -\frac{1}{(x^1)^2}$$

Similarly, $R^1_{313} = -1/(x^1)^2$. For the remainder, the partial-derivative terms all drop out, yielding

$$R^2_{323} = \Gamma^r_{33}\Gamma^2_{r2} - \Gamma^r_{32}\Gamma^2_{r3} = \Gamma^1_{33}\Gamma^2_{12} - 0 = -\frac{1}{(x^1)^2}$$

$$R^1_{213} = R^2_{123} = R^3_{132} = 0$$

Our basic set of nonzero terms is thus

(A) $R_{1212} = R_{1313} = R_{2323} = -1/(x^1)^4$

and, by Example 8.3,

(A) $G_{1212} = G_{1313} = G_{2323} = 1/(x^1)^4$

is a basic set of G_{ijkl}. Formula (8.10) or (8.12) now gives

$$\mathrm{K} = \frac{R_{1212}W_{1212} + R_{1313}W_{1313} + R_{2323}W_{2323}}{G_{1212}W_{1212} + G_{1313}W_{1313} + G_{2323}W_{2323}} = \frac{[-(x^1)^{-4}](W_{1212} + W_{1313} + W_{2323})}{[(x^1)^{-4}](W_{1212} + W_{1313} + W_{2323})} = -1$$

It is seen that this Riemannian space is more than just isotropic; it is a space of *constant curvature*.

THE RICCI TENSOR

8.13 For the metric of Example 8.4, calculate (*a*) R_{ij}, (*b*) R^i_j, (*c*) R.

(*a*) From $R_{ij} = R^k_{ijk} = R^1_{ij1} + R^2_{ij2} + R^3_{ij3}$ and the fact that $g_{ij} = 0$ for $i \neq j$, it follows that

$$R_{ij} = g^{11}R_{1ij1} + g^{22}R_{2ij2} + g^{33}R_{3ij3} \tag{1}$$

where $g^{11} = 1$, $g^{22} = 1/2x^1$, $g^{33} = 1/2x^2$. Now, a basic set of the R_{ijkl} was computed as

$$R_{1221}(= -R_{1212}) = -\frac{1}{2x^1} \qquad R_{2332}(= -R_{2323}) = -\frac{1}{2x^2} \qquad R_{3123}(= -R_{3132}) = -\frac{1}{2x^1}$$

and the only other nonzero components of the form R_{aija} generated by these are

$$R_{2112} = -\frac{1}{2x^1} \qquad R_{3223} = -\frac{1}{2x^2} \qquad R_{3213} = -\frac{1}{2x^1}$$

Hence, the nonzero R_{ij} may be read off from (1) as

$$R_{11} = g^{22}R_{2112} = -\frac{1}{4(x^1)^2}$$

$$R_{22} = g^{11}R_{1221} + g^{33}R_{3223} = -\frac{1}{2x^1} - \frac{1}{4(x^2)^2}$$

$$R_{33} = g^{22}R_{2332} = -\frac{1}{4x^1x^2}$$

$$R_{12} = g^{33}R_{3123} = -\frac{1}{4x^1x^2} = g^{33}R_{3213} = R_{21}$$

(*b*) $$R^i_j = g^{ik}R_{kj} = g^{ii}R_{ij} \quad \text{(no summation on } i\text{)}$$

(*c*) $$R = R^1_1 + R^2_2 + R^3_3 = g^{11}R_{11} + g^{22}R_{22} + g^{33}R_{33}$$

$$= (1)\left[-\frac{1}{4(x^1)^2}\right] + \left(\frac{1}{2x^1}\right)\left[-\frac{1}{2x^1} - \frac{1}{4(x^2)^2}\right] + \left(\frac{1}{2x^2}\right)\left(-\frac{1}{4x^1x^2}\right) = -\frac{x^1 + 2(x^2)^2}{(2x^1x^2)^2}$$

8.14 Derive (*8.16*) from (*8.14*).

Formula (*8.14*) involves two summations of the form Γ^s_{is}. By (*6.4*) and (*6.1b*),

$$\Gamma^s_{is} = g^{sr}\Gamma_{isr} = \frac{1}{2}\, g^{sr}(-g_{isr} + g_{sri} + g_{ris}) = -\frac{1}{2}\, g^{sr}g_{sir} + \frac{1}{2}\, g^{sr}g_{sri} + \frac{1}{2}\, g^{rs}g_{sir}$$

$$= \frac{1}{2}\, g^{sr}g_{rsi} \equiv \frac{1}{2}\, g^{sr}\frac{\partial g_{rs}}{\partial x^i} = \frac{\partial}{\partial x^i}(\ln \sqrt{|g|})$$

where Lemma 8.5 was used in the last step. Now substitute in (*8.14*):

$$R_{ij} = \frac{\partial^2(\ln\sqrt{|g|})}{\partial x^i \partial x^j} - \frac{\partial \Gamma^s_{ij}}{\partial x^s} + \Gamma^r_{is}\Gamma^s_{rj} - \Gamma^r_{ij}\frac{\partial(\ln\sqrt{|g|})}{\partial x^r}$$

$$= \frac{\partial^2(\ln\sqrt{|g|})}{\partial x^i \partial x^j} - \left(\frac{1}{\sqrt{|g|}}\sqrt{|g|}\,\frac{\partial \Gamma^s_{ij}}{\partial x^s} + \frac{1}{\sqrt{|g|}}\frac{\partial(\sqrt{|g|})}{\partial x^s}\Gamma^s_{ij}\right) + \Gamma^r_{is}\Gamma^s_{rj}$$

$$= \frac{\partial^2(\ln\sqrt{|g|})}{\partial x^i \partial x^j} - \frac{1}{\sqrt{|g|}}\frac{\partial}{\partial x^s}(\sqrt{|g|}\,\Gamma^s_{ij}) + \Gamma^r_{is}\Gamma^s_{rj}$$

Supplementary Problems

8.15 The *absolute partial derivatives* of a tensor $\mathbf{T} = (T^{i\,\cdots}_{j\,\cdots})$ defined on a 2-manifold $\mathcal{M}$: $x^i = x^i(u, v)$ are defined as

$$\frac{\delta \mathbf{T}}{\delta u} \equiv \left(T^{i\,\cdots}_{j\,\cdots,k}\,\frac{\partial x^k}{\partial u}\right) \qquad \text{and} \qquad \frac{\delta \mathbf{T}}{\delta v} \equiv \left(T^{i\,\cdots}_{j\,\cdots,k}\,\frac{\partial x^k}{\partial v}\right)$$

Since $(\partial x^i/\partial u)$ and $(\partial x^i/\partial v)$ are vectors, the inner products produce a pair of tensors of the same type and order as **T**; thus the operation of absolute partial differentiation may be repeated indefinitely. Prove that if (V^i) is any contravariant vector defined on $\mathcal{M}$,

$$\frac{\delta}{\delta u}\left(\frac{\delta V^i}{\delta v}\right) - \frac{\delta}{\delta v}\left(\frac{\delta V^i}{\delta u}\right) = R^i_{skl}V^s\,\frac{\partial x^k}{\partial u}\,\frac{\partial x^l}{\partial v}$$

[*Hint*: Expand the left side and use Problem 8.16.]

8.16 Prove that for any vector (V^i), $V^i_{,kl} - V^i_{,lk} = -R^i_{skl}V^s$.

8.17 For an arbitrary second-order contravariant tensor (T^{ij}), show that

$$T^{ij}_{,kl} - T^{ij}_{,lk} = -R^i_{skl}T^{sj} - R^j_{skl}T^{is}$$

[*Hint*: Lower superscripts and use Problem 8.3.]

8.18 For an arbitrary mixed tensor (T^i_j) show that

$$T^i_{j,kl} - T^i_{j,lk} = -R^i_{skl}T^s_j + R^s_{jkl}T^i_s$$

8.19 Verify the symmetry properties (*8.6*) for the G_{ijkl} [see (*8.7*)] and for the W_{ijkl} [see (*8.9*)].

8.20 Derive (*8.5*) from (*8.4*). [*Hint*: It is helpful to adopt the notation g_{ijkl} for $\partial^2 g_{ij}/\partial x^k \partial x^l$.]

8.21 List the independent (nonzero) components of R_{ijkl} when $n = 4$ and verify Theorem 8.2 for this case.

8.22 Calculate the Riemannian curvature K for the metric $\varepsilon\, ds^2 = (dx^1)^2 - 2x^1(dx^2)^2$.

8.23 Confirm that $\mathrm{K}=0$ for the Euclidean metric of polar coordinates,

$$ds^2 = (dx^1)^2 + (x^1\, dx^2)^2$$

(*a*) by a calculation; (*b*) by noting that K is an invariant.

8.24 Rework Example 8.4 for the pairs (*a*) $\mathbf{U}_{(1)} = (1,0,1)$, $\mathbf{V}_{(1)} = (1,1,1)$ and (*b*) $\mathbf{U}_{(2)} = (0,1,0)$, $\mathbf{V}_{(2)} = (2,1,2)$. (*c*) Explain why the answers should be the same for (*a*) and (*b*).

8.25 Let the surface of the 3-sphere of radius a be metrized by setting $x^1 = a$ in spherical coordinates and then allowing x^1, x^2 to replace x^2, x^3, respectively:

$$ds^2 = a^2(dx^1)^2 + (a \sin x^1)^2(dx^2)^2$$

Determine K for this non-Euclidean $\mathbf{R}^2$.

8.26 If the metric for Riemannian $\mathbf{R}^3$ is given by

$$g_{11} = f(x^2) \qquad g_{22} = g(x^2) \qquad g_{33} = h(x^2)$$

and $g_{ij} = 0$ for $i \neq j$, write explicit formulas for (*a*) $\mathrm{K}(x^2; \mathbf{U}, \mathbf{V})$, (*b*) R.

8.27 Specialize the results of Problem 8.26 to the case $f(x^2) \equiv g(x^2) \equiv h(x^2)$.

8.28 Find the isotropic points for the Riemannian metric

$$ds^2 = (\ln x^2)(dx^1)^2 + (\ln x^2)(dx^2)^2 + (\ln x^2)(dx^3)^2 \qquad (x^2 > 1)$$

and find the curvature K at those points. [*Hint*: Use Problem 8.27.]

8.29 Show that $\mathbf{R}^3$ under the metric

$$g_{11} = e^{x^2} \qquad g_{22} = 1 \qquad g_{33} = e^{x^2} \qquad g_{ij} = 0 \quad (i \neq j)$$

has constant Riemannian curvature with all points isotropic, and find that curvature.

8.30 Show that in a Riemannian 2-space [for which (*8.11*) holds]: (*a*) $R_{ij} = -g_{ij}\mathrm{K}$, (*b*) $R^i_j = -\delta^i_j\mathrm{K}$, and (*c*) $R = -2\mathrm{K}$.

8.31 Calculate the Ricci tensor R_{ij} for Problem 8.13 using (*8.16*), and compare your answers with those obtained earlier.

8.32 Use Problem 8.30 to calculate the Ricci tensors of both kinds and the curvature invariant for the spherical metric of Problem 8.25.

8.33 Calculate the Ricci tensors of both kinds and the curvature invariant for the (hyperbolic) metric of Problem 8.12. [*Hint*: Problem 8.27 can be used to good advantage here.]

8.34 Prove that for any tensor (T^{ij}), symmetric or not, $T^{ij}_{,ij} = T^{ij}_{,ji}$. [*Hint*: Use Problem 8.17 and the symmetry of the Ricci tensor.]

8.35 Is identical vanishing equivalent for the Riemannian curvature and the Ricci curvature invariant? Can you find an example where one is zero everywhere but not the other?

8.36 Is constancy in space equivalent for the two curvatures K and R?

Chapter 9

Spaces of Constant Curvature; Normal Coordinates

9.1 ZERO CURVATURE AND THE EUCLIDEAN METRIC

A fundamental question has run unanswered through preceding chapters: How can one tell whether a given metrization of $\mathbf{R}^n$ is Euclidean or not? To be sure that the meaning of "Euclidean" is clear, let us make the formal

Definition 1: A Riemannian metric $\mathbf{g} = (g_{ij})$, specified in a coordinate system (x^i), is the *Euclidean metric* if, under some permissible coordinate transformation (*3.1*), $\bar{\mathbf{g}} = (\delta_{ij})$.

Now, a coordinate system $(\bar{x}^i)$ in which $\bar{g}_{ij} = \delta_{ij}$ is (by Definition 1 of Chapter 3) a rectangular system. Hence our question amounts to: Does a given Riemannian space admit rectangular coordinates or does it not?

Suppose that a rectangular system $(\bar{x}^i)$ does exist. Then $\bar{\mathrm{K}} = 0$, since all Christoffel symbols vanish in $(\bar{x}^i)$. But Riemannian curvature is an invariant, so that $\mathrm{K} = 0$ in the original coordinates (x^i) as well. Moreover, by invariance,

$$g_{ij}U^iU^j = \bar{U}^i\bar{U}^i \geqq 0$$

Thus, the necessity part of the following theorem is immediate.

Theorem 9.1: A Riemannian metric (g_{ij}) is the Euclidean metric if and only if the Riemannian curvature K is zero at all points and the metric is positive definite.

To prove the sufficiency portion, we set up a system of first-order partial differential equations for n rectangular coordinates $\bar{x}^i$ as functions of the given coordinates x^j $(j = 1, 2, \ldots, n)$. The system that immediately comes to mind (Theorem 5.2) is $G = J^TJ$, or

$$\frac{\partial \bar{x}^k}{\partial x^i}\frac{\partial \bar{x}^k}{\partial x^j} = g_{ij}(x^1, x^2, \ldots, x^n) \tag{9.1}$$

But (*9.1*) is generally intractable because of its nonlinearity. Instead, we select the *linear* system that results when barred and unbarred coordinates are interchanged in (*6.6*) and then the $\bar{\Gamma}^i_{jk}$ are equated to zero:

$$\frac{\partial^2 \bar{x}^k}{\partial x^i \partial x^j} = \Gamma^r_{ij}(\mathbf{x})\frac{\partial \bar{x}^k}{\partial x^r} \tag{9.2}$$

Setting $w \equiv \bar{x}^k$ and $u_i \equiv \partial \bar{x}^k/\partial x^i$ yields the desired first-order system

$$\begin{aligned} \frac{\partial w}{\partial x^i} &= u_i \\ \frac{\partial u_i}{\partial x^j} &= \Gamma^r_{ij}u_r \end{aligned} \tag{9.3}$$

EXAMPLE 9.1 It is proved in Problems 9.7 and 9.8 that when $\mathrm{K} \equiv 0$, (*9.3*) is solvable for a coordinate system $(\bar{x}^k)$ for which all $\bar{g}_{ij}$ are constants (i.e., all $\bar{\Gamma}^i_{jk} = 0$); from these coordinates, rectangular coordinates can be reached, provided (g_{ij}) is positive definite. To make these results plausible, consider the two-dimensional metric

$$g_{11} = 1 \qquad g_{12} = g_{21} = 0 \qquad g_{22} = (x^2)^2$$

This metric is obviously positive definite and, because the only nonvanishing Christoffel symbol is $\Gamma^2_{22} = 1/x^2$, it has $R_{1212} = 0 = \mathrm{K}$. It is possible to solve (*9.1*) directly for the corresponding cartesian coordinates, and then to verify that that solution is contained in the general solution to (*9.3*).

Introduce the notation

$$f_1 \equiv \frac{\partial \bar{x}^1}{\partial x^1} \qquad f_2 \equiv \frac{\partial \bar{x}^1}{\partial x^2} \qquad f_3 \equiv \frac{\partial \bar{x}^2}{\partial x^1} \qquad f_4 \equiv \frac{\partial \bar{x}^2}{\partial x^2} \tag{1}$$

whereby (*9.1*) becomes the algebraic system

$$\begin{aligned} f_1^2 + f_3^2 &= 1 \\ f_1 f_2 + f_3 f_4 &= 0 \\ f_2^2 + f_4^2 &= (x^2)^2 \end{aligned} \tag{2}$$

System (*2*) can be solved for three of the f_i in terms of the fourth—say, f_1:

$$f_1 = f_1 \qquad f_2 = x^2\sqrt{1-f_1^2} \qquad f_3 = -\sqrt{1-f_1^2} \qquad f_4 = x^2 f_1 \tag{3}$$

Now (*1*) becomes two simple first-order systems in $\bar{x}^1$ alone and $\bar{x}^2$ alone:

$$\text{I:} \begin{cases} \dfrac{\partial \bar{x}^1}{\partial x^1} = f_1 \\ \dfrac{\partial \bar{x}^1}{\partial x^2} = x^2\sqrt{1-f_1^2} \end{cases} \qquad \text{and} \qquad \text{II:} \begin{cases} \dfrac{\partial \bar{x}^2}{\partial x^1} = -\sqrt{1-f_1^2} \\ \dfrac{\partial \bar{x}^2}{\partial x^2} = x^2 f_1 \end{cases}$$

The unknown function f_1 is determined by the requirements that the two equations I and the two equations II both be *compatible*:

$$\frac{\partial f_1}{\partial x^2} = \frac{\partial}{\partial x^1}\left(x^2\sqrt{1-f_1^2}\right) \qquad \text{and} \qquad \frac{\partial}{\partial x^2}\left(-\sqrt{1-f_1^2}\right) = \frac{\partial}{\partial x^1}\left(x^2 f_1\right)$$

The only function satisfying these two compatibility conditions is

$$f_1 = \text{const.} = \cos\phi$$

and I and II immediately integrate to give

$$\begin{aligned} \bar{x}^1 &= x^1\cos\phi + \frac{1}{2}(x^2)^2\sin\phi + c \\ \bar{x}^2 &= -x^1\sin\phi + \frac{1}{2}(x^2)^2\cos\phi + d \end{aligned} \tag{4}$$

We are, of course, free to set $\phi = c = d = 0$ in (*4*).

Turning to (*9.3*), we have to solve

$$(1) \quad \frac{\partial w}{\partial x^1} = u_1, \quad \frac{\partial w}{\partial x^2} = u_2$$

$$(2) \quad \frac{\partial u_1}{\partial x^1} = 0, \quad \frac{\partial u_1}{\partial x^2} = 0 \qquad (3) \quad \frac{\partial u_2}{\partial x^1} = 0, \quad \frac{\partial u_2}{\partial x^2} = u_2\Gamma^2_{22} = \frac{u_2}{x^2}$$

Note that these equations include their own compatibility conditions! For instance, the second equation (2) and the first equation (3) ensure the compatibility of the two equations (1). The fact that system (*9.3*) is automatically compatible whenever K = 0 is crucial to the proof of Theorem 9.1. Integrating the above equations in the order (3)–(2)–(1), we get:

$$w = a_1 x^1 + a_2 (x^2)^2 + a_3 \qquad (a_1, a_2, a_3 = \text{const.})$$

or, replacing the index k,

$$\bar{x}^k = a_1^k x^1 + a_2^k (x^2)^2 + a_3^k \qquad (a_i^k = \text{const.}) \tag{5}$$

As announced, (*5*) includes (*4*).

For subsequent use, the following compatibility theorem for quasilinear systems [which include linear systems such as (*9.3*)] is stated here, without proof:

Theorem 9.2: The quasilinear first-order system

$$\frac{\partial u_\lambda}{\partial x^j} = F_{\lambda j}(u_0, u_1, \ldots, u_m, x^1, x^2, \ldots, x^n) \qquad (\lambda = 0, 1, \ldots, m;\ j = 1, 2, \ldots, n)$$

where the functions $F_{\lambda j}$ are of differentiability class C^1, has a nontrivial solution for the u_λ, bounded over some region of $\mathbf{R}^n$, if and only if

$$\frac{\partial F_{\lambda j}}{\partial u_\nu} F_{\nu k} + \frac{\partial F_{\lambda j}}{\partial x^k} = \frac{\partial F_{\lambda k}}{\partial u_\nu} F_{\nu j} + \frac{\partial F_{\lambda k}}{\partial x^j} \qquad (\lambda = 0, 1, \ldots, m;\ 1 \leqq j < k \leqq n)$$

[The ν-summations run from 0 to m.]

9.2 FLAT RIEMANNIAN SPACES

A Riemannian space, or the determining metric, is termed *flat* if there is a transformation of coordinates $\bar{x}^i = \bar{x}^i(\mathbf{x})$ that puts the metric into the *standard form*

$$\varepsilon\, ds^2 = \varepsilon_1(d\bar{x}^1)^2 + \varepsilon_2(d\bar{x}^2)^2 + \cdots + \varepsilon_n(d\bar{x}^n)^2 \tag{9.4}$$

where $\varepsilon_i = \pm 1$ for each i. This condition generalizes the concept of the Euclidean metric. The essential distinction between the two concepts revolves about positive-definiteness; the analogue to Theorem 9.1 with positive-definiteness removed is:

Theorem 9.3: A Riemannian space is flat if and only if $\mathrm{K} = 0$ at all points.

Corollary 9.4: If $\mathrm{K} = 0$, then $R = 0$.

Proof: If $\mathrm{K} = 0$, then by Theorem 9.3 the space is flat, and hence the $\bar{g}_{ij}$ are constant for some coordinate system $(\bar{x}^i)$. It follows that all $\bar{\Gamma}_{ijk}$, $\bar{\Gamma}^i_{jk}$, $\bar{R}^i_{jkl}$, $\bar{R}_{ij}$, and $\bar{R}^i_j$ vanish. Therefore, $\bar{R} = \bar{R}^i_i = 0$, and since Ricci curvature is invariant, $R = 0$.

Remark 1: Problem 8.35 shows that the converse of Corollary 9.4 does not hold.

EXAMPLE 9.2 Consider the Riemannian metric

$$\varepsilon\, ds^2 = (dx^1)^2 + 4(x^2)^2(dx^2)^2 + 4(x^3)^2(dx^3)^2 - 4(x^4)^2(dx^4)^2$$

(*a*) Calculate the Riemannian curvature. (*b*) Find a solution of system (*9.3*) from which it may be inferred that the space is flat.

(*a*) Using Problem 6.4, we find as the nonvanishing Christoffel symbols

$$\Gamma^2_{22} = \frac{1}{x^2} \qquad \Gamma^3_{33} = \frac{1}{x^3} \qquad \Gamma^4_{44} = \frac{1}{x^4}$$

Because $\Gamma^i_{jk} = 0$ unless $i = j = k$, the partial-derivative terms drop out of (*8.2*), leaving

$$R^i_{jkl} = \Gamma^r_{jl}\Gamma^i_{rk} - \Gamma^r_{jk}\Gamma^i_{rl} = \Gamma^i_{ii}\Gamma^i_{ii} - \Gamma^i_{ii}\Gamma^i_{ii} = 0 \quad \text{(not summed)}$$

which in turn implies that $R_{ijkl} = 0$ and $\mathrm{K} = 0$.

(*b*) For the above-calculated Christoffel symbols

$$\frac{\partial u_1}{\partial x^1} = 0 \qquad \frac{\partial u_2}{\partial x^2} = \frac{u_2}{x^2} \qquad \frac{\partial u_3}{\partial x^3} = \frac{u_3}{x^3} \qquad \frac{\partial u_4}{\partial x^4} = \frac{u_4}{x^4}$$

with $\partial u_i/\partial x_j = 0$ for $i \neq j$. Integrating,

$$u_1 = f_1(x^2, x^3, x^4) \qquad u_2 = x^2 f_2(x^1, x^3, x^4) \qquad u_3 = x^3 f_3(x^1, x^2, x^4) \qquad u_4 = x^4 f_4(x^1, x^2, x^3)$$

for arbitrary functions f_i. But the remaining equations (*9.3*), $\partial w/\partial x^i = u_i$, give rise to the compatibility relations

$$\frac{\partial u_i}{\partial x^j} = \frac{\partial u_j}{\partial x^i}$$

which are satisfied only if $f_i = c_i = \text{const}$. Therefore,

$$w = a_1 x^1 + a_2(x^2)^2 + a_3(x^3)^2 + a_4(x^4)^2 + a_5$$

and the transformation must be of the general form

$$\bar{x}^k = a_1^k x^1 + a_2^k (x^2)^2 + a_3^k (x^3)^2 + a_4^k (x^4)^2 + a_5^k \qquad (a_i^k \text{ constants})$$

We wish to specialize the constants so that the covariant law $G = J^T \bar{G} J$ will hold, with $\bar{G}$ corresponding to (*9.4*). As a preliminary guess, set

$$[a_i^k]_{45} = \begin{bmatrix} b_1 & 0 & 0 & 0 & 0 \\ 0 & b_2 & 0 & 0 & 0 \\ 0 & 0 & b_3 & 0 & 0 \\ 0 & 0 & 0 & b_4 & 0 \end{bmatrix}$$

so that the covariant law becomes

$$\begin{bmatrix} 1 & & & \\ & 4(x^2)^2 & & \\ & & 4(x^3)^2 & \\ & & & -4(x^4)^2 \end{bmatrix} = \begin{bmatrix} b_1 & & & \\ & 2b_2x^2 & & \\ & & 2b_3x^3 & \\ & & & 2b_4x^4 \end{bmatrix} \begin{bmatrix} \varepsilon_1 & & & \\ & \varepsilon_2 & & \\ & & \varepsilon_3 & \\ & & & \varepsilon_4 \end{bmatrix} \begin{bmatrix} b_1 & & & \\ & 2b_2x^2 & & \\ & & 2b_3x^3 & \\ & & & 2b_4x^4 \end{bmatrix}$$

By inspection, the choice $b_1 = b_2 = b_3 = b_4 = 1$ will render $\varepsilon_1 = \varepsilon_2 = \varepsilon_3 = -\varepsilon_4 = 1$.

In connection with (*9.4*) there is an interesting theorem (Sylvester's law of inertia). Define as the *signature* of a flat metric (g_{ij}) the ordered n-tuple

$$(\operatorname{sgn} \varepsilon_1, \operatorname{sgn} \varepsilon_2, \ldots, \operatorname{sgn} \varepsilon_n)$$

composed of the signs of the coefficients in the standard form (i.e., the signs of $\bar{g}_{11}, \ldots, \bar{g}_{nn}$).

Theorem 9.5: The signature of a flat metric is uniquely determined up to order.

9.3 NORMAL COORDINATES

It is possible to introduce local, quasirectangular coordinates in Riemannian space the use of which greatly simplifies the proofs of certain complicated tensor identities.

Let O denote an arbitrary point of $\mathbf{R}^n$, and $\mathbf{p} = (p^i)$ an arbitrary direction (unit vector) at O. *Assuming a positive-definite metric*, consider the differential equations for geodesics,

$$\frac{d^2x^i}{ds^2} + \Gamma^i_{jk} \frac{dx^j}{ds} \frac{dx^k}{ds} = 0 \tag{9.5}$$

[cf. (*7.13*)], along with initial conditions

$$\left. \frac{dx^i}{ds} \right|_{s=0} = p^i \tag{9.6}$$

Here the arc-length parameter is chosen to make $s = 0$ at O.

Remark 2: Under an *indefinite* metric, there could exist directions at O in which arc length could not be defined; see, e.g., Problem 7.22. There would then be no hope of satisfying (*9.6*) with (p^i) arbitrary.

It can be shown that for a given $\mathbf{p}$, the system (*9.5*)–(*9.6*) has a unique solution; moreover, for each point P in some neighborhood $\mathcal{N}$ of O, there is a unique choice of direction $\mathbf{p}$ at O such that the solution curve $x^i = x^i(s)$ (a geodesic) passes through P. Accordingly, for each P in $\mathcal{N}$, take as the coordinates of P

$$y^i = sp^i \tag{9.7}$$

where s is the distance along the geodesic from O to P. The numbers (y^i) are called the *normal coordinates* (or *geodesic* or *Riemannian coordinates*) of P.

EXAMPLE 9.3 Show that if the Riemannian metric $ds^2 = g_{ij}\,dx^i\,dx^j$ for $\mathbf{R}^2$ is Euclidean and there is a point O

at which $g_{12} = 0$, then normal coordinates (y^i) with origin O are constant multiples of (z^i), for some rectangular coordinate system (z^i).

Because $g_{12} = 0$ at point O, the vectors $\mathbf{T} = (1/\sqrt{g_{11}}, 0)$ and $\mathbf{S} = (0, 1/\sqrt{g_{22}})$ are, at O, an orthonormal pair. The space, being Euclidean, admits a rectangular coordinate system; in particular, a system (z^i) with origin O and unit vectors **T** and **S** (Fig. 9-1). Again because the space is Euclidean, the straight line segment OP is the unique geodesic connecting O with the arbitrary point P. With $s = \overline{OP}$ and **p** the direction vector of OP, we have the vector equation

$$z^1\mathbf{T} + z^2\mathbf{S} = s\mathbf{p}$$

or componentwise,

$$z^1\left(\frac{1}{\sqrt{g_{11}}}\right) = sp^1 \equiv y^1 \qquad \text{and} \qquad z^2\left(\frac{1}{\sqrt{g_{22}}}\right) = sp^2 \equiv y^2$$

QED.

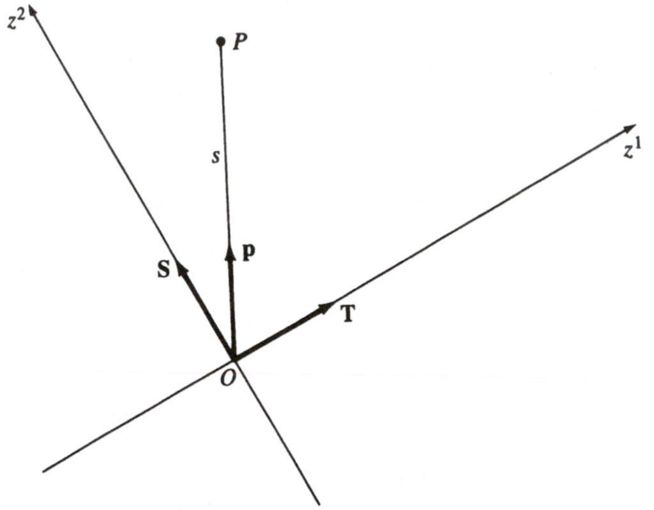

Fig. 9-1

The chief value of Riemannian coordinates resides in the following theorem (Problem 9.10).

Theorem 9.6: If the metric tensor (g_{ij}) is positive definite, then, at the origin of a Riemannian coordinate system (y^i), all $\partial g_{ij}/\partial y^k$, $\partial g^{ij}/\partial y^k$, Γ_{ijk}, and Γ^i_{jk} are zero.

Remark 3: Recall that neither the partial derivatives of the metric tensor nor the Christoffel symbols are tensorial. Thus, their (y^i)-representations can vanish at O without their (x^i)-representations doing so. For instance, because the transformation between (x^i) and (y^i) has $J = I$ at O, *(6.5)* gives:

$$\Gamma^i_{jk}(\mathbf{x})\big|_O = \frac{\partial^2 y^i}{\partial x^j \partial x^k}\bigg|_O$$

The right side is generally nonzero, unless the coordinate transformation happens to be linear.

EXAMPLE 9.4 Prove *Bianchi's first identity*, $R_{ijkl} + R_{iklj} + R_{iljk} = 0$.

Theorem 9.6 implies that at O, the origin of normal coordinates,

$$R_{ijkl} = \frac{\partial \Gamma_{jli}}{\partial y^k} - \frac{\partial \Gamma_{jki}}{\partial y^l}$$

If we use the notation Γ_{ijkl} for $\partial \Gamma_{ijk}/\partial y^l$, for arbitrary i, j, k, l, then

$$R_{ijkl} = \Gamma_{jlik} - \Gamma_{jkil}$$
$$R_{iklj} = \Gamma_{kjil} - \Gamma_{klij}$$
$$R_{iljk} = \Gamma_{lkij} - \Gamma_{ljik}$$

On summing these three relations and observing the cancellations which take place, we see that the desired identity holds at O in the coordinates (y^i). This *tensor* identity must therefore remain valid at O in the alias coordinates (x^i). But O is any point of $\mathbf{R}^n$, and the proof is complete.

EXAMPLE 9.5 Prove *Bianchi's second identity*,

$$R_{ijkl,u} + R_{ijlu,k} + R_{ijuk,l} = 0 \tag{9.8}$$

Working with the Riemann tensor of the second kind, we have, at the origin O of normal coordinates,

$$R^i_{jkl,u} = \frac{\partial R^i_{jkl}}{\partial y^u} = \frac{\partial}{\partial y^u}\left(\frac{\partial \Gamma^i_{jl}}{\partial y^k} - \frac{\partial \Gamma^i_{jk}}{\partial y^l} + \Gamma^r_{jl}\Gamma^i_{rk} - \Gamma^r_{jk}\Gamma^i_{rl}\right)$$
$$= \Gamma^i_{jlku} - \Gamma^i_{jklu}$$

since terms like $(\partial \Gamma^r_{jl}/\partial y^u)\Gamma^i_{rk}$ vanish along with the Γ^i_{rk} at O. From this, permutation of subscripts yields

$$R^i_{jkl,u} + R^i_{jlu,k} + R^i_{juk,l} = 0$$

at O, and the validity of (*9.8*) at O follows from the fact that covariant differentiation commutes with the lowering of a superscript (Problem 6.11). We conclude, as in Example 9.4, that (*9.8*) holds generally in (x^i).

A positive definite metric has been tacitly assumed, both here and in Example 9.4. The assumption can be dropped; see Problem 9.13.

9.4 SCHUR'S THEOREM

From Chapter 8 it is known that although every point of Riemannian two-space is isotropic, the curvature ($= R_{1212}/g$) can still vary from one isotropic point to the next. However, Problems 8.11, 8.12, 8.28, and 8.29 suggest that a different situation prevails in $\mathbf{R}^3$. To prove the general theorem, known as *Schur's theorem*, it is necessary to establish a preliminary result, a generalization of (*8.11*).

Lemma 9.7: At an isotropic point of $\mathbf{R}^n$ the Riemannian curvature is given by

$$\mathrm{K} = \frac{R_{abcd}}{g_{ac}g_{bd} - g_{ad}g_{bc}} \equiv \frac{R_{abcd}}{G_{abcd}} \tag{9.9}$$

for any specific subscript string such that $G_{abcd} \neq 0$. [If $G_{abcd} = 0$, then $R_{abcd} = 0$ also.]

For a proof, see Problem 9.8.

Theorem 9.8 (*Schur's Theorem*): If all points in some neighborhood $\mathcal{N}$ in a Riemannian $\mathbf{R}^n$ are isotropic and $n \geqq 3$, then K is constant throughout that neighborhood.

For a proof, see Problem 9.14.

9.5 THE EINSTEIN TENSOR

The *Einstein tensor* is defined in terms of the Ricci tensor R_{ij} and the curvature invariant R (Section 8.4):

$$G^i_j \equiv R^i_j - \frac{1}{2}\,\delta^i_j R \tag{9.10}$$

It is clear that (G^i_j) is in fact a mixed tensor of order two.

As a direct generalization of the notion of the divergence of a vector field $\mathbf{V} = (V^i)$ relative to rectangular coordinates (x^i),

$$\operatorname{div}\mathbf{V} = \frac{\partial V^1}{\partial x^1} + \frac{\partial V^2}{\partial x^2} + \cdots + \frac{\partial V^n}{\partial x^n} \equiv \frac{\partial V^r}{\partial x^r}$$

we define the *divergence* of the general tensor $\mathbf{T} = (T^{i_1 i_2 \ldots i_k \ldots i_p}_{j_1 j_2 \ldots j_q})$ with respect to its kth contravariant index to be the tensor

$$\operatorname{div}\mathbf{T} \equiv (T^{i_1 i_2 \ldots r \ldots i_p}_{j_1 j_2 \ldots j_q, r}) \tag{9.11}$$

In Problem 9.15 is proved

Theorem 9.9: For any Riemannian metric, the divergence of the Einstein tensor is zero at all points.

Solved Problems

ZERO CURVATURE AND THE EUCLIDEAN METRIC

9.1 Test the compatibility conditions (Theorem 9.2) for the system

$$\frac{\partial u_0}{\partial x^1} = \frac{u_0}{x^1} \qquad \frac{\partial u_0}{\partial x^2} = 2x^2 u_0 \tag{1}$$

If it is compatible, solve the system.

In the notation of Theorem 9.2, there is only the condition corresponding to $\lambda = 0, j = 1, k = 2$ to be satisfied.

$$\frac{\partial F_{01}}{\partial u_0} F_{02} + \frac{\partial F_{01}}{\partial x^2} \stackrel{?}{=} \frac{\partial F_{02}}{\partial u_0} F_{01} + \frac{\partial F_{02}}{\partial x^1}$$

$$\frac{\partial}{\partial u_0}\left(\frac{u_0}{x^1}\right) \cdot 2x^2 u_0 + \frac{\partial}{\partial x^2}\left(\frac{u_0}{x^1}\right) \stackrel{?}{=} \frac{\partial}{\partial u_0}(2x^2 u_0) \cdot \frac{u_0}{x^1} + \frac{\partial}{\partial x^1}(2x^2 u_0)$$

$$\frac{2x^2 u_0}{x^1} = \frac{2x^2 u_0}{x^1}$$

Therefore, the system is compatible. The first equation (*1*) integrates to $u_0 = x^1\,\phi(x^2)$; the second equation then gives

$$x^1\phi' = 2x^2 x^1 \phi \qquad \text{whence} \qquad \phi = c\exp(x^2)^2$$

Hence the solution of (*1*) is $u_0 = cx^1 \exp(x^2)^2$.

9.2 Show that $\mathbf{R}^3$ under the metric $ds^2 = [(x^1)^2 + (x^2)^2](dx^1)^2 + [(x^1)^2 + (x^2)^2](dx^2)^2 + (dx^3)^2$ is Euclidean.

This metric has $g_{33} = \text{const.}$, and g_{11} and g_{22} independent of x^3. Problem 6.4 then shows that $\Gamma^i_{jk} = 0$ whenever i, j, or k equals 3; consequently, of the six independent components of the Riemann tensor, only R_{1212} is possibly nonzero. But (from Problem 6.4), with $z \equiv (x^1)^2 + (x^2)^2$,

$$\Gamma^1_{11} = \frac{x^1}{z} \qquad \Gamma^1_{12} = \Gamma^1_{21} = \frac{x^2}{z} \qquad \Gamma^1_{22} = -\frac{x^1}{z}$$

$$\Gamma^2_{11} = -\frac{x^2}{z} \qquad \Gamma^2_{12} = \Gamma^2_{21} = \frac{x^1}{z} \qquad \Gamma^2_{22} = \frac{x^2}{z}$$

so that

$$R^1_{212} = \frac{\partial \Gamma^1_{22}}{\partial x^1} - \frac{\partial \Gamma^1_{21}}{\partial x^2} + \Gamma^1_{22}\Gamma^1_{11} + \Gamma^2_{22}\Gamma^1_{21} - \Gamma^1_{21}\Gamma^1_{12} - \Gamma^2_{21}\Gamma^1_{22}$$

$$= \frac{-z + x^1(2x^1)}{z^2} - \frac{z - x^2(2x^2)}{z^2} + \left(-\frac{x^1}{z}\right)\left(\frac{x^1}{z}\right) + \frac{x^2}{z}\left(\frac{x^2}{z}\right) - \frac{x^2}{z}\left(\frac{x^2}{z}\right) - \frac{x^1}{z}\left(-\frac{x^1}{z}\right) = 0$$

Consequently, $R_{1212} = 0 = K$. As the metric is clearly positive definite, Theorem 9.1 implies that the space is Euclidean.

9.3 For the Euclidean space of Problem 9.2, exhibit a transformation from the given coordinate system (x^i) to a rectangular system $(\bar{x}^i)$.

Using the Christoffel symbols as calculated in Problem 9.2, we obtain from (*9.3*) the following system for the u_i:

$$\frac{\partial u_1}{\partial x^1} = \frac{x^1 u_1 - x^2 u_2}{z} \qquad \frac{\partial u_1}{\partial x^2} = \frac{x^2 u_1 + x^1 u_2}{z} \qquad \frac{\partial u_1}{\partial x^3} = 0 \tag{1}$$

$$\frac{\partial u_2}{\partial x^1} = \frac{x^2 u_1 + x^1 u_2}{z} \qquad \frac{\partial u_2}{\partial x^2} = \frac{-x^1 u_1 + x^2 u_2}{z} \qquad \frac{\partial u_2}{\partial x^3} = 0 \tag{2}$$

$$\frac{\partial u_3}{\partial x^1} = 0 \qquad \frac{\partial u_3}{\partial x^2} = 0 \qquad \frac{\partial u_3}{\partial x^3} = 0 \tag{3}$$

Thus u_1 and u_2 are functions of x^1, x^2 alone, and $u_3 = \text{const}$. Since the g_{ij} are all polynomials of degree 2 in x^1, x^2, use the method of undetermined coefficients, assuming polynomial forms

$$u_i = a_i(x^1)^2 + b_i x^1 x^2 + c_i(x^2)^2 + d_i x^1 + e_i x^2 + f_i \qquad (i = 1, 2)$$

The (compatibility) relation $\partial u_1/\partial x^2 = \partial u_2/\partial x^1$ implied by the second equation (*1*) and the first equation (*2*) requires

$$b_1 = 2a_2 \qquad 2c_1 = b_2 \qquad e_1 = d_2$$

Similarly, $\partial u_1/\partial x^1 = -\partial u_2/\partial x^2$ implies

$$2a_1 = -b_2 \qquad b_1 = -2c_2 \qquad d_1 = -e_2$$

Using the first equation (*1*), or $z(\partial u_1/\partial x^1) = x^1 u_1 - x^2 u_2$, we get:

$$a_1 = 0 \qquad a_2 = 0 \qquad c_1 = b_2 \qquad b_1 = -c_2 \qquad d_1 = -e_2 \qquad f_1 = 0 = -f_2$$

It follows that $b_1 = b_2 = c_1 = c_2 = 0$, and therefore (renotating d_1 and e_1)

$$u_1 = ax^1 + bx^2 \qquad u_2 = bx^1 - ax^2 \qquad u_3 = c$$

[*Note*: This solution of (*1*)–(*2*)–(*3*) may be obtained by the *method of characteristics*, without any prior assumptions.]

The first equations (*9.3*),

$$\frac{\partial w}{\partial x^1} = ax^1 + bx^2 \qquad \frac{\partial w}{\partial x^2} = bx^1 - ax^2 \qquad \frac{\partial w}{\partial x^3} = c$$

may now be integrated to give

$$w = \frac{a}{2}(x^1)^2 + bx^1x^2 - \frac{a}{2}(x^2)^2 + cx^3 + d$$

or, replacing $\bar{x}^k$ and corresponding superscripts, and with $d = 0$,

$$\bar{x}^k = \frac{a^k}{2}(x^1)^2 + b^k x^1 x^2 - \frac{a^k}{2}(x^2)^2 + c^k(x^3)^2$$

It is clear that we may take $c^1 = c^2 = 0 = a^3 = b^3$ and $c^3 = 1$:

$$\bar{x}^1 = \frac{1}{2} a^1(x^1)^2 + b^1x^1x^2 - \frac{1}{2} a^1(x^2)^2$$
$$\bar{x}^2 = \frac{1}{2} a^2(x^1)^2 + b^2x^1x^2 - \frac{1}{2} a^2(x^2)^2$$
$$\bar{x}^3 = x^3$$

The Jacobian matrix is

$$J = \begin{bmatrix} a^1x^1 + b^1x^2 & b^1x^1 - a^1x^2 & 0 \\ a^2x^1 + b^2x^2 & b^2x^1 - a^2x^2 & 0 \\ 0 & 0 & 1 \end{bmatrix}$$

Since $J^TJ = G$, we must have

$$(a^1)^2 + (a^2)^2 = 1 \qquad a^1b^1 + a^2b^2 = 0 \qquad (b^1)^2 + (b^2)^2 = 1$$

so take $a^1 = 0$, $a^2 = 1$, $b^2 = 0$, $b^1 = 1$. The transformation is, finally,

$$\bar{x}^1 = x^1x^2 \qquad \bar{x}^2 = \frac{1}{2} [(x^1)^2 - (x^2)^2] \qquad \bar{x}^3 = x^3$$

FLAT RIEMANNIAN SPACES

9.4 Determine whether the following metric is flat and/or Euclidean:

$$\varepsilon\, ds^2 = (dx^1)^2 - (x^2)^2(dx^2)^2 \qquad (n = 2)$$

Since the metric is not positive definite, it cannot be Euclidean. To determine flatness, it suffices to examine $R_{1212} = g_{11}R^1_{212}$. But Problem 6.4 shows that $R^1_{212} = 0$; hence the space is flat.

9.5 Show that if the metric tensor is constant, the space is flat and the coordinate transformation $\bar{x} = Ax$, where A is a rank-n matrix of eigenvectors of $G = (g_{ij})$, diagonalizes the metric (i.e., $\bar{g}_{ij} = 0$ if $i \neq j$).

Since all partial derivatives of g_{ij} are zero, all Christoffel symbols will vanish and all $R_{ijkl} = 0$, making $K = 0$. Thus, by Theorem 9.3, the space is flat. By Chapters 2 and 3, if $\bar{x} = Ax$, then $J = A$ and

$$G = J^T\bar{G}J = A^T\bar{G}A$$

However, since G is real and symmetric, its eigenvectors form an orthogonal matrix which we now choose as A, with

$$AGA^{-1} = AGA^T = D \quad \text{(diagonal matrix of eigenvalues of } G\text{)}$$

Hence, $\bar{G} = AGA^T = D$ QED.

9.6 Find the signature of the flat metric

$$\varepsilon\, ds^2 = 4(dx^1)^2 + 5(dx^2)^2 - 2(dx^3)^2 + 2(dx^4)^2 - 4\, dx^2\, dx^3 - 4\, dx^2\, dx^4 - 10\, dx^3\, dx^4$$

In view of Problem 9.5, it suffices to find the eigenvalues λ of $G = (g_{ij})$. The characteristic equation is

$$|G - \lambda I| = \begin{vmatrix} 4-\lambda & 0 & 0 & 0 \\ 0 & 5-\lambda & -2 & -2 \\ 0 & -2 & -2-\lambda & -5 \\ 0 & -2 & -5 & 2-\lambda \end{vmatrix}$$

$$= (4-\lambda)\begin{vmatrix} 5-\lambda & -2 & -2 \\ -2 & -2-\lambda & -5 \\ -2 & -5 & 2-\lambda \end{vmatrix} = -(4-\lambda)\begin{vmatrix} 5-\lambda & 2 & 0 \\ -2 & 2+\lambda & -3+\lambda \\ -2 & 5 & 7-\lambda \end{vmatrix}$$

$$= -(4-\lambda)[(5-\lambda)(29-\lambda^2) + 8(5-\lambda)] = -(4-\lambda)(5-\lambda)(37-\lambda^2) = 0$$

from which the eigenvalues are $\lambda = +4, +5, +\sqrt{37}, -\sqrt{37}$. This means that there is a transformation which changes the metric into the form

$$\varepsilon\, ds^2 = 4(dx^1)^2 + 5(dx^2)^2 + \sqrt{37}(dx^3)^2 - \sqrt{37}(dx^4)^2 = (d\bar{x}^1)^2 + (d\bar{x}^2)^2 + (d\bar{x}^3)^2 - (d\bar{x}^4)^2$$

with the obvious change of coordinates. Hence, the signature is $(+\,+\,+\,-)$, or some permutation thereof (Theorem 9.5).

9.7 Show that the conditions $R_{ijkl} = 0$ are sufficient for the compatibility of (*9.3*).

In the notation of Theorem 9.2, (*9.3*) takes the form (with $m = n$)

$$\lambda = 0 \qquad \frac{\partial u_0}{\partial x^j} = F_{0j} \equiv u_j$$

$$\lambda > 0 \qquad \frac{\partial u_\lambda}{\partial x^j} = F_{\lambda j} \equiv u_r \Gamma^r_{\lambda j}(\mathbf{x})$$

The corresponding compatibility conditions are

$$\lambda = 0 \qquad \delta^\nu_j u_r \Gamma^r_{\nu k} = \delta^\nu_k u_r \Gamma^r_{\nu j}$$

or $u_r \Gamma^r_{jk} = u_r \Gamma^r_{kj}$, which holds trivially, and

$$\lambda > 0 \qquad \delta^\nu_r \Gamma^r_{\lambda j} u_s \Gamma^s_{\nu k} + u_r \frac{\partial \Gamma^r_{\lambda j}}{\partial x^k} = \delta^\nu_r \Gamma^r_{\lambda k} u_s \Gamma^s_{\nu j} + u_r \frac{\partial \Gamma^r_{\lambda k}}{\partial x^j}$$

which rearranges to

$$\underbrace{\left(\frac{\partial \Gamma^r_{\lambda j}}{\partial x^k} - \frac{\partial \Gamma^r_{\lambda k}}{\partial x^j} + \Gamma^s_{\lambda j}\Gamma^r_{sk} - \Gamma^s_{\lambda k}\Gamma^r_{sj} \right)}_{R^r_{\lambda kj}} u_r = 0$$

Thus, $R_{r\lambda kj} = 0$ forces $R^r_{\lambda kj} = 0$ and compatibility.

9.8 Prove Lemma 9.7.

As (R_{ijkl}) and (G_{ijkl}) are tensors [see Example 8.3] and K is an invariant,

$$(T_{ijkl}) \equiv (R_{ijkl} - \mathrm{K} G_{ijkl})$$

is a tensor of the same type and order. It must be proved that all $T_{ijkl} = 0$ at an isotropic point P. Since K is independent of direction at P, so are the T_{ijkl}; and (*8.7*) gives

$$T_{ijkl} U^i V^j U^k V^l = 0 \qquad (T_{ijkl} = T_{ijkl}(P)) \tag{1}$$

If we define the second-order tensor $(S_{ik}) \equiv (T_{ijkl} V^j V^l)$, we find that $S_{ik} = S_{ki}$, and by (*1*), $S_{ik} U^i U^k = 0$ at P for any (U^i). It follows that all $S_{ik} = 0$ at P. Now set $V^i = \delta^i_a$. Then, at P,

$$0 = S_{ik} = T_{ijkl}\delta^j_a \delta^l_a = T_{iaka}$$

for arbitrary (fixed) index a. Next set $V^i = \delta^i_a + \delta^i_b$ for arbitrary fixed indices a and b:

$$0 = T_{ijkl} V^j V^l = T_{ijkl}(\delta^j_a + \delta^j_b)(\delta^l_a + \delta^l_b) = T_{iaka} + T_{iakb} + T_{ibka} + T_{ibkb}$$

or $T_{iakb} + T_{ibka} = 0$. Therefore, since T_{ijkl} obeys the same symmetry laws as R_{ijkl} and G_{ijkl},

$$T_{ijkl} - T_{iljk} = 0 \tag{2}$$

$$T_{ijkl} + T_{iklj} + T_{iljk} = 0 \tag{3}$$

Adding (*2*) and (*3*),

$$2T_{ijkl} + T_{iklj} = 0 \tag{4}$$

But, from (*2*), $T_{iklj} = T_{ijkl}$, so that (*4*) implies $T_{ijkl} = 0$, as desired.

9.9 Prove Theorems 9.1 and 9.3.

We already know that if the space is either Euclidean or flat, $K \equiv 0$. Suppose, conversely, that $K \equiv 0$; then every point is isotropic, and Lemma 9.7 implies that all R_{ijkl} vanish. It then follows from Problem 9.7 that there exists a coordinate system $(\bar{x}^i)$ for which $\bar{\Gamma}^i_{jk} = 0$ or $\bar{g}_{ij} = \text{const.}$ By Problem 9.5, there exists another coordinate system, (y^i), in which the metric takes the form (for real constants a_i)

$$\varepsilon\, ds^2 = \varepsilon_1 a_1^2 (dy^1)^2 + \varepsilon_2 a_2^2 (dy^2)^3 + \cdots + \varepsilon_n a_n^2 (dy^n)^2$$

The transformation $\bar{y}^1 = a_1 y^1$, $\bar{y}^2 = a_2 y^2, \ldots, \bar{y}^n = a_n y^n$ now reduces the metric to

$$\varepsilon\, ds^2 = \varepsilon_1 (d\bar{y}^1)^2 + \varepsilon_2 (d\bar{y}^2)^2 + \cdots + \varepsilon_n (d\bar{y}^n)^2 \tag{1}$$

and the space is flat. This proves Theorem 9.3. If the given metric is positive definite, then in (*1*), $\varepsilon_i = 1$ for each i. In this case the metric is Euclidean, proving Theorem 9.1.

NORMAL COORDINATES

9.10 Prove Theorem 9.6.

If (y^i) are normal coordinates, then the geodesic through O and any point P in some neighborhood $\mathcal{N}$ of O has the parametric form

$$y^i = sp^i \qquad (p^i = \text{const.})$$

This geodesic thus obeys the differential equations

$$\frac{dy^i}{ds} = p^i \qquad \text{and} \qquad \frac{d^2y^i}{ds^2} = 0$$

But it must also satisfy (*9.5*), $\delta \mathbf{T}/\delta s = 0$, in the coordinates (y^i):

$$\frac{d^2y^i}{ds^2} + \Gamma^i_{jk} \frac{dy^j}{ds}\frac{dy^k}{ds} = 0$$

Thus, by substitution, $\Gamma^i_{jk} p^j p^k = 0$ for all directions (p^i) at O. But Γ^i_{jk} is symmetric for each i; hence, $\Gamma^i_{jk} = 0$ at O for all i, j, k. Also, $\Gamma_{ijk} = g_{kr}\Gamma^r_{ij} = 0$; hence, $\partial g_{ij}/\partial y^k = 0$ at O, by (*6.2*). Finally, since $g^{ij}g_{jr} = \delta^i_r$, the product rule for differentiation yields $\partial g^{ij}/\partial y^k = 0$ at O.

9.11 Prove that at the origin of a Riemannian coordinate system (y^i),

$$\frac{\partial \Gamma^i_{ji}}{\partial y^k} = \frac{\partial \Gamma^i_{ki}}{\partial y^j} \qquad \text{(all } j \text{ and } k\text{; summed on } i\text{)}$$

Since Γ^i_{jk} and $\partial g^{ij}/\partial y^k$ all vanish at the origin O of the Riemannian coordinate system,

$$\frac{\partial \Gamma^i_{ji}}{\partial y^k} = \frac{\partial}{\partial y^k}(g^{ir}\Gamma_{jir}) = g^{ir}\frac{\partial}{\partial y^k}\left[\frac{1}{2}(-g_{jir} + g_{irj} + g_{rji})\right] = \frac{1}{2}\, g^{ir}(-g_{jirk} + g_{irjk} + g_{rjik}) \tag{1}$$

at O, with $g_{ijkl} \equiv \partial^2 g_{ij}/\partial y^k \partial y^l$. But, since $g^{ir} = g^{ri}$,

$$g^{ir}g_{jirk} = g^{ri}g_{jirk} = g^{ir}g_{jrik} = g^{ir}g_{rjik}$$

and (*1*) becomes

$$\frac{\partial \Gamma^i_{ji}}{\partial y^k} = \frac{1}{2}\, g^{ir}g_{irjk} = \frac{1}{2}\, g^{ir}g_{irkj} = \frac{\partial \Gamma^i_{ki}}{\partial y^j}$$

9.12 Prove the identity $R_{ijkl,u} + R_{iljk,u} = R_{ikul,j} + R_{ikju,l}$.

Covariant differentiation of Bianchi's first identity, (*8.6*), gives $R_{ijkl,u} + R_{iklj,u} + R_{iljk,u} = 0$. Then the second identity, (*9.8*), yields

$$R_{ijkl,u} + R_{iljk,u} = -R_{iklj,u} = R_{ikju,l} + R_{ikul,j}$$

9.13 Show that Bianchi's identities remain valid under an indefinite metric.

One can appeal to the topological fact that, at a given point P of $\mathbf{R}^n$, the directions for which a given metric (g_{ij}) is indefinite span, at worst, a hyperplane. Hence, normal coordinates are possible along geodesics whose tangent vectors (p^i) at P do not lie in the hyperplane; Problem 9.10 gives $\Gamma^i_{jk}p^jp^k = 0$ for these directions. But the Γ^i_{jk} are continuous, and any direction *in* the hyperplane is the limit of a sequence of directions not in the hyperplane. It follows that $\Gamma^i_{jk}p^ip^j = 0$ for all (p^i), yielding Theorem 9.6 and the Bianchi identities.

SCHUR'S THEOREM

9.14 Prove Schur's theorem (Theorem 9.8).

By Lemma 9.7, $R_{ijkl} = G_{ijkl}\mathrm{K}$ throughout $\mathcal{N}$. Take the covariant derivative of both sides with respect to x^u, then permute indices ($G_{ijkl,u} = 0$ because $g_{ij,u} = 0$ in general):

$$R_{ijkl,u} = G_{ijkl}\mathrm{K}_{,u} \qquad R_{ijlu,k} = G_{ijlu}\mathrm{K}_{,k} \qquad R_{ijuk,l} = G_{ijuk}\mathrm{K}_{,l}$$

Add the three equations and apply (*9.8*):

$$G_{ijkl}\mathrm{K}_{,u} + G_{ijlu}\mathrm{K}_{,k} + G_{ijuk}\mathrm{K}_{,l} = 0 \qquad (1)$$

Multiply both sides of (*1*) by $g^{ik}g^{jl}$ and sum. Since

$$\begin{aligned} g^{ik}g^{jl}G_{ijkl} &= g^{ik}g^{jl}(g_{ik}g_{jl} - g_{il}g_{jk}) = \delta^k_k\delta^l_l - \delta^k_l\delta^l_k = n^2 - n \\ g^{ik}g^{jl}G_{ijlu} &= g^{ik}g^{jl}(g_{il}g_{ju} - g_{iu}g_{jl}) = \delta^k_l\delta^l_u - \delta^k_u\delta^l_l = \delta^k_u - n\delta^k_u \\ g^{ik}g^{jl}G_{ijuk} &= g^{ik}g^{jl}(g_{iu}g_{jk} - g_{ik}g_{ju}) = \delta^k_u\delta^l_k - \delta^k_k\delta^l_u = \delta^l_u - n\delta^l_u \end{aligned}$$

that summation yields the relation

$$\begin{aligned} 0 &= (n^2 - n)\mathrm{K}_{,u} + (\delta^k_u - n\delta^k_u)\mathrm{K}_{,k} + (\delta^l_u - n\delta^l_u)\mathrm{K}_{,l} \\ &= (n^2 - n)\mathrm{K}_{,u} + (1 - n)\mathrm{K}_{,u} + (1 - n)\mathrm{K}_{,u} = (n - 2)(n - 1)\mathrm{K}_{,u} \end{aligned}$$

For $n \geqq 3$, $\mathrm{K}_{,u} = \partial\mathrm{K}/\partial x^u = 0$. Since u was arbitrary, K must be constant over $\mathcal{N}$. QED

THE EINSTEIN TENSOR

9.15 Prove Theorem 9.9.

We must prove that $G^r_{i,r} = 0$. Multiply both sides of (*9.8*) by $g^{il}g^{jk}$ and sum:

$$\begin{aligned} 0 &= g^{il}g^{jk}R_{ijkl,u} - g^{il}g^{jk}R_{ijul,k} - g^{il}g^{jk}R_{jiuk,l} \\ &= g^{jk}R^l_{jkl,u} - g^{jk}R^l_{jul,k} - g^{il}R^k_{iuk,l} = g^{jk}R_{jk,u} - g^{jk}R_{ju,k} - g^{il}R_{iu,l} \\ &= R^k_{k,u} - R^k_{u,k} - R^l_{u,l} = 2\left(\frac{1}{2}R_{,u} - R^k_{u,k}\right) \end{aligned}$$

or, changing u to i and k to r, $\frac{1}{2}\delta^r_i R_{,r} - R^r_{i,r} = 0$. But, by Problem 6.32, $\delta^r_{i,j} = 0$ for all i, j, r; hence,

$$\left(R^r_i - \frac{1}{2}\delta^r_i R\right)_{,r} = 0 \qquad \text{or} \qquad G^r_{i,r} = 0$$

9.16 Show that G_{ij}, the associated Einstein tensor obtained by lowering the index i in G^i_j, is symmetric.

By definition,

$$G_{ij} = g_{ik}G^k_j = g_{ik}\left(R^k_j - \frac{1}{2}\delta^k_j R\right) = R_{ij} - \frac{1}{2}g_{ij}R$$

which is obviously symmetric (by symmetry of the Ricci tensor).

Supplementary Problems

9.17 Solve, if compatible, the system $\partial u_\lambda / \partial x^j = F_{\lambda j}$, with

(*a*) $$F_{01} = x^2/2u_0 \qquad F_{02} = x^1/2u_0$$

(*b*) $$F_{01} = u_0 x^1 \qquad F_{02} = u_1 x^2 \qquad F_{11} = u_0 x^1 \qquad F_{12} = u_1 x^2$$

9.18 Verify that $ds^2 = (dx^1)^2 + (x^1)^2(dx^2)^2$ represents the Euclidean metric (in polar coordinates).

9.19 Consider the metric $\varepsilon\, ds^2 = (dx^1)^2 - (x^1\, dx^2)^2 - (x^1\, dx^3)^2$. Show that $R_{2323} = -(x^1)^2$ and that, therefore, the space is not flat.

9.20 Determine whether the following metric is flat and/or Euclidean:

$$\varepsilon\, ds^2 = (dx^1)^2 - (x^1)^2(dx^2)^2 \qquad (n = 2)$$

9.21 Determine whether the following metric is flat and/or Euclidean:

$$ds^2 = (dx^1)^2 + (x^3)^2(dx^2)^2 + (dx^3)^2$$

9.22 Find the signature of the metric for $\mathbf{R}^3$ given by

$$\varepsilon\, ds^2 = 2(dx^1)^2 + 2(dx^2)^2 + 5(dx^3)^2 - 8\, dx^1\, dx^2 - 4\, dx^1\, dx^3 - dx^2\, dx^3$$

9.23 Prove that $R^i_{ijk} = 0$. [*Hint*: Use the first of (*8.6*).]

9.24 Use Problem 9.11 to obtain a simplified proof for Problem 8.34.

9.25 Show that the *Einstein invariant*, $G \equiv G^i_i$, vanishes if the space is flat. [*Hint*: Use Corollary 9.4.]

9.26 In the general theory of relativity one encounters the *Schwarzschild metric*,

$$\varepsilon\, ds^2 = e^{\varphi}(dx^1)^2 + (x^1)^2[(dx^2)^2 + (\sin^2 x^2)(dx^3)^2] - e^{\psi}(dx^4)^2$$

where both φ and ψ are functions of x^1 and x^4 only. Calculate the nonzero components of the Einstein tensor. [*Hint*: Use the method of Problem 8.13, with $n = 4$. Also helpful is the observation that when $i \neq j$, $G^i_j = R^i_j$, and for each fixed index $i = j = \alpha$, $G^\alpha_\alpha = -\frac{1}{2}R^\alpha_\alpha - \frac{1}{2}R^i_i$, summed on $i \neq \alpha$.]

Chapter 10

Tensors in Euclidean Geometry

10.1 INTRODUCTION

There exists a startling correlation between formulas of *differential geometry*, developed to answer questions about curves and surfaces in Euclidean 3-space, and tensor identities previously introduced to handle changes of coordinate systems. Differential geometry was used to great advantage by Einstein in his development of relativity.

The metric will be assumed to be the Euclidean metric, and to emphasize this fact we shall designate the space by $\mathbf{E}^3$, which means $\mathbf{R}^3$ with the metric

$$ds^2 = (dx^1)^2 + (dx^2)^2 + (dx^3)^2$$

Moreover, we shall use the familiar notation (x, y, z) in place of (x^1, x^2, x^3).

10.2 CURVE THEORY; THE MOVING FRAME

A *curve* $\mathscr{C}$ in $\mathbf{E}^3$ is the image of a class C^3 mapping, $\mathbf{r}$, from an interval $\mathscr{I}$ of real numbers into $\mathbf{E}^3$, as indicated in Fig. 10-1. The image of the real number t in $\mathscr{I}$ will be denoted

$$\mathbf{r}(t) \equiv (x(t), y(t), z(t)) \tag{10.1}$$

a vector field of class C^3.

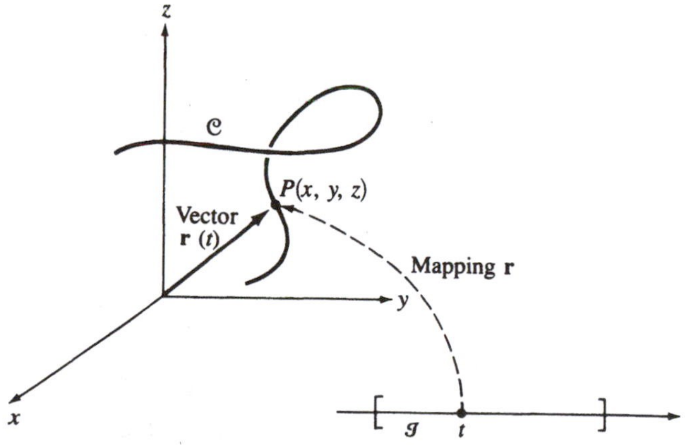

Fig. 10-1

Regular Curves

The *tangent vector* of $\mathscr{C}$ is given by

$$\frac{d\mathbf{r}}{dt} \equiv \dot{\mathbf{r}} = \left(\frac{dx}{dt}, \frac{dy}{dt}, \frac{dz}{dt}\right) \tag{10.2}$$

$\mathscr{C}$ is said to be *regular* if $\dot{\mathbf{r}}(t) \neq \mathbf{0}$ for each t in $\mathscr{I}$.

Remark 1: This corresponds to the definition of regularity, given in Section 7.3, in the case of a positive definite metric.

EXAMPLE 10.1 An *elliptical helix* (Fig. 10-2) is a helix lying on an elliptical cylinder $x^2/a^2 + y^2/b^2 = 1$ in xyz-space; it is given by $\mathscr{C} \ : \ x = a\cos t, \quad y = b\sin t, \quad z = ct,$ with $\mathscr{I}$ the entire real line. The *pitch* is defined as the number c. If $a = b$, the helix is called *circular*, with *radius* a.

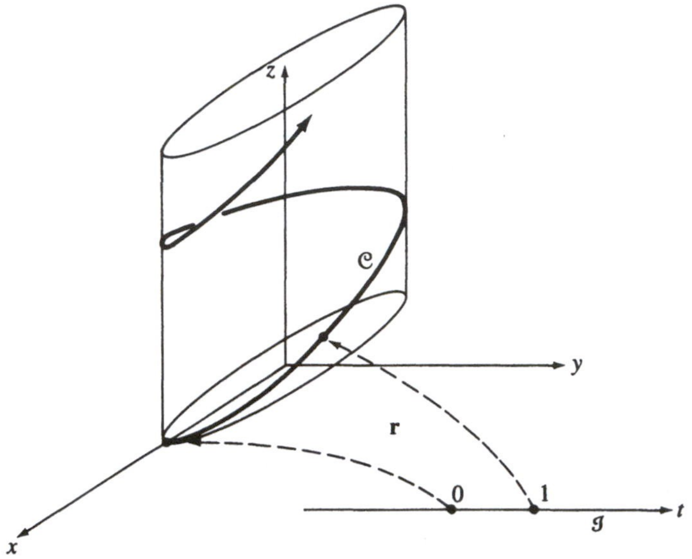

Fig. 10-2

EXAMPLE 10.2 The space curve $\mathscr{C} \ : \ x = t, \quad y = at^2, \quad z = bt^3 \qquad (\mathscr{I} = \mathbf{R})$ captures the salient local features of all curves; it is known as the *twisted cubic*. As indicated in Fig. 10-3, the projection of $\mathscr{C}$ in the xy-plane is a parabola, $y = ax^2$; its projection in the xz-plane is a standard cubic curve, $z = bx^3$; in the yz-plane, the semicubical parabola $(y/a)^3 = (z/b)^2$.

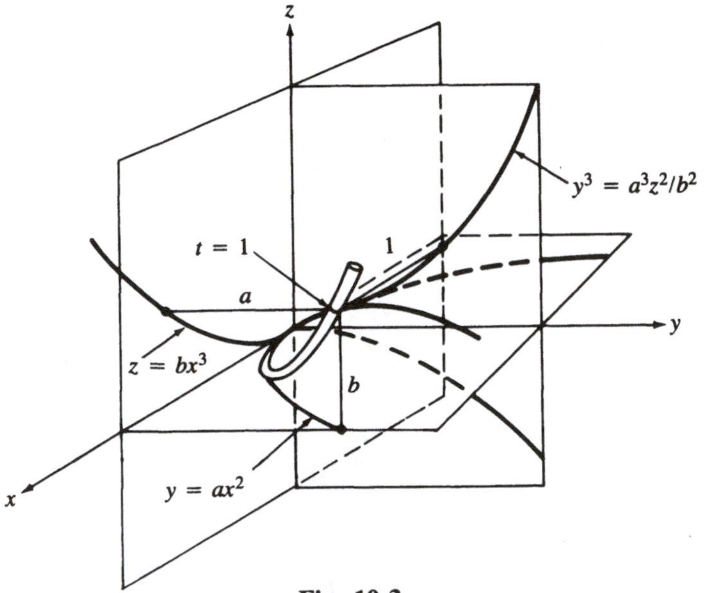

Fig. 10-3

Arc Length

Since the Euclidean metric is positive definite, every regular curve has an arc-length parameterization $\mathbf{r} = \mathbf{r}(s)$, such that

$$s = \int_a^t \left\| \frac{d\mathbf{r}}{du} \right\| du \qquad \text{or} \qquad \frac{ds}{dt} = \|\dot{\mathbf{r}}\| \tag{10.3}$$

(The dot, as in $\dot{\mathbf{r}}$, is used to denote differentiation with respect to t, and a prime, as in $\mathbf{r}'$, denotes differentiation with respect to s.) The mapping $t \rightarrow s$ defined by (*10.3*) has the inverse relation $s \rightarrow t$ given explicitly by $t = \varphi(s)$, where φ is also differentiable:

$$\frac{dt}{ds} = \varphi'(s) = \frac{1}{\|\dot{\mathbf{r}}\|} \tag{10.4}$$

The Moving Frame

Three vectors of fundamental importance to curve theory will now be discussed. Two of them were introduced in Chapter 7: the *unit tangent vector*—the (unique) vector

$$\mathbf{T} \equiv \mathbf{r}' = \left(\frac{dx}{ds}, \frac{dy}{ds}, \frac{dz}{ds} \right)$$

—and the *unit principal normal*—any unit, class C^1 vector $\mathbf{N}$ that is orthogonal to $\mathbf{T}$ and is parallel to $\mathbf{T}'$ wherever $\mathbf{T}' \neq \mathbf{0}$. The *binormal vector* associated with a curve is the unit vector $\mathbf{B} \equiv \mathbf{T} \times \mathbf{N}$ [for the cross product, see (*2.10*)]; $\mathbf{B}$ is uniquely determined once $\mathbf{N}$ has been chosen.

Not all regular curves have a principal normal vector (see Problem 10.1). However, it was proved in Problem 7.14 that all *planar* curves possess a principal normal, of the form

$$\mathbf{N} = (-\sin\theta, \cos\theta, 0) \qquad (\text{plane } z = 0)$$

if $\mathbf{T} = (\cos\theta, \sin\theta, 0)$. The following result provides further information.

Theorem 10.1: Every planar curve has a principal normal vector. If a space curve has a principal normal vector, that vector lies in the plane of the curve for any nonstraight planar segment of the curve. Along any straight-line segment, the principal normal can be chosen as any class C^1 vector orthogonal to the unit tangent vector.

At each point of $\mathscr{C}$ where $\mathbf{N}$ can be defined, the mutually orthogonal triplet of unit vectors $\mathbf{T}$, $\mathbf{N}$, $\mathbf{B}$ constitutes a right-handed system of basis elements for $\mathbf{E}^3$. This triad, which changes continuously along $\mathscr{C}$ (Fig. 10-4), is often called the *moving frame* or *moving triad*; the plane of $\mathbf{T}$ and $\mathbf{N}$ is known as the *osculating plane*.

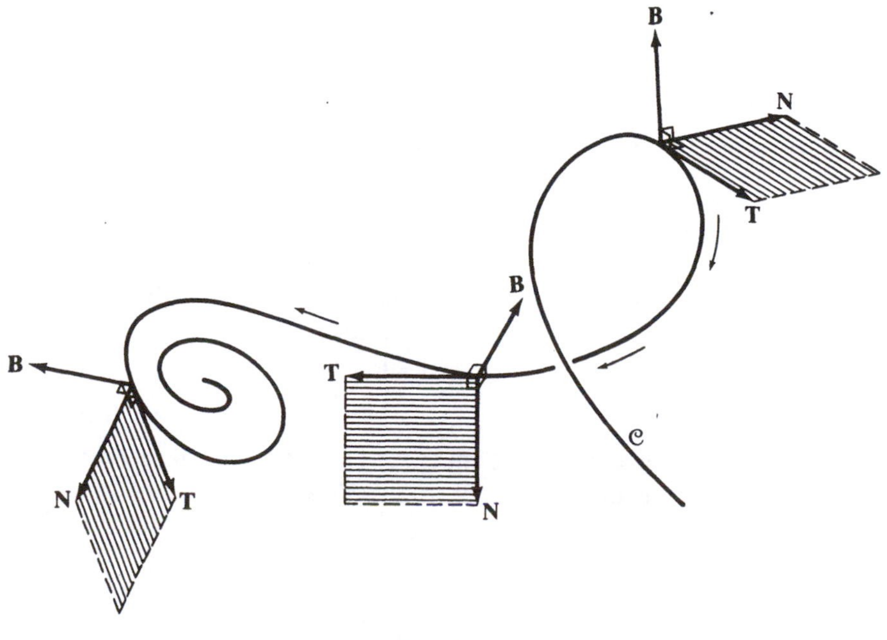

Fig. 10-4

The moving frame has been defined for the arc-length parameterization. When it is necessary to use the original parameter t instead, the following expressions may be established (Problem 10.4) for any point at which $\dot{\mathbf{r}} \neq \mathbf{0}$ and $\dot{\mathbf{r}} \times \ddot{\mathbf{r}} \neq \mathbf{0}$:

$$\mathbf{T} = \frac{\dot{\mathbf{r}}}{\|\dot{\mathbf{r}}\|} \qquad \mathbf{N} = \varepsilon\,\frac{(\dot{\mathbf{r}}\dot{\mathbf{r}})\ddot{\mathbf{r}} - (\dot{\mathbf{r}}\ddot{\mathbf{r}})\dot{\mathbf{r}}}{\|\dot{\mathbf{r}}\|\,\|\dot{\mathbf{r}} \times \ddot{\mathbf{r}}\|} \qquad \mathbf{B} = \varepsilon\,\frac{\dot{\mathbf{r}} \times \ddot{\mathbf{r}}}{\|\dot{\mathbf{r}} \times \ddot{\mathbf{r}}\|} \tag{10.5}$$

Here, $\varepsilon = \pm 1$, the choice of sign depending on the choice of $\mathbf{N}$ as a class C^1 vector.

10.3 CURVATURE AND TORSION

Two important numbers, or more accurately, scalar fields, are associated with space curves.

Definition 1: The *curvature* κ and *torsion* τ of a curve $\mathscr{C}$: $\mathbf{r} = \mathbf{r}(s)$ in $\mathbf{E}^3$ are, respectively, the real numbers

$$\kappa \equiv \mathbf{N}\mathbf{T}' \qquad \text{and} \qquad \tau \equiv -\mathbf{N}\mathbf{B}' \tag{10.6}$$

The sign of κ will depend on that chosen for $\mathbf{N}$; however, since $\mathbf{B}$ and $\mathbf{B}'$ change in sign together with $\mathbf{N}$, τ is uniquely determined.

It follows (cf. Problem 7.13) that the absolute values of curvature and torsion are given by

$$\kappa_0 \equiv |\kappa| = \|\mathbf{T}'\| \qquad \text{and} \qquad \tau_0 \equiv |\tau| = \|\mathbf{B}'\| \tag{10.7}$$

Thus, κ_0 measures the absolute rate of change of the unit tangent vector and the amount of "bending" a curve possesses at any given point, while τ_0 measures the absolute rate of change of the binormal and the tendency of the curve to "twist" out of its osculating plane at each point. The significance of negative values for κ and τ will become apparent later.

Remark 2: It can be shown that the two functions $\kappa = \kappa(s)$ and $\tau = \tau(s)$ determine the curve $\mathscr{C}$ up to a rigid motion in $\mathbf{E}^3$.

In the t-parameterization of $\mathscr{C}$, we have (Problem 10.7):

$$\kappa = \frac{\varepsilon\|\dot{\mathbf{r}} \times \ddot{\mathbf{r}}\|}{\|\dot{\mathbf{r}}\|^3} \qquad \text{and} \qquad \tau = \frac{\det[\dot{\mathbf{r}}\,\ddot{\mathbf{r}}\,\dddot{\mathbf{r}}]}{\|\dot{\mathbf{r}} \times \ddot{\mathbf{r}}\|^2} \tag{10.8}$$

where $\varepsilon = \pm 1$ and $[\dot{\mathbf{r}}\,\ddot{\mathbf{r}}\,\dddot{\mathbf{r}}]$ represents the 3×3 matrix having as row vectors $\dot{\mathbf{r}}$, $\ddot{\mathbf{r}}$, and $\dddot{\mathbf{r}}$. [Recall the identity

$$\mathbf{a} \cdot (\mathbf{b} \times \mathbf{c}) = \det[\mathbf{a}\,\mathbf{b}\,\mathbf{c}]$$

for the *triple scalar product* of three vectors.]

Serret–Frenet Formulas

The derivatives of the vectors composing the moving triad are given by

$$\begin{aligned} \mathbf{T}' &= \kappa\mathbf{N} \\ \mathbf{N}' &= -\kappa\mathbf{T} + \tau\mathbf{B} \\ \mathbf{B}' &= -\tau\mathbf{N} \end{aligned} \qquad \text{or} \qquad \begin{bmatrix} \mathbf{T} \\ \mathbf{N} \\ \mathbf{B} \end{bmatrix}' = \begin{bmatrix} 0 & \kappa & 0 \\ -\kappa & 0 & \tau \\ 0 & -\tau & 0 \end{bmatrix} \begin{bmatrix} \mathbf{T} \\ \mathbf{N} \\ \mathbf{B} \end{bmatrix} \tag{10.9}$$

Note the skew-symmetry of the coefficient matrix. The first of these formulas was established in Problem 7.13; the other two are derived in Problem 10.8.

10.4 REGULAR SURFACES

Surfaces are generally encountered in the calculus in the form $z = F(x, y)$; that is, as graphs of two-variable functions in three-dimensional space. Here, however, it is more convenient to adopt the

Definition 2: A *surface* $\mathcal{S}$ in $\mathbf{E}^3$ is the image of a C^3 vector function,

$$\mathbf{r}(x^1, x^2) = (f(x^1, x^2), g(x^1, x^2), h(x^1, x^2))$$

which maps some region $\mathscr{V}$ of $\mathbf{E}^2$ into $\mathbf{E}^3$.

(See Fig. 10-5; in general, primes will designate objects in the parameter plane (x^i) corresponding to those on the surface in xyz-space.) The coordinate breakdown of the mapping $\mathbf{r}$,

$$x = f(x^1, x^2) \qquad y = g(x^1, x^2) \qquad z = h(x^1, x^2) \tag{10.10}$$

is called the *Gaussian form* or *representation* of $\mathcal{S}$.

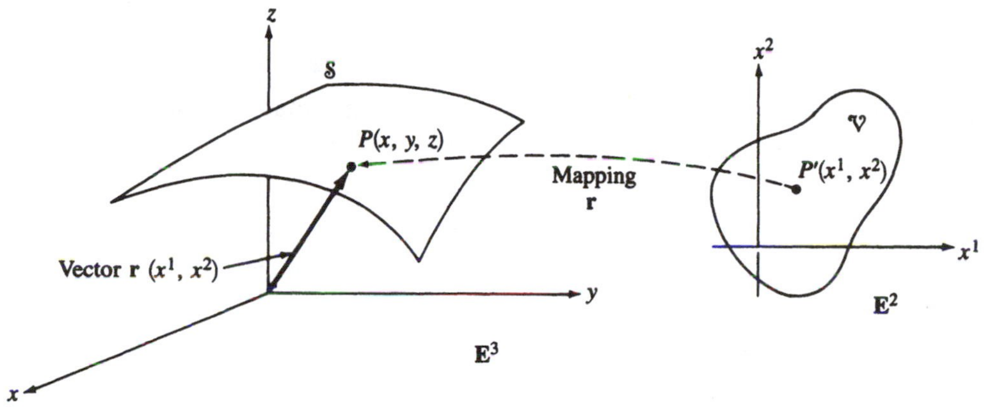

Fig. 10-5

Point P is a *regular point* of $\mathcal{S}$ if

$$\frac{\partial \mathbf{r}}{\partial x^1} \times \frac{\partial \mathbf{r}}{\partial x^2} \equiv \begin{vmatrix} \mathbf{i} & \mathbf{j} & \mathbf{k} \\ \dfrac{\partial f}{\partial x^1} & \dfrac{\partial g}{\partial x^1} & \dfrac{\partial h}{\partial x^1} \\ \dfrac{\partial f}{\partial x^2} & \dfrac{\partial g}{\partial x^2} & \dfrac{\partial h}{\partial x^2} \end{vmatrix} \neq \mathbf{0} \tag{10.11}$$

at P'; otherwise, P is a *singular point*. If every point of $\mathcal{S}$ is a regular point, then $\mathcal{S}$ is a *regular surface*.

Remark 3: Condition (*10.11*) is tantamount to the linear independence of the two vectors $(\partial\mathbf{r}/\partial x^1)_P$ and $(\partial\mathbf{r}/\partial x^2)_P$. Equivalently, and of more geometrical interest, the condition ensures that every *curve* in $\mathcal{S}$ through P which we take to be the image under $\mathbf{r}$ of a regular curve in $\mathscr{V}$ through P', is, in a neighborhood of P, *regular* in the sense of Section 10.2.

EXAMPLE 10.3 For a C^3 function F, show that the graph $z = F(x, y)$ is a regular surface.

The surface has the Gaussian representation

$$x = x^1 \qquad y = x^2 \qquad z = F(x^1, x^2)$$

and thus

$$\frac{\partial \mathbf{r}}{\partial x^1} \times \frac{\partial \mathbf{r}}{\partial x^2} = \begin{vmatrix} \mathbf{i} & \mathbf{j} & \mathbf{k} \\ 1 & 0 & \partial F/\partial x^1 \\ 0 & 1 & \partial F/\partial x^2 \end{vmatrix} = \left(-\frac{\partial F}{\partial x^1}, -\frac{\partial F}{\partial x^2}, 1\right) \neq \mathbf{0}$$

at an arbitrary surface point P. (This would be true if F were merely class C^1.)

Subscript Notation for Partial Derivatives

From now on, write

$$\frac{\partial \mathbf{r}}{\partial x^1} \equiv \mathbf{r}_1 \qquad \frac{\partial \mathbf{r}}{\partial x^2} \equiv \mathbf{r}_2 \qquad \frac{\partial^2 \mathbf{r}}{\partial x^1 \partial x^1} = \mathbf{r}_{11} \qquad \text{etc.}$$

so that, e.g., (*10.11*) takes the compact form $\mathbf{r}_1 \times \mathbf{r}_2 \neq \mathbf{0}$.

10.5 PARAMETRIC LINES; TANGENT SPACE

Let (x^i) be taken as coordinates—for the moment, *rectangular* coordinates—in the parameter plane $\mathbf{E}^2$, yielding two (orthogonal) families of coordinate lines:

$$\begin{cases} x^1 = t \\ x^2 = d \end{cases} \qquad \text{and} \qquad \begin{cases} x^1 = c \\ x^2 = \sigma \end{cases}$$

If (c, d) runs over $\mathcal{V}$ (the pre-image of surface $\mathcal{S}$), then the images under $\mathbf{r}$ of these two families are the two sets of *parametric lines* (or *coordinate curves*) on $\mathcal{S}$:

$$\underbrace{\mathbf{r} = \mathbf{r}(t, d) \equiv \mathbf{p}(t)}_{x^1\text{-curves}} \qquad \underbrace{\mathbf{r} = \mathbf{r}(c, \sigma) \equiv \mathbf{q}(\sigma)}_{x^2\text{-curves}}$$

Figure 10-6 suggests that the net of parametric lines is orthogonal also. This is not, of course, true in general. In fact, since the tangent fields to the x^1-curves and the x^2-curves are respectively $d\mathbf{p}/dt = \mathbf{r}_1$ and $d\mathbf{q}/d\sigma = \mathbf{r}_2$, the net is orthogonal if and only if $\mathbf{r}_1\mathbf{r}_2 = 0$ at every point of $\mathcal{S}$.

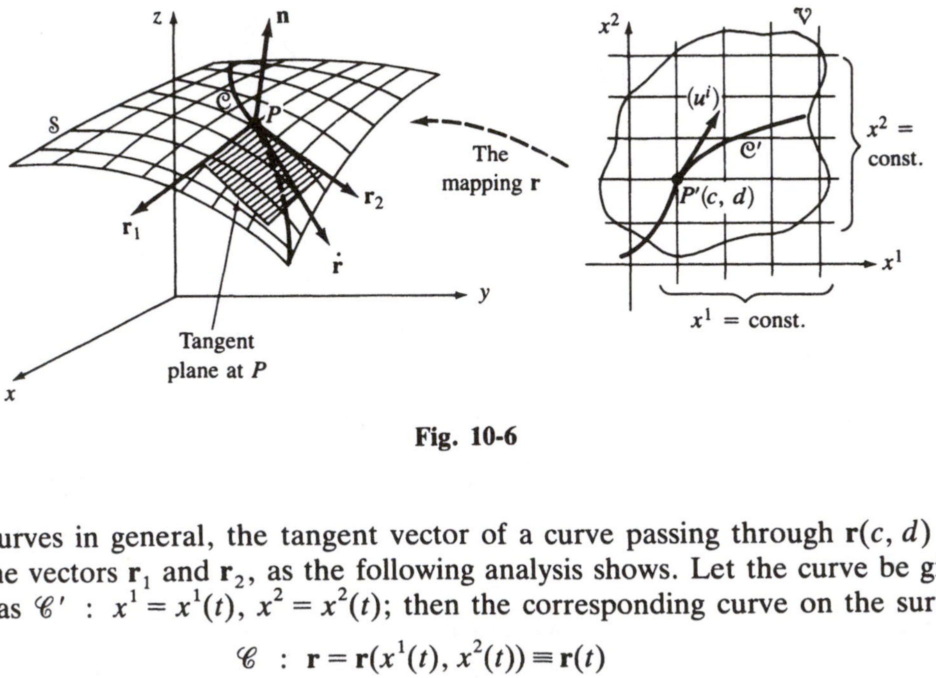

Fig. 10-6

For surface curves in general, the tangent vector of a curve passing through $\mathbf{r}(c, d)$ is a linear combination of the vectors $\mathbf{r}_1$ and $\mathbf{r}_2$, as the following analysis shows. Let the curve be given in the parameter plane as $\mathscr{C}' : x^1 = x^1(t),\ x^2 = x^2(t)$; then the corresponding curve on the surface is

$$\mathscr{C} : \mathbf{r} = \mathbf{r}(x^1(t), x^2(t)) \equiv \mathbf{r}(t)$$

with tangent vector

$$\dot{\mathbf{r}} = \frac{\partial \mathbf{r}}{\partial x^1}\frac{dx^1}{dt} + \frac{\partial \mathbf{r}}{\partial x^2}\frac{dx^2}{dt} \equiv u^1\mathbf{r}_1 + u^2\mathbf{r}_2 \equiv u^i\mathbf{r}_i \tag{10.12}$$

Here, $u^1 \equiv dx^1/dt$, $u^2 \equiv dx^2/dt$, so that the vector (u^i) in the parameter plane is the tangent to $\mathscr{C}'$ at P' (see Fig. 10-6).

Definition 3: The collection of linear combinations of the vectors $\mathbf{r}_1(P)$ and $\mathbf{r}_2(P)$ is called the *tangent space* of $\mathscr{S}$ at P. The *unit surface normal* is the unit vector $\mathbf{n}$ in the direction of $\mathbf{r}_1 \times \mathbf{r}_2$:

$$\mathbf{n} = \frac{1}{E}(\mathbf{r}_1 \times \mathbf{r}_2) \qquad (E \equiv \|\mathbf{r}_1 \times \mathbf{r}_2\| > 0) \tag{10.13}$$

The geometric realization of the tangent space is obviously the *tangent plane at P*, and the surface normal can be identified with a line segment through P perpendicular to this tangent plane; that is, orthogonal to the surface at P, as indicated in Fig. 10-6.

To summarize this whole affair, the linearly independent (by regularity) triad of vectors $\mathbf{r}_1$, $\mathbf{r}_2$, $\mathbf{n}$ forms a moving frame for the surface, as shown in Fig. 10-7, much in the manner that a moving triad exists for a regular curve having a principal normal.

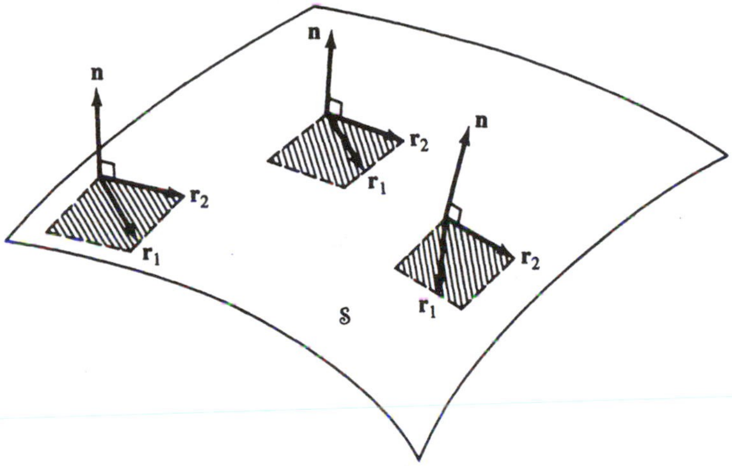

Fig. 10-7

10.6 FIRST FUNDAMENTAL FORM

Consider a curve on the regular surface $\mathscr{S}$: $\mathbf{r} = \mathbf{r}(x^1, x^2)$ given by $\mathscr{C}$: $\mathbf{r} = \mathbf{r}(x^1(t), x^2(t)) \equiv \mathbf{r}(t)$, with pre-image $\mathscr{C}'$: $x^i = x^i(t)$ in the parameter plane. Using (*10.12*) and recalling that the (Euclidean) inner product is distributive over linear combinations of vectors, arc length along $\mathscr{C}$ is calculated as

$$\left(\frac{ds}{dt}\right)^2 = \|\dot{\mathbf{r}}\|^2 = \dot{\mathbf{r}}\dot{\mathbf{r}} = (u^i\mathbf{r}_i)(u^j\mathbf{r}_j) \equiv g_{ij}u^iu^j \tag{10.14a}$$

where we define

$$g_{ij} = \mathbf{r}_i\mathbf{r}_j \qquad (1 \leqq i, j \leqq 2) \tag{10.15}$$

and, as above, $u^i = dx^i/dt$. In the equivalent differential form,

$$ds^2 = g_{ij}\,dx^i\,dx^j \equiv \mathrm{I} \tag{10.14b}$$

the arc-length formula is known as the *First Fundamental Form* (abbreviated FFF) of the surface $\mathscr{S}$. In view of (*10.12*) and the regularity of $\mathscr{S}$, $\|\dot{\mathbf{r}}\| = 0$ if and only if $u^1 = u^2 = 0$; this proves

Lemma 10.2: The FFF of a regular surface is positive definite.

Lemma 10.2 implies that $g \equiv \det(g_{ij}) > 0$; in fact, we can use Lagrange's identity,

$$(\mathbf{r}_1 \times \mathbf{r}_2)^2 = (\mathbf{r}_1^2)(\mathbf{r}_2^2) - (\mathbf{r}_1\mathbf{r}_2)^2$$

to establish that

$$g = E^2 \tag{10.16}$$

cf. (*10.13*).

EXAMPLE 10.4 Compute the FFF for the right helicoid (Fig. 10-8),

$$\mathbf{r} = (x^1 \cos x^2, x^1 \sin x^2, ax^2)$$

We have:

$$\mathbf{r}_1 = (\cos x^2, \sin x^2, 0) \qquad \mathbf{r}_2 = (-x^1 \sin x^2, x^1 \cos x^2, a)$$

whence

$$g_{11} = \mathbf{r}_1^2 = \cos^2 x^2 + \sin^2 x^2 + 0^2 = 1$$
$$g_{12} = \mathbf{r}_1\mathbf{r}_2 = (\cos x^2)(-x^1 \sin x^2) + (\sin x^2)(x^1 \cos x^2) = 0$$
$$g_{22} = \mathbf{r}_2^2 = (-x^1)^2(\sin^2 x^2) + (x^1)^2(\cos^2 x^2) + a^2 = (x^1)^2 + a^2$$

and

$$\mathrm{I} = (dx^1)^2 + [(x^1)^2 + a^2](dx^2)^2$$

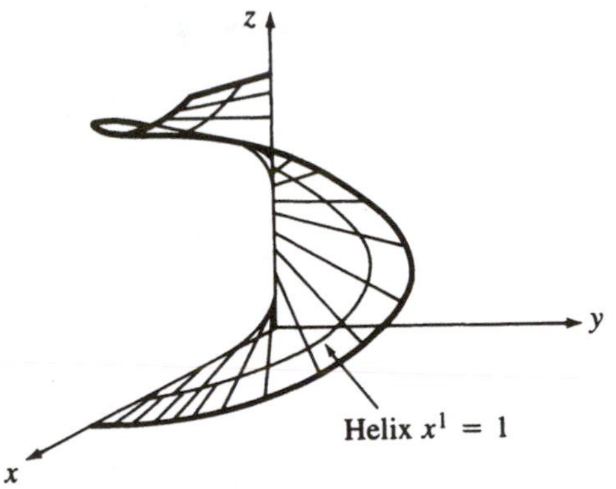

Fig. 10-8

Along with the FFF, tensor calculus enters the picture. For the *intrinsic properties* of a particular surface $\mathscr{S}$ in $\mathbf{E}^3$ (the properties defined by measurements of distance *on* the surface) are all implicit in (*10.14b*), which can be interpreted as a particular Riemannian metrization of the parameter plane. Thus, the study of intrinsic properties of surfaces becomes the tensor analysis of Riemannian metrics in $\mathbf{R}^2$—*and this may be conducted without any reference to* $\mathbf{E}^3$ *whatever.* Observe that the metrics under consideration will all be positive definite (Lemma 10.2) but not necessarily Euclidean (see Theorem 9.1). Accordingly, we shall drop the designation $\mathbf{E}^2$ for the parameter plane, which shall henceforth be referred to general coordinates (x^i).

EXAMPLE 10.5 The metric for $\mathbf{R}^2$ corresponding to the right helicoid (Example 10.4) is non-Euclidean, as is demonstrated in Problem 10.27. Now the parameters x^1 and x^2, which are actual polar coordinates in the xy-plane of $\mathbf{E}^3$ (see Fig. 10-8), formally keep that significance when the plane is considered abstractly as parameter space. This is an instance of the formal use of a familiar coordinate system in a non-Euclidean space, as mentioned in Section 3.1.

Unit Tangent Vector

If $\mathscr{C}\ :\ \mathbf{r} = \mathbf{r}(x^1(t), x^2(t))$ is any curve on $\mathscr{S}$, then by (*10.12*) and (*10.14a*),

$$\mathbf{T} = \frac{\dot{\mathbf{r}}}{\|\dot{\mathbf{r}}\|} = \frac{u^i\mathbf{r}_i}{\sqrt{g_{jk}u^ju^k}} \tag{10.17}$$

Angle Between Two Curves

Let $\mathscr{C}_1$ and $\mathscr{C}_2$ be two intersecting curves on $\mathscr{S}$ that correspond to $x^i = \phi^i(t)$ and $x^i = \psi^i(\sigma)$ $(i = 1, 2)$ in the parameter plane. Writing $u^i \equiv d\phi^i/dt$ and $v^i \equiv d\psi^i/d\sigma$, we have for the angle θ between $\mathbf{T}_1$ of $\mathscr{C}_1$ and $\mathbf{T}_2$ of $\mathscr{C}_2$:

$$\cos\theta = \mathbf{T}_1\mathbf{T}_2 = \frac{u^i\mathbf{r}_i}{\sqrt{g_{pq}u^pu^q}}\cdot\frac{v^j\mathbf{r}_j}{\sqrt{g_{rs}v^rv^s}} = \frac{g_{ij}u^iv^j}{\sqrt{g_{pq}u^pu^q}\sqrt{g_{rs}v^rv^s}} \tag{10.18}$$

Compare (*5.11*).

Theorem 10.3: The angle between the two parametric lines through a surface point is

$$\cos\theta = \frac{g_{12}}{\sqrt{g_{11}}\sqrt{g_{22}}} \tag{10.19}$$

Corollary 10.4: The two families of parametric lines form an orthogonal net if and only if $g_{12} = 0$ at every point of $\mathscr{S}$.

10.7 GEODESICS ON A SURFACE

A further link with tensors is provided by the concept of geodesics for regular surfaces. One can intuitively imagine stretching a string between two points on a surface, and pulling it tight: on a sphere this would lead to a great circular arc, and on a right circular cylinder, a helical arc. Since from our point of view the surface is disregarded and (g_{ij}) is taken as a metric for the parameter plane, the problem has already been worked out (Section 7.6).

Relative to the FFF of $\mathscr{S}$, define the Christoffel symbols through formulas (*6.1*) and (*6.4*), $n = 2$. [Problem 10.48 gives an equivalent "extrinsic" definition, in terms of the vector $\mathbf{r}$.] Then a *geodesic* on $\mathscr{S}$ is any curve $\mathbf{r} = \mathbf{r}(x^1(t), x^2(t))$ in the surface whose pre-image in the parameter $\mathbf{R}^2$ satisfies the system of differential equations (*7.11*)–(*7.12*); if $t = s =$ arc length, the governing system is (*7.13*). [Remember that the (non-Euclidean) distance measured by s in $\mathbf{R}^2$ is the Euclidean distance along the geodesic as a curve in $\mathbf{E}^3$.]

Similarly, harking back to Section 6.5, the *intrinsic curvature* of a curve $\mathscr{C}$ in $\mathscr{S}$ is the function

$$\tilde{\kappa}(s) = \sqrt{g_{ij}b^ib^j} \tag{10.20}$$

—cf. (*6.12*)—where the *intrinsic curvature vector* (b^i) (in $\mathbf{R}^2$) is given by (*6.11*).

Remark 4: Intrinsic curvature can be shown to be the instantaneous rate of change of the angle between the tangent vector of $\mathscr{C}$ and another vector in the tangent space that is "transported parallelly" along the curve. Here, the term "parallel" refers to a certain generalization of Euclidean parallelism (see Problem 10.22).

Theorem 10.5: A curve on a surface is a geodesic if and only if its intrinsic curvature $\tilde{\kappa}$ is identically zero.

In contrast to the above intrinsic characterization of geodesics, there is an interesting and useful extrinsic characterization, proved as Problem 10.18. It adds a visual dimension that often allows the immediate identification of a geodesic.

Theorem 10.6: A curve on a regular surface is a geodesic if and only if a principal normal $\mathbf{N}$ of the curve can be chosen that coincides with the surface normal $\mathbf{n}$ at all points along the curve.

10.8 SECOND FUNDAMENTAL FORM

By taking the dot product of the surface normal with the second partial derivatives of **r** with respect to x^1 and x^2,

$$f_{ij} \equiv \mathbf{n}\mathbf{r}_{ij} \tag{10.21}$$

we generate the (symmetric) coefficients of the *Second Fundamental Form* (SFF) of a surface:

$$f_{ij}\,dx^i\,dx^j \equiv \mathrm{II} \tag{10.22}$$

Curvature of a Normal Section

If $\mathscr{F}$ is a plane containing the surface normal **n** at some point P of $\mathscr{S}$ (Fig. 10-9), the curvature of the *normal section* of $\mathscr{S}$ (the curve of intersection of $\mathscr{S}$ and $\mathscr{F}$), denoted $\mathscr{C}_{\mathscr{F}}$, is given at point P by the formula

$$\kappa_{\mathscr{F}} = \frac{f_{ij}u^iu^j}{g_{kl}u^ku^l} = \frac{\mathrm{II}}{\mathrm{I}} \tag{10.23}$$

where $(u^i) = (dx^i/dt)$ gives the direction, at P', of the curve corresponding to $C_{\mathscr{F}}$ in the parameter plane; see Problem 10.23.

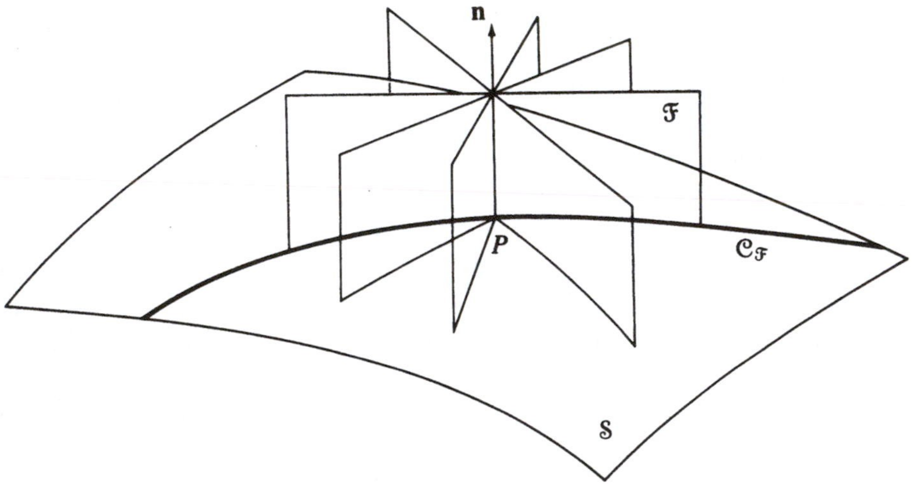

Fig. 10-9

As $\mathscr{F}$ rotates about **n**, the curvature $\kappa_{\mathscr{F}}$ of $\mathscr{C}_{\mathscr{F}}$ at P is periodic and will reach an absolute maximum and an absolute minimum; let

$$\max \kappa_{\mathscr{F}} \equiv \kappa_1 \qquad \min \kappa_{\mathscr{F}} \equiv \kappa_2 \tag{10.24}$$

The two section curves having these two extremal curvatures are called *principal curves*, and their directions are the *principal directions*. If $\kappa_1 = \kappa_2$ at P, all the normal sections at P have the same curvature and unique principal directions do not exist. (In this case, P is called an *umbilical point* of the surface.)

Surface Curvature

Two measures of the curvature of a surface $\mathscr{S}$ are commonly used.

Definition 4: The *Gaussian curvature* of $\mathscr{S}$ at point P is the number $\mathrm{K} = \kappa_1\kappa_2$; the *mean curvature* is the number $\mathrm{H} = \kappa_1 + \kappa_2$.

It will be proved in Problem 10.25 that the extreme curvatures κ_1 and κ_2 are the roots of the following quadratic equation in λ:

$$(g_{11}g_{22} - g_{12}^2)\lambda^2 - (f_{11}g_{22} + f_{22}g_{11} - 2f_{12}g_{12})\lambda + (f_{11}f_{22} - f_{12}^2) = 0 \tag{10.25}$$

The relations between the roots and the coefficients of a polynomial equation then give:

$$\mathrm{K} = \frac{f_{11}f_{22} - f_{12}^2}{g_{11}g_{22} - g_{12}^2} \qquad \mathrm{H} = \frac{f_{11}g_{22} + f_{22}g_{11} - 2f_{12}g_{12}}{g_{11}g_{22} - g_{12}^2} \tag{10.26}$$

10.9 STRUCTURE FORMULAS FOR SURFACES

Two fundamental sets of relationships involve the parts of the moving triad of a surface, $(\mathbf{r}_1, \mathbf{r}_2, \mathbf{n})$.

Equations of Weingarten

Since $\mathbf{n}^2 = 1$, $\partial(\mathbf{n}^2)/\partial x^i = 2\mathbf{n}\mathbf{n}_i = 0$ $(i = 1, 2)$. Hence, for each i, $\mathbf{n}_i$ lies in the tangent space: $\mathbf{n}_i = u_i^1\mathbf{r}_1 + u_i^2\mathbf{r}_2$, for certain scalars u_i^k. Similarly, from orthogonality,

$$0 = (\mathbf{n}\mathbf{r}_i)_j = \mathbf{n}_j\mathbf{r}_i + \mathbf{n}\mathbf{r}_{ij} \qquad \text{or} \qquad \mathbf{n}_j\mathbf{r}_i = -f_{ij}$$

It follows (Problem 10.28) that

$$\mathbf{n}_i = -g^{jk}f_{ij}\mathbf{r}_k \tag{10.27a}$$

for $i = 1, 2$. From the explicit form of the inverse metric matrix (g^{ij}), (*10.27a*) may be spelled out as follows using the symmetry of g_{ij} and f_{ij}:

$$\begin{aligned} \mathbf{n}_1 &= \frac{g_{12}f_{12} - g_{22}f_{11}}{g}\mathbf{r}_1 + \frac{g_{12}f_{11} - g_{11}f_{12}}{g}\mathbf{r}_2 \\ \mathbf{n}_2 &= \frac{g_{12}f_{22} - g_{22}f_{12}}{g}\mathbf{r}_1 + \frac{g_{12}f_{12} - g_{11}f_{22}}{g}\mathbf{r}_2 \end{aligned} \tag{10.27b}$$

Equations of Gauss

Since $(\mathbf{r}_1, \mathbf{r}_2, \mathbf{n})$ is a basis for $\mathbf{E}^3$, we can write $\mathbf{r}_{ij} = u_{ij}^1\mathbf{r}_1 + u_{ij}^2\mathbf{r}_2 + u_{ij}^3\mathbf{n}$. Evaluation of the coefficients (Problem 10.29) leads to

$$\mathbf{r}_{ij} = \Gamma_{ij}^k\mathbf{r}_k + f_{ij}\mathbf{n} \tag{10.28}$$

An Identity Between FFF and SFF

Since $\mathbf{r}_{ijk} = \mathbf{r}_{ikj}$, (*10.28*) implies $(\Gamma_{ij}^s\mathbf{r}_s + f_{ij}\mathbf{n})_k = (\Gamma_{ik}^s\mathbf{r}_s + f_{ik}\mathbf{n})_j$, or

$$(\Gamma_{ij}^s)_k\mathbf{r}_s + \Gamma_{ij}^s\mathbf{r}_{sk} + f_{ijk}\mathbf{n} + f_{ij}\mathbf{n}_k = (\Gamma_{ik}^s)_j\mathbf{r}_s + \Gamma_{ik}^s\mathbf{r}_{sj} + f_{ikj}\mathbf{n} + f_{ik}\mathbf{n}_j$$

Dot both sides with $\mathbf{r}_l$ and use the definition $\mathbf{r}_l\mathbf{r}_s \equiv g_{ls}$ and the relations $\mathbf{r}_l\mathbf{r}_{sk} = \Gamma_{skl}$ (Problem 10.48) and $\mathbf{r}_l\mathbf{n} = 0$:

$$(\Gamma_{ij}^s)_k g_{sl} + \Gamma_{ij}^s\Gamma_{skl} + f_{ij}\mathbf{n}_k\mathbf{r}_l = (\Gamma_{ik}^s)_j g_{sl} + \Gamma_{ik}^s\Gamma_{sjl} + f_{ik}\mathbf{n}_j\mathbf{r}_l$$

Now substitute for the $\mathbf{n}_i$ from (*10.27a*) and use $\mathbf{r}_t\mathbf{r}_l \equiv g_{tl}$ and $g^{st}g_{tl} = \delta_l^s$ to simplify the result:

$$-f_{ij}f_{kl} + f_{ik}f_{jl} = g_{sl}\left(\frac{\partial\Gamma_{ik}^s}{\partial x^j} - \frac{\partial\Gamma_{ij}^s}{\partial x^k} + \Gamma_{ik}^r\Gamma_{rj}^s - \Gamma_{ij}^r\Gamma_{rk}^s\right)$$

Finally, introducing the Riemann tensor via (*8.2*) and (*8.3*), we obtain

$$R_{ijkl} = f_{ik}f_{jl} - f_{il}f_{jk} \tag{10.29}$$

The left member of (*10.29*) depends only on the coefficients of I together with their first and second derivatives; the right member depends only on the coefficients of II. This essential *compatibility relation* between the two fundamental forms must hold at every point of a regular surface.

The 'Most Excellent Theorem' of Gauss

By (*10.26*) and (*10.29*),

$$\mathrm{K} = \frac{f_{11}f_{22} - f_{12}^2}{g_{11}g_{22} - g_{12}^2} = \frac{R_{1212}}{g} \tag{10.30}$$

Thus, the numerator of K can be derived entirely from the FFF. Since the denominator is also obviously from the FFF, we have:

Theorem 10.7 (*Theorema Egregium*): The Gaussian curvature is an intrinsic property of a surface, depending only on the First Fundamental Form and its derivatives.

Remark 5: The motive for the definition (*8.7*) of Riemannian curvature is now apparent.

10.10 ISOMETRIES

The practical question of whether inhabitants of a fog-enshrouded planet could, solely by measuring distances on the surface of the planet, determine its curvature, is answered in the affirmative by Theorem 10.7. A further important conclusion can be drawn.

Suppose that two surfaces, $\mathscr{S}^{(1)}$: $\mathbf{r}^{(1)} = \mathbf{r}^{(1)}(x^1, x^2)$ and $\mathscr{S}^{(2)}$: $\mathbf{r}^{(2)} = \mathbf{r}^{(2)}(x^1, x^2)$, are defined over the same region $\mathscr{V}$ of the plane and that the First Fundamental Forms agree on $\mathscr{V}$. This will obviously set up a correspondence between $\mathscr{S}^{(1)}$ and $\mathscr{S}^{(2)}$ in $\mathbf{E}^3$ that is bijective between small patches (induced by the neighborhoods of $\mathscr{V}$ over which both $\mathbf{r}^{(i)}$ are bijective) of the two surfaces. This correspondence is called a *local isometry* between $\mathscr{S}^{(1)}$ and $\mathscr{S}^{(2)}$ because the two surfaces are, patch for patch, metrically identical. But then (Theorem 10.7) the Gaussian curvatures $\mathrm{K}^{(1)}$ and $\mathrm{K}^{(2)}$ must be equal at corresponding points.

Theorem 10.8: If two surfaces are locally isometric, their Gaussian curvatures are identical.

In the case of *constant* Gaussian curvature K, *Beltrami's theorem* tells us that there is a parameterization for $\mathscr{S}$ for which the FFF takes on the form:

$$\begin{aligned} ds^2 &= a^2(dx^1)^2 + (a^2 \sinh^2 x^1)(dx^2)^2 && \text{if } \mathrm{K} = -1/a^2 < 0 \\ ds^2 &= (dx^1)^2 + (dx^2)^2 && \text{if } \mathrm{K} = 0 \\ ds^2 &= a^2(dx^1)^2 + (a^2 \sin^2 x^1)(dx^2)^2 && \text{if } \mathrm{K} = 1/a^2 > 0 \end{aligned}$$

EXAMPLE 10.6 The plane and the sphere are surfaces of constant zero curvature and constant positive curvature, respectively. For a surface of constant negative curvature, see Problem 10.49.

Beltrami's theorem implies a partial converse of Theorem 10.8:

Theorem 10.9 (*Minding's Theorem*): If two surfaces are of the same constant Gaussian curvature, they are locally isometric.

Remark 6: A proof of Theorem 10.9 for zero curvature was given in Problem 9.9.

Solved Problems

CURVE THEORY; THE MOVING FRAME

10.1 The curve

$$\mathscr{C}\ :\ \begin{cases} x = t \\ y = t^4 \\ z = 0 \end{cases} \quad (t<0) \qquad \begin{cases} x = t \\ y = 0 \\ z = t^4 \end{cases} \quad (t \geqq 0)$$

lies partly in the xy-plane and partly in the xz-plane (Fig. 10-10). Show that it is regular of class C^3, but that it possesses no principal normal vector.

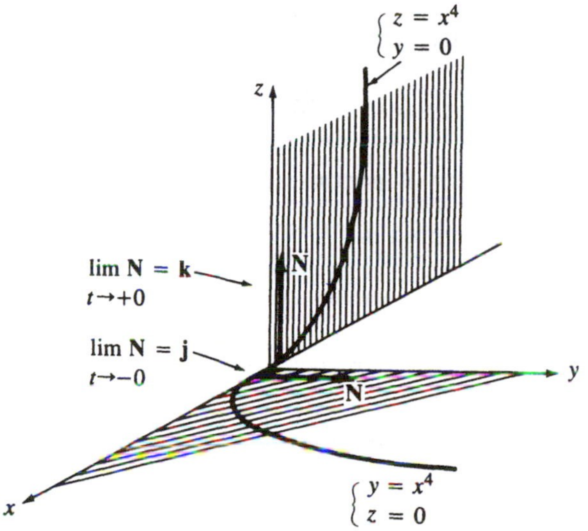

Fig. 10-10

The component functions for $\mathbf{r}(t)$ are

$$x(t) = t \qquad y(t) = \begin{cases} t^4 & t<0 \\ 0 & t \geqq 0 \end{cases} \qquad z(t) = \begin{cases} 0 & t<0 \\ t^4 & t \geqq 0 \end{cases}$$

When $t<0$, $\dot{y}(t) = 4t^3$. As $t \to 0$,

$$\lim_{t\to -0} \frac{y(t)-y(0)}{t-0} = \lim_{t\to -0} \frac{t^4}{t} = 0 \qquad \lim_{t\to +0} \frac{y(t)-y(0)}{t-0} = \lim_{t\to +0} \frac{0}{t} = 0$$

hence, $y(t)$ is differentiable at $t=0$. Clearly, $\dot{y}(t) = 0$ for $t>0$. A similar analysis applies to $z(t)$. Hence:

$$\dot{y}(t) = \begin{cases} 4t^3 & t<0 \\ 0 & t \geqq 0 \end{cases} \qquad \dot{z}(t) = \begin{cases} 0 & t<0 \\ 4t^3 & t \geqq 0 \end{cases}$$

which are continuous functions. Continuing the analysis up to the third derivatives:

$$\dddot{y}(t) = \begin{cases} 24t & t<0 \\ 0 & t \geqq 0 \end{cases} \qquad \dddot{z}(t) = \begin{cases} 0 & t<0 \\ 24t & t \geqq 0 \end{cases}$$

Hence, $x(t)$ being differentiable to all orders, $\mathbf{r}(t)$ is of class C^3. Furthermore, because $\dot{x}(t) \equiv 1$, $\dot{\mathbf{r}}(t) \neq 0$ for all t and $\mathscr{C}$ is regular. However, the principal normal, which exists for the separate parts of $\mathscr{C}$ (lying in the xy-plane for $t<0$ and in the xz-plane for $t>0$), cannot possibly be continuous at $t=0$, let alone differentiable. Hence, $\mathscr{C}$ does not possess a principal normal.

10.2 (*a*) Describe the curve $\mathbf{r} = (\cos t, \sin t, \tan^{-1} t)$, where $0 \leqq t$ and where the principal value of the arctangent is understood. (*b*) Find the arc length between the points $\mathbf{r}(0)$ and $\mathbf{r}(1)$.

(*a*) This is a form of the circular helix, except that the pitch z/t decreases with increasing t. The curve lies on the right circular cylinder $x^2 + y^2 = 1$; beginning at $(1, 0, 0)$, it winds around the cylinder and approaches the circle $x^2 + y^2 = 1$, $z = \pi/2$ asymptotically as $t \to \infty$.

(*b*) $$\dot{\mathbf{r}} = \left(-\sin t, \cos t, \frac{1}{t^2+1}\right) \qquad \text{or} \qquad \frac{ds}{dt} = \sqrt{\sin^2 t + \cos^2 t + \frac{1}{(t^2+1)^2}}$$

A numerical method of integration is required. Using Simpson's rule on a programmable calculator, one obtains

$$L = \int_0^1 \frac{\sqrt{t^4 + 2t^2 + 2}}{t^2 + 1}\, dt \approx 1.27797806$$

10.3 Find the moving frame for the curve

$$\mathscr{C}\ :\ \mathbf{r} = \left(\frac{3 - 3t^3}{5}, \frac{4 + 4t^3}{5}, 3t\right) \qquad (t \text{ real})$$

Show that the binormal vector $\mathbf{B}$ is constant, so that the curve is actually planar.

Making the calculations required in (*10.5*):

$$\dot{\mathbf{r}} = \left(\frac{-9t^2}{5}, \frac{12t^2}{5}, 3\right) \qquad \|\dot{\mathbf{r}}\| = \sqrt{\frac{81}{25}t^4 + \frac{144}{25}t^4 + 9} = 3\sqrt{t^4 + 1}$$

and $$\mathbf{T} = \frac{(-9t^2/5, 12t^2/5, 3)}{3\sqrt{t^4+1}} = \frac{(-3t^2, 4t^2, 5)}{5\sqrt{t^4+1}}$$

$$\ddot{\mathbf{r}} = \left(\frac{-18t}{5}, \frac{24t}{5}, 0\right)$$

$$\dot{\mathbf{r}} \times \ddot{\mathbf{r}} = \begin{vmatrix} \mathbf{i} & \mathbf{j} & \mathbf{k} \\ -\frac{9}{5}t^2 & \frac{12}{5}t^2 & 3 \\ -\frac{18t}{5} & \frac{24t}{5} & 0 \end{vmatrix} = \left(-\frac{72t}{5}, -\frac{54t}{5}, 0\right) = \frac{-18t}{5}(4, 3, 0)$$

$$\|\dot{\mathbf{r}} \times \ddot{\mathbf{r}}\| = \frac{18|t|}{5}\sqrt{4^2 + 3^2 + 0^2} = 18\,|t|$$

$$(\dot{\mathbf{r}}\dot{\mathbf{r}})\ddot{\mathbf{r}} = (9t^4 + 9)\left(-\frac{18}{5}t, \frac{24}{5}t, 0\right) = \left(-\frac{162}{5}t^5 - \frac{162}{5}t, \frac{216}{5}t^5 + \frac{216}{5}t, 0\right)$$

$$(\dot{\mathbf{r}}\ddot{\mathbf{r}})\dot{\mathbf{r}} = \left(\frac{9\cdot 18}{25}t^3 + \frac{24\cdot 12}{25}t^3 + 0\right)\left(-\frac{9}{5}t^2, \frac{12}{5}t^2, 3\right) = \left(-\frac{162}{5}t^5, \frac{216}{5}t^5, 54t^3\right)$$

$$(\dot{\mathbf{r}}\dot{\mathbf{r}})\ddot{\mathbf{r}} - (\dot{\mathbf{r}}\ddot{\mathbf{r}})\dot{\mathbf{r}} = \left(-\frac{162}{5}t, \frac{216}{5}t, -54t^3\right) = 18t\left(-\frac{9}{5}, \frac{12}{5}, -3t^2\right)$$

and $$\mathbf{N} = \varepsilon\,\frac{-18t(9/5, -12/5, 3t^2)}{(3\sqrt{t^4+1})(18|t|)} = -\frac{\varepsilon t}{|t|}\,\frac{(3, -4, 5t^2)}{5\sqrt{t^4+1}}$$

Now choose $\varepsilon = +1$ when $t < 0$ and -1 otherwise, making

$$\mathbf{N} = \frac{(3, -4, 5t^2)}{5\sqrt{t^4+1}} \qquad \mathbf{B} = \varepsilon\,\frac{(-18t/5)(4, 3, 0)}{18\,|t|} = \frac{-\varepsilon t}{|t|}\left(\frac{4}{5}, \frac{3}{5}, 0\right) = \left(\frac{4}{5}, \frac{3}{5}, 0\right)$$

10.4 Establish the general formulas (*10.5*) for the moving frame of the curve $\mathscr{C}\ :\ \mathbf{r} = \mathbf{r}(t)$, with an arbitrary parameter t.

By definition,

$$\frac{ds}{dt} = \|\dot{\mathbf{r}}\| \equiv (\dot{\mathbf{r}}\dot{\mathbf{r}})^{1/2} \qquad \text{or} \qquad \frac{dt}{ds} = \|\dot{\mathbf{r}}\|^{-1}$$

and we have at once for the unit tangent vector

$$\mathbf{T} = \mathbf{r}' = \dot{\mathbf{r}}\,\frac{dt}{ds} = \frac{\dot{\mathbf{r}}}{\|\dot{\mathbf{r}}\|}$$

To obtain a principal normal, first calculate

$$\frac{d}{dt}\,\|\dot{\mathbf{r}}\| \equiv \frac{d}{dt}\,(\dot{\mathbf{r}}\dot{\mathbf{r}})^{1/2} = \frac{1}{2}\,(\dot{\mathbf{r}}\dot{\mathbf{r}})^{-1/2}(\ddot{\mathbf{r}}\dot{\mathbf{r}} + \dot{\mathbf{r}}\ddot{\mathbf{r}}) = \frac{\dot{\mathbf{r}}\ddot{\mathbf{r}}}{\|\dot{\mathbf{r}}\|}$$

(Note the general formula $d\|\mathbf{u}\|/dt = \mathbf{u}\dot{\mathbf{u}}/\|\mathbf{u}\|$.) Hence,

$$\frac{d}{dt}\,\|\dot{\mathbf{r}}\|^{-1} = -\|\dot{\mathbf{r}}\|^{-2}\,\frac{d}{dt}\,\|\dot{\mathbf{r}}\| = -\frac{\dot{\mathbf{r}}\ddot{\mathbf{r}}}{\|\dot{\mathbf{r}}\|^3}$$

and

$$\dot{\mathbf{T}} = \ddot{\mathbf{r}}\,\frac{dt}{ds} + \dot{\mathbf{r}}\,\frac{d}{dt}\,\|\dot{\mathbf{r}}\|^{-1} = \frac{\ddot{\mathbf{r}}}{\|\dot{\mathbf{r}}\|} - \frac{(\dot{\mathbf{r}}\ddot{\mathbf{r}})\dot{\mathbf{r}}}{\|\dot{\mathbf{r}}\|^3} = \frac{\|\dot{\mathbf{r}}\|^2\ddot{\mathbf{r}} - (\dot{\mathbf{r}}\ddot{\mathbf{r}})\dot{\mathbf{r}}}{\|\dot{\mathbf{r}}\|^3}$$

$$\mathbf{T}' = \dot{\mathbf{T}}\,\frac{dt}{ds} = \frac{(\dot{\mathbf{r}}\dot{\mathbf{r}})\ddot{\mathbf{r}} - (\dot{\mathbf{r}}\ddot{\mathbf{r}})\dot{\mathbf{r}}}{\|\dot{\mathbf{r}}\|^4} = -\frac{\dot{\mathbf{r}} \times (\dot{\mathbf{r}} \times \ddot{\mathbf{r}})}{\|\dot{\mathbf{r}}\|^4}$$

where the last step used the vector identify $\mathbf{u} \times (\mathbf{v} \times \mathbf{w}) = (\mathbf{u}\mathbf{w})\mathbf{v} - (\mathbf{u}\mathbf{v})\mathbf{w}$. It follows that $\mathbf{N}$ can be constructed by normalizing the vector

$$\mathbf{N}^* \equiv -\dot{\mathbf{r}} \times (\dot{\mathbf{r}} \times \ddot{\mathbf{r}})$$

Since $\dot{\mathbf{r}}$ and $\dot{\mathbf{r}} \times \ddot{\mathbf{r}}$ are orthogonal, $\|\mathbf{N}^*\| = \|\dot{\mathbf{r}}\|\,\|\dot{\mathbf{r}} \times \ddot{\mathbf{r}}\|$ and so, provided $\dot{\mathbf{r}} \times \ddot{\mathbf{r}} \neq \mathbf{0}$,

$$\mathbf{N} = \varepsilon\,\frac{(\dot{\mathbf{r}} \times \ddot{\mathbf{r}}) \times \dot{\mathbf{r}}}{\|\dot{\mathbf{r}} \times \ddot{\mathbf{r}}\|\,\|\dot{\mathbf{r}}\|} = \varepsilon\,\frac{(\dot{\mathbf{r}}\dot{\mathbf{r}})\ddot{\mathbf{r}} - (\dot{\mathbf{r}}\ddot{\mathbf{r}})\dot{\mathbf{r}}}{\|\dot{\mathbf{r}}\|\,\|\dot{\mathbf{r}} \times \ddot{\mathbf{r}}\|}$$

Finally, for the binormal vector, with $\mathbf{v} \equiv \dot{\mathbf{r}} \times \ddot{\mathbf{r}} \neq \mathbf{0}$,

$$\mathbf{B} = \mathbf{T} \times \mathbf{N} = \frac{\dot{\mathbf{r}}}{\|\dot{\mathbf{r}}\|} \times \varepsilon\,\frac{\mathbf{v} \times \dot{\mathbf{r}}}{\|\mathbf{v}\|\,\|\dot{\mathbf{r}}\|} = \varepsilon\,\frac{(\dot{\mathbf{r}}\dot{\mathbf{r}})\mathbf{v} - (\dot{\mathbf{r}}\mathbf{v})\dot{\mathbf{r}}}{\|\dot{\mathbf{r}}\|^2\|\mathbf{v}\|} = \varepsilon\,\frac{\|\dot{\mathbf{r}}\|^2\mathbf{v} - (0)\dot{\mathbf{r}}}{\|\dot{\mathbf{r}}\|^2\|\mathbf{v}\|} = \varepsilon\,\frac{\mathbf{v}}{\|\mathbf{v}\|}$$

CURVATURE AND TORSION

10.5 Find the curvature and the torsion of the circular helix

$$\mathbf{r} = \left(a\cos\frac{s}{c},\ a\sin\frac{s}{c},\ \frac{bs}{c}\right) \qquad (c = \sqrt{a^2 + b^2})$$

where s is arc length.

By differentiation with respect to arc length,

$$\mathbf{T} = \mathbf{r}' = \left(-\frac{a}{c}\sin\frac{s}{c},\ \frac{a}{c}\cos\frac{s}{c},\ \frac{b}{c}\right) \qquad \mathbf{T}' = \left(-\frac{a}{c^2}\cos\frac{s}{c},\ -\frac{a}{c^2}\sin\frac{s}{c},\ 0\right)$$

Normalizing $\mathbf{T}'$, choose

$$\mathbf{N} = \left(-\cos\frac{s}{c},\ -\sin\frac{s}{c},\ 0\right)$$

and, correspondingly,

$$\mathbf{B} = \mathbf{T} \times \mathbf{N} = \begin{vmatrix} \mathbf{i} & \mathbf{j} & \mathbf{k} \\ -\frac{a}{c}\sin\frac{s}{c} & \frac{a}{c}\cos\frac{s}{c} & \frac{b}{c} \\ -\cos\frac{s}{c} & -\sin\frac{s}{c} & 0 \end{vmatrix} = \left(\frac{b}{c}\sin\frac{s}{c},\ -\frac{b}{c}\cos\frac{s}{c},\ \frac{a}{c}\right)$$

$$\mathbf{B}' = \left(\frac{b}{c^2}\cos\frac{s}{c}, \frac{b}{c^2}\sin\frac{s}{c}, 0\right)$$

Then, by (*10.6*),

$$\kappa = \frac{a}{c^2}\cos^2\frac{s}{c} + \frac{a}{c^2}\sin^2\frac{s}{c} + 0^2 = \frac{a}{c^2} \qquad \tau = \frac{b}{c^2}\cos^2\frac{s}{c} + \frac{b}{c^2}\sin^2\frac{s}{c} + 0^2 = \frac{b}{c^2}$$

[If we introduce the "time" parameter $t = cs$, we then have:

$$\frac{dz}{dt} = \frac{b}{c^2} = \tau$$

i.e. the rate at which the helix rises out of the xy-plane (its osculating plane at $t = 0$) is given by its (constant) torsion.]

10.6 Find the curvature and torsion of the curve $\mathbf{r} = (t^2 + t\sqrt{2}, t^2 - t\sqrt{2}, 2t^3/3)$ (t real).

Use the formulas (*10.8*):

$$\dot{\mathbf{r}} = (2t + \sqrt{2}, 2t - \sqrt{2}, 2t^2) \qquad \|\dot{\mathbf{r}}\| = \sqrt{(2t+\sqrt{2})^2 + (2t-\sqrt{2})^2 + 4t^4} = 2(t^2+1)$$

$$\ddot{\mathbf{r}} = (2, 2, 4t) \qquad \dddot{\mathbf{r}} = (0, 0, 4)$$

$$\dot{\mathbf{r}} \times \ddot{\mathbf{r}} = \begin{vmatrix} \mathbf{i} & \mathbf{j} & \mathbf{k} \\ 2t+\sqrt{2} & 2t-\sqrt{2} & 2t^2 \\ 2 & 2 & 4t \end{vmatrix} = 4(t^2 - t\sqrt{2}, -(t^2 + t\sqrt{2}), \sqrt{2})$$

$$(\dot{\mathbf{r}} \times \ddot{\mathbf{r}})^2 = 16[(t^2 - t\sqrt{2})^2 + (t^2 + t\sqrt{2})^2 + 2] = 32(t^2+1)^2$$

$$\det[\dot{\mathbf{r}}\,\ddot{\mathbf{r}}\,\dddot{\mathbf{r}}] = \dddot{\mathbf{r}} \cdot (\dot{\mathbf{r}} \times \ddot{\mathbf{r}}) = (0,0,4) \cdot 4(t^2 - t\sqrt{2}, -t^2 - t\sqrt{2}, \sqrt{2}) = 16\sqrt{2}$$

Hence

$$\kappa = \frac{\varepsilon\sqrt{32(t^2+1)^2}}{8(t^2+1)^3} = \frac{\varepsilon}{\sqrt{2}\,(t^2+1)^2} \qquad \tau = \frac{16\sqrt{2}}{32(t^2+1)^2} = \frac{1}{\sqrt{2}\,(t^2+1)^2}$$

10.7 Prove (*10.8*).

Using the results of Problem 10.4, we have

$$\kappa = \mathbf{N}\mathbf{T}' = \left(\varepsilon\,\frac{(\dot{\mathbf{r}} \times \ddot{\mathbf{r}}) \times \dot{\mathbf{r}}}{\|\dot{\mathbf{r}}\|\,\|\dot{\mathbf{r}} \times \ddot{\mathbf{r}}\|}\right) \cdot \left(-\frac{\dot{\mathbf{r}} \times (\dot{\mathbf{r}} \times \ddot{\mathbf{r}})}{\|\dot{\mathbf{r}}\|^4}\right) = \varepsilon\,\frac{\|\dot{\mathbf{r}} \times (\dot{\mathbf{r}} \times \ddot{\mathbf{r}})\|^2}{\|\dot{\mathbf{r}}\|^5\,\|\dot{\mathbf{r}} \times \ddot{\mathbf{r}}\|}$$

$$= \varepsilon\,\frac{\|\dot{\mathbf{r}}\|^2\,\|\dot{\mathbf{r}} \times \ddot{\mathbf{r}}\|^2 \sin^2(\pi/2)}{\|\dot{\mathbf{r}}\|^5\,|\dot{\mathbf{r}} \times \ddot{\mathbf{r}}\|} = \varepsilon\,\frac{\|\dot{\mathbf{r}} \times \ddot{\mathbf{r}}\|}{\|\dot{\mathbf{r}}\|^3}$$

The torsion requires the computation of $\mathbf{B}'$. By (*10.5*),

$$\varepsilon\mathbf{B} = \frac{\dot{\mathbf{r}} \times \ddot{\mathbf{r}}}{\|\dot{\mathbf{r}} \times \ddot{\mathbf{r}}\|} \equiv \frac{\mathbf{v}}{\|\mathbf{v}\|}$$

whence

$$\varepsilon\dot{\mathbf{B}} = \frac{d}{dt}\left(\frac{\mathbf{v}}{\|\mathbf{v}\|}\right) = \frac{1}{\|\mathbf{v}\|}\,\dot{\mathbf{v}} + \frac{d}{dt}\left(\frac{1}{\|\mathbf{v}\|}\right)\mathbf{v} = \frac{\dot{\mathbf{v}}}{\|\mathbf{v}\|} - \frac{(\mathbf{v}\dot{\mathbf{v}})\mathbf{v}}{\|\mathbf{v}\|^3} = \frac{\|\mathbf{v}\|^2\dot{\mathbf{v}} - (\mathbf{v}\dot{\mathbf{v}})\mathbf{v}}{\|\mathbf{v}\|^3}$$

But $\dot{\mathbf{v}} = d(\dot{\mathbf{r}} \times \ddot{\mathbf{r}})/dt = (\ddot{\mathbf{r}} \times \ddot{\mathbf{r}}) + (\dot{\mathbf{r}} \times \dddot{\mathbf{r}}) = \dot{\mathbf{r}} \times \dddot{\mathbf{r}}$; hence,

$$\varepsilon\mathbf{B}' = \frac{\varepsilon\dot{\mathbf{B}}}{\|\dot{\mathbf{r}}\|} = \frac{\|\dot{\mathbf{r}} \times \ddot{\mathbf{r}}\|^2(\dot{\mathbf{r}} \times \dddot{\mathbf{r}}) - [(\dot{\mathbf{r}} \times \ddot{\mathbf{r}})(\dot{\mathbf{r}} \times \dddot{\mathbf{r}})](\dot{\mathbf{r}} \times \ddot{\mathbf{r}})}{\|\dot{\mathbf{r}}\|\,\|\dot{\mathbf{r}} \times \ddot{\mathbf{r}}\|^3}$$

Dot this with

$$\varepsilon\mathbf{N} = \frac{(\dot{\mathbf{r}}\dot{\mathbf{r}})\ddot{\mathbf{r}} - (\dot{\mathbf{r}}\ddot{\mathbf{r}})\dot{\mathbf{r}}}{\|\dot{\mathbf{r}}\|\,\|\dot{\mathbf{r}} \times \ddot{\mathbf{r}}\|}$$

from (*10.5*), and use $\mathbf{u}\cdot(\mathbf{u}\times\mathbf{w})=0$:

$$-\tau = \mathbf{NB}' = \epsilon^2\,\mathbf{NB} = \frac{(\dot{\mathbf{r}}\ddot{\mathbf{r}}\,\|\dot{\mathbf{r}}\times\ddot{\mathbf{r}}\|^2[\dot{\mathbf{r}}\cdot(\dot{\mathbf{r}}\times\dddot{\mathbf{r}})] - 0 - 0 + 0}{\|\dot{\mathbf{r}}\|^2\,\|\dot{\mathbf{r}}\times\ddot{\mathbf{r}}\|^4} = \frac{\|\dot{\mathbf{r}}\|^2(-\det[\dot{\mathbf{r}}\,\ddot{\mathbf{r}}\,\dddot{\mathbf{r}}])}{\|\dot{\mathbf{r}}\|^2\,\|\dot{\mathbf{r}}\times\ddot{\mathbf{r}}\|^2}$$

or

$$\tau = \frac{\det[\dot{\mathbf{r}}\,\ddot{\mathbf{r}}\,\dddot{\mathbf{r}}]}{\|\dot{\mathbf{r}}\times\ddot{\mathbf{r}}\|^2}$$

10.8 Prove (*a*) $\mathbf{N}' = -\kappa\mathbf{T} + \tau\mathbf{B}$, (*b*) $\mathbf{B}' = -\tau\mathbf{N}$.

(*a*) Since $\mathbf{NN}=1$, $(\mathbf{NN})' = 2\mathbf{N}'\mathbf{N} = 0$ and $\mathbf{N}'$ is orthogonal to $\mathbf{N}$, which puts it in the place of $\mathbf{T}$ and $\mathbf{B}$. Therefore, for certain real λ and μ,

$$\mathbf{N}' = \lambda\mathbf{T} + \mu\mathbf{B} \tag{1}$$

Dot both sides by $\mathbf{T}$, then by $\mathbf{B}$, and use $\mathbf{TN}=0$, $\mathbf{T}'\mathbf{N}+\mathbf{TN}'=0$, $\mathbf{N}'\mathbf{B}+\mathbf{NB}'=0$, $\kappa=\mathbf{NT}'$, $\tau=-\mathbf{NB}'$:

$$\mathbf{TN}' = \lambda\mathbf{T}^2 + \mu\mathbf{TB} = \lambda \qquad\text{or}\qquad \lambda = -\mathbf{T}'\mathbf{N} = -\kappa$$
$$\mathbf{BN}' = \tau = \lambda\mathbf{BT} + \mu\mathbf{B}^2 = \mu$$

Substitution for λ and μ in (*1*) then yields the desired results.

(*b*) From $\mathbf{B} = \mathbf{T}\times\mathbf{N}$ and part (*a*),

$$\mathbf{B}' = \mathbf{T}'\times\mathbf{N} + \mathbf{T}\times\mathbf{N}' = (\kappa\mathbf{N})\times\mathbf{N} + \mathbf{T}\times(-\kappa\mathbf{T}+\tau\mathbf{B})$$
$$= 0 + 0 + \tau(\mathbf{T}\times\mathbf{B}) = \tau(-\mathbf{N}) = -\tau\mathbf{N}$$

10.9 Prove that if a curve has $\kappa' = 0$ at some point, then $\mathbf{N}''$ is orthogonal to $\mathbf{T}$ at that point.

From $\mathbf{N}' = -\kappa\mathbf{T} + \tau\mathbf{B}$, it follows that $\mathbf{N}'' = -\kappa'\mathbf{T} - \kappa\mathbf{T}' + \tau'\mathbf{B} + \tau\mathbf{B}'$. But $\kappa' = 0$; and from the Serret–Frenet formulas for $\mathbf{T}'$ and $\mathbf{B}'$ we obtain

$$\mathbf{N}'' = -\kappa(\kappa\mathbf{N}) + \tau'\mathbf{B} + \tau(-\tau\mathbf{N}) = (-\kappa^2 - \tau^2)\mathbf{N} + \tau'\mathbf{B}$$

As $\mathbf{N}''$ is in the plane of $\mathbf{N}$ and $\mathbf{B}$, it is orthogonal to $\mathbf{T}$.

SURFACES IN EUCLIDEAN SPACE

10.10 Show that a surface of revolution is regular and exhibit the unit surface normal.

The Gaussian form of a surface of revolution about the z-axis (Fig. 10-11) is

$$\mathbf{r} = (f(x^1)\cos x^2,\ f(x^1)\sin x^2,\ g(x^1)) \qquad (f(x^1) > 0)$$

so

$$\mathbf{r}_1 = (f'(x^1)\cos x^2,\ f'(x^1)\sin x^2,\ g'(x^1)) \qquad \mathbf{r}_2 = (-f(x^1)\sin x^2,\ f(x^1)\cos x^2,\ 0)$$

and $\mathbf{r}_1\times\mathbf{r}_2 = (-fg'\cos x^2,\ -fg'\sin x^2,\ ff'(\cos^2 x^2 + \sin^2 x^2))$, with norm

$$E = \sqrt{f^2g'^2\cos^2 x^2 + f^2g'^2\sin^2 x^2 + f^2f'^2} = f\sqrt{f'^2 + g'^2}$$

Now $f = f(x^1) \neq 0$; further, the generating curve is regular, which means that, with $t = x^1$, the tangent vector of that curve,

$$\left(\frac{dx}{dt},\ 0,\ \frac{dz}{dt}\right) = (f',\ 0,\ g')$$

is non-null and $f'^2 + g'^2 \neq 0$. Therefore, $E \neq 0$ and the surface is regular.

The unit surface normal is

$$\mathbf{n} = \frac{1}{E}(\mathbf{r}_1\times\mathbf{r}_2) = \left(-\frac{g'}{\sqrt{f'^2+g'^2}}\cos x^2,\ -\frac{g'}{\sqrt{f'^2+g'^2}}\sin x^2,\ \frac{f'}{\sqrt{f'^2+g'^2}}\right)$$

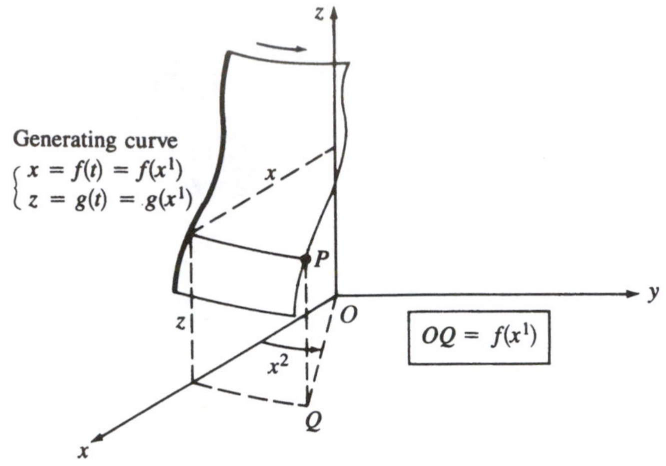

Fig. 10-11

10.11 Identify the x^1- and x^2-curves for the right helicoid (Example 10.4) and describe the behavior of the unit surface normal along an x^1-curve.

The x^1-curves ($x^2 = \text{const.}$) are given by

$$\mathbf{r} = (0, 0, ax^2) + x^1(\cos x^2, \sin x^2, 0) \qquad (x^1 \geqq 0)$$

thus, they are rays parallel to the xy-plane. The x^2-curves ($x^1 = \text{const.}$) are given by

$$\sqrt{x^2 + y^2} = x^1 \qquad z = ax^2$$

i.e., circular helices of radii x^1.

We have:

$$\mathbf{r}_1 = (\cos x^2, \sin x^2, 0) \qquad \mathbf{r}_2 = (-x^1 \sin x^2, x^1 \cos x^2, a)$$

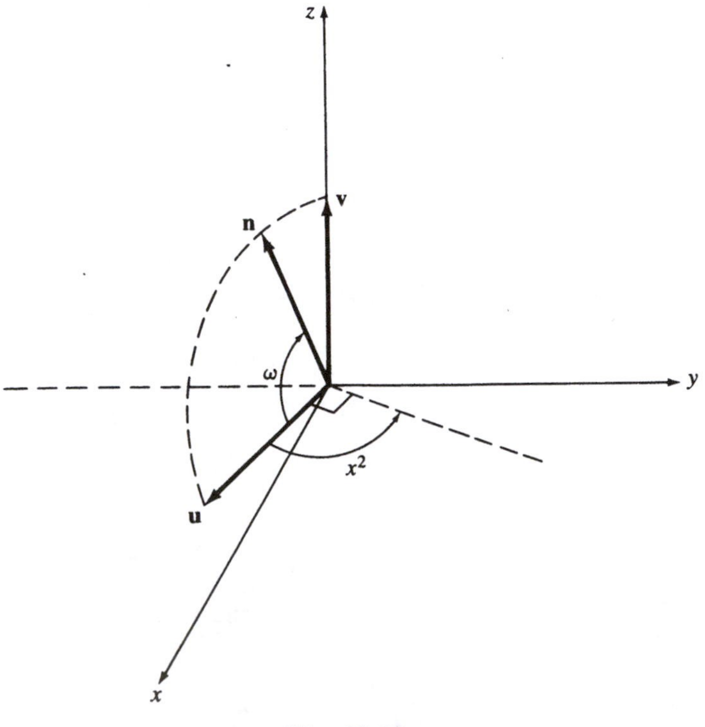

Fig. 10-12

$$\mathbf{r}_1 \times \mathbf{r}_2 = \begin{vmatrix} \mathbf{i} & \mathbf{j} & \mathbf{k} \\ \cos x^2 & \sin x^2 & 0 \\ -x^1 \sin x^2 & x^1 \cos x^2 & a \end{vmatrix} = (a \sin x^2, -a \cos x^2, x^1)$$

and
$$\mathbf{n} = \frac{\mathbf{r}_1 \times \mathbf{r}_2}{\|\mathbf{r}_1 \times \mathbf{r}_2\|} = \left(\frac{a \sin x^2}{\sqrt{a^2 + (x^1)^2}}, \frac{-a \cos x^2}{\sqrt{a^2 + (x^1)^2}}, \frac{x^1}{\sqrt{a^2 + (x^1)^2}} \right)$$
$$= (\cos \omega)\mathbf{u} + (\sin \omega)\mathbf{v}$$

where $\omega \equiv \tan^{-1}(x^1/a)$, $\mathbf{u} \equiv (\sin x^2, -\cos x^2, 0)$, $\mathbf{v} \equiv (0, 0, 1)$. On an $\mathbf{x}^1$-ray, $\mathbf{u}$ and $\mathbf{v}$ are fixed unit vectors, while ω increases from 0 to a $\pi/2$ as x^1 increases from 0 to ∞. Thus, $\mathbf{n}$ traces out a quarter-circle as the ray is described (see Fig. 10-12).

10.12 Find the FFF for any surface of revolution, and specialize to a right circular cone.

With $\mathbf{r}_1$ and $\mathbf{r}_2$ as obtained in Problem 10.10,

$$g_{11} = \mathbf{r}_1\mathbf{r}_1 = (f' \cos x^2)^2 + (f' \sin x^2)^2 + (g')^2 = f'^2 + g'^2$$
$$g_{12} = g_{21} = \mathbf{r}_1\mathbf{r}_2 = -f'f \cos x^2 \sin x^2 + f'f \sin x^2 \cos x^2 + (g')(0) = 0$$
$$g_{22} = \mathbf{r}_2\mathbf{r}_2 = (-f \sin x^2)^2 + (f \cos x^2)^2 + 0^2 = f^2$$

and

$$\mathrm{I} = (f'^2 + g'^2)(dx^1)^2 + f^2(dx^2)^2 \qquad (1)$$

For a right circular cone (Fig. 10-13), $f = x^1$ and $g = ax^1$; hence,

$$\mathrm{I} = (1 + a^2)(dx^1)^2 + (x^1)^2(dx^2)^2 \qquad (2)$$

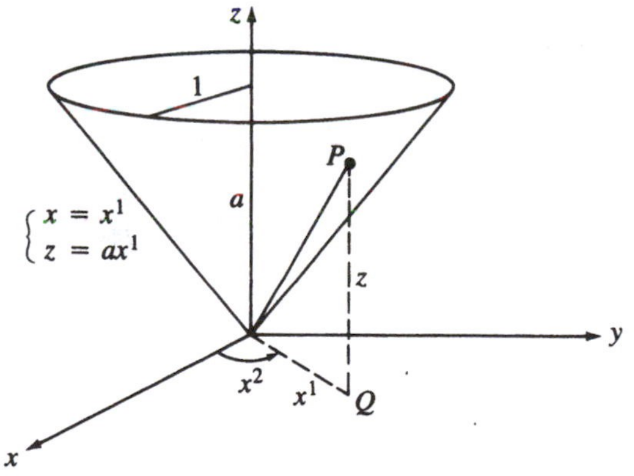

Fig. 10-13

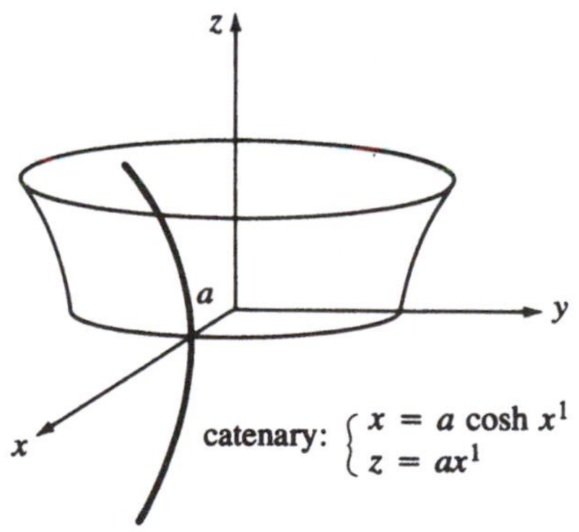

Fig. 10-14

10.13 Find the FFF for the catenoid (Fig. 10-14) and compute the length of the curve given by $x^1 = t$, $x^2 = t$ $\quad (0 \leqq t \leqq \ln(1 + \sqrt{2}))$.

Here $f(x^1) = a \cosh x^1$, $g(x^1) = ax^1$, and (1) of Problem 10.12 gives, along the curve,

$$\left(\frac{ds}{dt}\right)^2 = (a^2 \cosh^2 x^1)\left(\frac{dx^1}{dt}\right)^2 + (a^2 \cosh^2 x^1)\left(\frac{dx^2}{dt}\right)^2 = 2a^2 \cosh^2 t$$

and

$$L = a\sqrt{2} \int_0^{\ln(1+\sqrt{2})} \cosh t \, dt = a\sqrt{2} \sinh[\ln(1 + \sqrt{2})] = a\sqrt{2}$$

10.14 Let $\mathscr{C}_1$ and $\mathscr{C}_2$ be two curves on the right circular cone $\mathbf{r} = (x^1 \cos x^2, x^1 \sin x^2, 2x^1)$ whose pre-images in the parameter plane are

$$\mathscr{C}_1 : \begin{cases} x^1 = 3 - t \\ x^2 = t/2 \end{cases} \qquad \mathscr{C}_2 : \begin{cases} x^1 = \sigma > 0 \\ x^2 = \sigma^2 \end{cases}$$

At the point of intersection, find the angle between $\mathscr{C}_1$ and $\mathscr{C}_2$, and show that orthogonality in the x^1x^2-plane does not carry over to the cone.

The intersection point P' of the two pre-image curves is determined by the simultaneous equations

$$3 - t = \sigma \qquad \text{and} \qquad \frac{t}{2} = \sigma^2$$

which give $t = 2$, $\sigma = 1$, and $P' = (1, 1)$. Thus, the two tangent vectors at P' are:

$$(u^i) = \left(\frac{dx^i}{dt}\right)\bigg|_{t=2} = (-1, \tfrac{1}{2}) \qquad (v^i) = \left(\frac{dx^i}{d\sigma}\right)\bigg|_{\sigma=1} = (1, 2)$$

Considered in the Euclidean sense, (u^i) and (v^i) are orthogonal.

To express the angle between tangents at the image of P', we adopt the metric (*2*) of Problem 10.12 (with $a = 2$) and apply (*10.18*) for $x^1 = 1$, $x^2 = 1$:

$$\cos\theta = \frac{(1 + 2^2)(-1)(1) + (1)^2(\tfrac{1}{2})(2)}{D} = \frac{-4}{D} \neq 0$$

Therefore, the curves are not orthogonal at the image of P'.

10.15 Prove Theorem 10.3 and verify Corollary 10.4 geometrically for the right helicoid (Example 10.4) and for any surface of revolution (Problem 10.12).

The proof consists merely in taking $(u^i) = (1, 0)$ and $(v^i) = (0, 1)$ in (*10.18*). (Compare Problem 5.31.)

As is clear from Problem 10.11, the right helicoid is a *ruled surface*, generated by a half-line (an x^1-curve), pivoted on the z-axis, that rotates parallel to the xy-plane while the pivot point travels up the z-axis. A given point P of the generator thus describes a helical x^2-curve (Fig. 10-8), which is necessarily everywhere orthogonal to the generator (i.e., to the x^1-curves). As for surfaces of revolution, it is clear that the parameter lines that match the revolved planar curve (x^1-curves, or *meridians*) and the circles traced by individual points of the planar curve (x^2-curves, or *parallels of latitude*) are mutually orthogonal. By previous computations, $g_{12} = 0$ for both the helicoid and for the general surface of revolution.

10.16 Show that under a change of coordinates $x^1 = x^1(\bar{x}^1, \bar{x}^2)$, $x^2 = x^2(\bar{x}^1, \bar{x}^2)$ in the plane, the surface metric (g_{ij}) transforms as a second-order covariant tensor.

We have by substitution $\mathbf{r}(x^1, x^2) = \mathbf{r}(x^1(\bar{x}^1, \bar{x}^2), x^2(\bar{x}^1, \bar{x}^2)) \equiv \bar{\mathbf{r}}(\bar{x}^1, \bar{x}^2)$, the latter being the "new" parameterization for $\mathscr{S}$. To compute the metric under this parameterization, write (by the chain rule for partial derivatives and the bilinearity of the inner product)

$$\bar{g}_{ij} \equiv \bar{\mathbf{r}}_i \bar{\mathbf{r}}_j \equiv \frac{\partial \bar{\mathbf{r}}}{\partial \bar{x}^i}\frac{\partial \bar{\mathbf{r}}}{\partial \bar{x}^j} = \left(\frac{\partial \mathbf{r}}{\partial x^p}\frac{\partial x^p}{\partial \bar{x}^i}\right)\cdot\left(\frac{\partial \mathbf{r}}{\partial x^q}\frac{\partial x^q}{\partial \bar{x}^j}\right) = \left(\mathbf{r}_p \frac{\partial x^p}{\partial \bar{x}^i}\right)\cdot\left(\mathbf{r}_q \frac{\partial x^q}{\partial \bar{x}^j}\right)$$

$$= \mathbf{r}_p \mathbf{r}_q \frac{\partial x^p}{\partial \bar{x}^i}\frac{\partial x^q}{\partial \bar{x}^j} \equiv g_{pq}\frac{\partial x^p}{\partial \bar{x}^i}\frac{\partial x^q}{\partial \bar{x}^j}$$

which is the correct formula for tensor character.

GEODESICS

10.17 (*a*) Find the Christoffel symbols of the second kind for the sphere of radius a. (*b*) Verify that the great circles passing through the north and south poles (i.e., the x^1-curves) are geodesics.

(*a*) The FFF for the sphere of radius a may be calculated from Problem 10.12:

$$g_{11} = a^2 \qquad g_{12} = 0 = g_{21} \qquad g_{22} = a^2 \sin^2 x^1$$

The formulas from Problem 6.4 can be used, since (g_{ij}) is diagonal; the nonzero Christoffel symbols are found to be:

$$\Gamma^1_{22} = -\sin x^1 \cos x^1 \qquad \Gamma^2_{12} = \Gamma^2_{21} = \cot x^1$$

(*b*) We want to show that the family of curves $x^1 = t$, $x^2 = d = \text{const.}$ are integral curves of the differential system (*7.11*)–(*7.12*), which may be conveniently written as

$$\frac{d^2x^i}{dt^2} + \Gamma^i_{jk}\frac{dx^j}{dt}\frac{dx^k}{dt} = \frac{1}{2}\frac{dx^i}{dt}\left[\frac{d}{dt}\ln\left(g_{jk}\frac{dx^j}{dt}\frac{dx^k}{dt}\right)\right]$$

or, for the given metric,

$$i = 1 \qquad \frac{d^2x^1}{dt^2} - (\sin x^1 \cos x^1)\left(\frac{dx^2}{dt}\right)^2 = \frac{1}{2}\frac{dx^1}{dt}\left[\frac{d}{dt}\ln\left(a^2\left(\frac{dx^1}{dt}\right)^2 + (a^2\sin^2 x^1)\left(\frac{dx^2}{dt}\right)^2\right)\right]$$

$$i = 2 \qquad \frac{d^2x^2}{dt^2} + (2\cot x^1)\frac{dx^1}{dt}\frac{dx^2}{dt} = \frac{1}{2}\frac{dx^2}{dt}\left[\frac{d}{dt}\ln \mathrm{I}\right]$$

Since $dx^1/dt = 1$ and $dx^2/dt = 0$, both equations reduce to $0 = 0$, and the verification is complete.

10.18 Prove Theorem 10.6: A curve on a regular surface is a geodesic if and only if, by proper choice of the principal normal, $\mathbf{N} = \mathbf{n}$.

Let any curve on the surface be given by $\mathscr{C}\ :\ \mathbf{r} = \mathbf{r}(x^1(s), x^2(s))$, where s = arc length, Then,

$$\mathbf{T} = \mathbf{r}_i \frac{dx^i}{ds}$$

and the first formula (*10.9*) gives

$$\kappa\mathbf{N} = \mathbf{T}' = \frac{d^2x^i}{ds^2}\mathbf{r}_i + \frac{dx^i}{ds}\left(\frac{\partial \mathbf{r}_i}{\partial x^j}\frac{dx^j}{ds}\right) \equiv \frac{d^2x^i}{ds^2}\mathbf{r}_i + \frac{dx^i}{ds}\frac{dx^j}{ds}\mathbf{r}_{ij} \tag{1}$$

Dot both sides of (*1*) by the vector $\mathbf{r}_k$ and use the result of Problem 10.48:

$$\kappa\mathbf{r}_k\mathbf{N} = \frac{d^2x^i}{ds^2}\mathbf{r}_i\mathbf{r}_k + \frac{dx^i}{ds}\frac{dx^j}{ds}\mathbf{r}_{ij}\mathbf{r}_k \equiv \frac{d^2x^i}{ds^2}g_{ik} + \frac{dx^i}{ds}\frac{dx^j}{ds}\Gamma_{ijk} \tag{2}$$

Multiply both sides of (*2*) by g^{kl} and sum on k:

$$g^{kl}\kappa\mathbf{r}_k\mathbf{N} = \frac{d^2x^i}{ds^2}\delta^l_i + \frac{dx^i}{ds}\frac{dx^j}{ds}g^{kl}\Gamma_{ijk} = \frac{d^2x^l}{ds^2} + \Gamma^l_{ij}\frac{dx^i}{ds}\frac{dx^j}{ds} \tag{3}$$

Now if $\mathscr{C}$ is a geodesic, the right side of (*3*) vanishes, and this implies that $\kappa\mathbf{r}_k\mathbf{N} = 0$ for $k = 1, 2$. If $\kappa \neq 0$, then $\mathbf{r}_1\mathbf{N} = 0 = \mathbf{r}_2\mathbf{N}$; so that $\mathbf{N}$ is orthogonal to both $\mathbf{r}_1$ and $\mathbf{r}_2$ (thus to the tangent plane). Therefore, but for orientation, $\mathbf{N} = \mathbf{n}$. If $\kappa = 0$ at some point P and there is a sequence of points along the curve approaching P for which $\kappa \neq 0$, then, by continuity, $\mathbf{N} = \mathbf{n}$. Otherwise, $\kappa = 0$ on an interval and the curve is a straight line on that interval, in which case its principal normal $\mathbf{N}$ can be chosen to agree with $\mathbf{n}$. Conversely, if the curve has the property that $\mathbf{N} = \mathbf{n}$ at all points, then $\mathbf{r}_k\mathbf{N} = \mathbf{r}_k\mathbf{n} = 0$ and the left side of (*3*) vanishes, showing that the pre-image of $\mathscr{C}$ satisfies the differential equations for a geodesic.

10.19 Apply Theorem 10.6 to the plane sections of a torus.

In Fig. 10-15 is shown a torus and various examples of plane sections. In (*a*), an elliptical-shaped vertical section, the section cannot be a geodesic because the surface normal at P does not lie in the plane of the curve (which contains the curve normal). In (*b*), a horizontal section that is a circle, again the surface normals do not lie in the plane, and this section is not a geodesic. The circles shown in (*c*) and (*d*) are geodesics, since the surface normal will coincide with a correctly chosen principal normal of the curve.

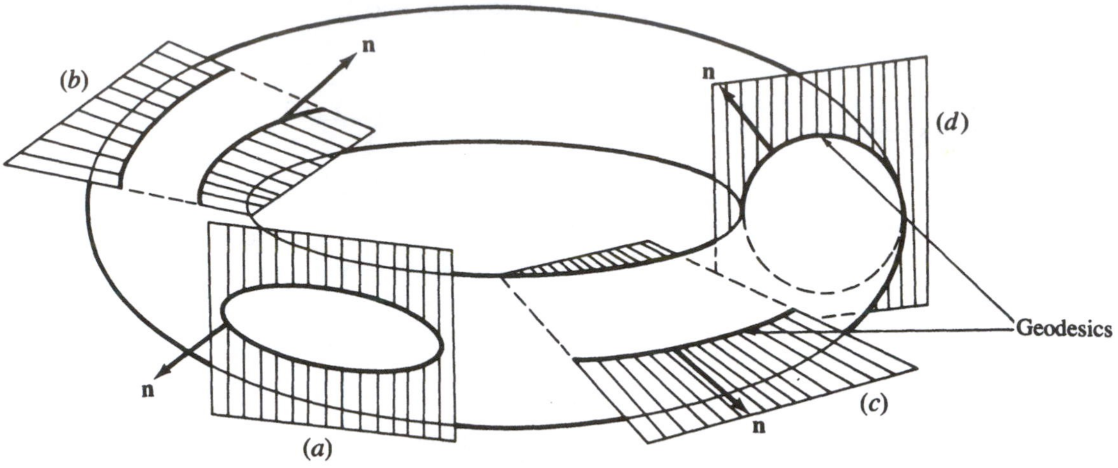

Fig. 10-15

10.20 Demonstrate that the intrinsic curvature $\tilde{\kappa}$ of a curve on a surface can be different from its curvature κ as a curve in $\mathbf{E}^3$.

One example is a circle on a sphere of radius a. If the circle also has radius a, it is a great circle and, hence, a geodesic with zero intrinsic curvature. But its curvature as a (planar) curve in $\mathbf{E}^3$ is $1/a$. Another example is the circular helix: its curvature is nonzero as a curve in $\mathbf{E}^3$, but as a geodesic on a circular cylinder its intrinsic curvature is zero.

SECOND FUNDAMENTAL FORM

10.21 (*a*) Find the SFF for the right circular cone of Problem 10.12. (*b*) At the point $P(1, 0, a)$, calculate the curvature of the normal section having the direction $\mathbf{j}$ at P (see Fig. 10-16).

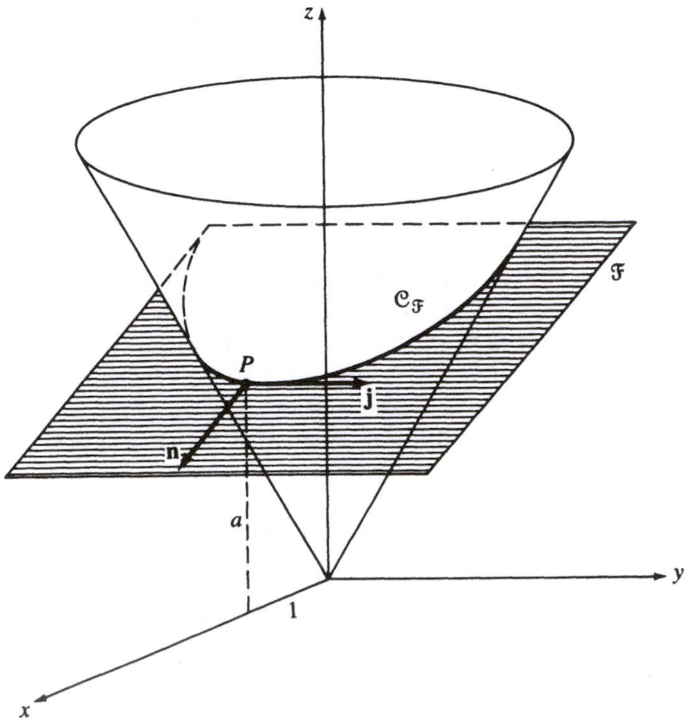

Fig. 10-16

(*a*) From $\mathbf{r} = (x^1 \cos x^2, x^1 \sin x^2, ax^1)$ $(x^1 > 0)$, we obtain:

$$\mathbf{r}_1 = (\cos x^2, \sin x^2, a) \qquad \mathbf{r}_2 = (-x^1 \sin x^2, x^1 \cos x^2, 0)$$
$$\mathbf{r}_{11} = (0, 0, 0) \qquad \mathbf{r}_{12} = \mathbf{r}_{21} = (-\sin x^2, \cos x^2, 0) \qquad \mathbf{r}_{22} = (-x^1 \cos x^2, -x^1 \sin x^2, 0)$$

and by Problem 10.10, $\mathbf{n} = (a^2+1)^{-1/2}(-a \cos x^2, -a \sin x^2, 1)$. The coefficients in II are thus $f_{11} \equiv \mathbf{n}\mathbf{r}_{11} = 0$, $f_{12} = f_{21} \equiv \mathbf{n}\mathbf{r}_{12} = 0$, $f_{22} \equiv \mathbf{n}\mathbf{r}_{22} = (a^2+1)^{-1/2} ax^1$.

(*b*) The direction $\mathbf{j}$ at P corresponds to the direction (u^1, u^2) at $P' = (1, 0)$ in the parameter plane, where

$$\mathbf{j} = u^1\mathbf{r}_1(P) + u^2\mathbf{r}_2(P)$$
$$(0,1,0) = u^1(1,0,a) + u^2(0,1,0)$$
$$(0,1,0) = (u^1, u^2, au^1)$$

Thus, $u^1 = 0$ and $u^2 = 1$. Appropriating I from Problem 10.12, we have

$$\kappa_{\mathscr{F}} = \frac{\mathrm{II}(u^1, u^2)}{\mathrm{I}(u^1, u^2)} = \frac{f_{22}(P)(u^2)^2}{(a^2+1)(u^1)^2 + (1)^2(u^2)^2} = f_{22}(P) = \frac{a}{\sqrt{a^2+1}}$$

10.22 Develop geometrically a notion of "parallel transport" of a vector along a curve $\mathscr{C}$ on a regular surface $\mathscr{S}$.

Imagine $\mathscr{C}$, and with it $\mathscr{S}$, as being rolled without slipping onto a fixed plane $\mathscr{F}$, in such fashion that the point of contact is aways on $\mathscr{C}$ and the tangent plane to $\mathscr{S}$ at the point of contacts always coincides with $\mathscr{F}$. This maps $\mathscr{C}$ to a (planar) curve $\mathscr{C}^*$ in $\mathscr{F}$ that has the same arc-length parameter and the same tangent vector. Then, any vector in $\mathscr{F}$ that is attached to the point of contact and that remains parallel to itself (in the ordinary Euclidean sense) as the contact point describes $\mathscr{C}^*$ may—under the inverse mapping $\mathscr{C}^* \to \mathscr{C}$—be considered as undergoing *parallel transport along* $\mathscr{C}$. In general, parallel transport of a given vector around a closed curve on a surface *does not* reproduce the initial vector.

10.23 Prove (*10.23*).

Start with the formula for the unit tangent vector of any curve $\mathscr{C}$ on $\mathscr{S}$:

$$\mathbf{T} = \frac{u^i\mathbf{r}_i}{\sqrt{g_{kl}u^k u^l}} \qquad \left(u^i \equiv \frac{dx^i}{dt}\right)$$

Then

$$\dot{\mathbf{T}} = \frac{d}{dt}\left(\frac{u^i}{\sqrt{g_{kl}u^k u^l}}\right)\mathbf{r}_i + \frac{u^i}{\sqrt{g_{kl}u^k u^l}}\dot{\mathbf{r}}_i = Q^i\mathbf{r}_i + \frac{u^i}{\sqrt{g_{kl}u^k u^l}}\mathbf{r}_{ij}u^j \tag{1}$$

where Q^i is an abbreviation for the scalar coefficient of $\mathbf{r}_i$. Now the Frenet formula gives $\kappa\mathbf{N} = \mathbf{T}' = \dot{\mathbf{T}}/\sqrt{g_{kl}u^k u^l}$; together with (*1*), this yields:

$$\kappa\mathbf{N} = \frac{Q_i}{\sqrt{g_{kl}u^k u^l}}\mathbf{r}_i + \frac{u^i u^j}{g_{kl}u^k u^l}\mathbf{r}_{ij} \tag{2}$$

Dot both sides of (*2*) with $\mathbf{n}$ (the surface normal) and use the fact that $\mathbf{r}_i\mathbf{n} = 0$ for each i:

$$\kappa\mathbf{n}\mathbf{N} = \frac{u^i u^j}{g_{kl}u^k u^l}\mathbf{r}_{ij}\mathbf{n} \equiv \frac{u^i u^j}{g_{kl}u^k u^l} f_{ij} \tag{3}$$

If $\mathscr{C}$ is a normal section $\mathscr{C}_{\mathscr{F}}$ at P, and κ, $\mathbf{N}$, and the right side of (*3*) are all evaluated at P, then $\kappa = \kappa_{\mathscr{F}}$, $\mathbf{n}\mathbf{N} = \mathbf{n}^2 = 1$, and (*3*) becomes the desired expression

$$\kappa_{\mathscr{F}} = \frac{f_{ij}u^i u^j}{g_{kl}u^k u^l}$$

CURVATURE OF SURFACES

10.24 Show that the maximum and minimum values of the function

$$F(\mathbf{u}) = \frac{a_{ij}u^i u^j}{b_{kl}u^k u^l} = \frac{\mathbf{u}^T A\mathbf{u}}{\mathbf{u}^T B\mathbf{u}}$$

where $A = [a_{ij}]_{22}$, $B = [b_{ij}]_{22}$, $\mathbf{u} = (u^1, u^2) \neq (0, 0)$, with B positive definite, are the two roots of the quadratic equation in λ

$$\det(A - \lambda B) \equiv \begin{vmatrix} a_{11} - \lambda b_{11} & a_{12} - \lambda b_{12} \\ a_{21} - \lambda b_{21} & a_{22} - \lambda b_{22} \end{vmatrix} = 0 \qquad (1)$$

(hence, eigenvalues of $B^{-1}A$), and that the extreme values of F occur for vectors $\mathbf{u}$ satisfying $(A - xB)\mathbf{u} = \mathbf{0}$, where x takes on the two eigenvalues of $B^{-1}A$ (hence, eigenvectors of $B^{-1}A$).

We may assume without loss of generality that A and B are symmetric. Let $\mathscr{G}$ be any simple closed curve in the u^1u^2-plane having the origin in its interior. The Weierstrass theorem guarantees that $F(\mathbf{u})$ asumes a largest value on $\mathscr{G}$; say, $F(\mathbf{w}) = M$. Because F is constant on rays emanating from the origin ($F(\lambda\mathbf{u}) = F(\mathbf{u})$ for any $\lambda \neq 0$), the absolute maximum on $\mathscr{G}$ is both an absolute and a relative maximum in the u^1u^2-plane; hence, the gradient of F must vanish at $\mathbf{w}$. We have:

$$\frac{\partial F(\mathbf{u})}{\partial u^p} = \frac{(b_{kl}u^k u^l)(2a_{pj}u^j) - (a_{ij}u^i u^j)(2b_{pl}u^l)}{(b_{kl}u^k u^l)^2} = \frac{2}{b_{kl}u^k u^l}[a_{pj}u^j - F(\mathbf{u})(b_{pl}u^l)]$$

or

$$\nabla F(\mathbf{u}) = \frac{2}{\mathbf{u}^T B\mathbf{u}}[A\mathbf{u} - F(\mathbf{u})\, B\mathbf{u}]$$

Therefore, $A\mathbf{w} - M\,B\mathbf{w} = \mathbf{0}$, which shows (i) that M is an eigenvalue of $B^{-1}A$ and thus is a root of the characteristic equation (*1*); (ii) that $\mathbf{w}$ is an eigenvector belonging to M.

A like consideration of the minimum value, m, of F on $\mathscr{G}$ leads to the other eigenvalue and associated eigenvector.

10.25 Prove that the extreme normal curvatures κ_1, κ_2 are the two roots of the quadratic equation (*10.25*).

In Problem 10.24, take $a_{ij} = f_{ij}$ and $b_{kl} = g_{kl}$; expand (*1*) to obtain (*10.25*).

10.26 Prove that the two normal section curves through P on $\mathscr{S}$ giving rise to max $\kappa_{\mathscr{F}} = \kappa_1$ and min $\kappa_{\mathscr{F}} = \kappa_2$ are orthogonal when $\kappa_1 \neq \kappa_2$ (that is, when P is not an umbilical point on $\mathscr{S}$).

Let us prove the general result, in the notation of Problem 10.24. We have:

$$A\mathbf{w} - M\,B\mathbf{w} = \mathbf{0} \qquad A\mathbf{v} - m\,B\mathbf{v} = \mathbf{0}$$

With the inner product of column vectors defined as $\mathbf{p}\cdot\mathbf{q} \equiv \mathbf{p}^T B\mathbf{q}$, multiply the first equation by $\mathbf{v}^T$ and the second by $\mathbf{w}^T$, and subtract:

$$(m - M)\mathbf{v}\cdot\mathbf{w} = 0$$

Hence, if $m \neq M$, $\mathbf{v}$ and $\mathbf{w}$ are orthogonal.

10.27 Calculate K and H for the right helicoid. Show that as $x^1 \to \infty$, K tends to zero (the surface becomes "flatter" as the distance from its axis increases without bound).

From Problem 10.11 and Example 10.4,

$$\mathbf{n} = \frac{1}{\sqrt{(x^1)^2 + a^2}}(a \sin x^2, -a\cos x^2, x^1)$$

$$\mathbf{r}_{11} = (0, 0, 0) \qquad \mathbf{r}_{12} = \mathbf{r}_{21} = (-\sin x^2, \cos x^2, 0) \qquad \mathbf{r}_{22} = (-x^1 \cos x^2, -x^1 \sin x^2, 0)$$

so that

$$f_{11} = \mathbf{n}\mathbf{r}_{11} = 0 \qquad f_{12} = f_{21} = -a/\sqrt{(x^1)^2 + a^2} \qquad f_{22} = 0$$

and

$$\mathrm{K} = \frac{f_{11}f_{22} - f_{12}^2}{g_{11}g_{22} - g_{12}^2} = \frac{0 - a^2/[(x^1)^2 + a^2]}{[(x^1)^2 + a^2]} = -\frac{a^2}{[(x^1)^2 + a^2]^2} \to 0 \quad \text{as } x^1 \to \infty$$

$$\mathrm{H} = \frac{f_{11}g_{22} + f_{22}g_{11} - 2f_{12}g_{12}}{g_{11}g_{22} - g_{12}^2} = \frac{0 + 0 - 2(0)}{g} = 0$$

STRUCTURE FORMULAS; ISOMETRIES

10.28 Complete the proof of *(10.27a)*.

The relations $\mathbf{n}_i = u_i^k \mathbf{r}_k$ and $\mathbf{n}_j \mathbf{r}_i = -f_{ij}$ imply

$$-f_{ij} = u_j^k g_{ki}$$

Multiply both sides by g^{is} and sum over i, obtaining $u_j^s = -g^{ls} f_{lj}$. Hence,

$$\mathbf{n}_i = u_i^k \mathbf{r}_k = -g^{lk} f_{li} \mathbf{r}_k = -g^{lk} f_{il} \mathbf{r}_k$$

10.29 Prove *(10.28)*.

Let the equation $\mathbf{r}_{ij} = u_{ij}^1 \mathbf{r}_1 + u_{ij}^2 \mathbf{r}_2 + u_{ij}^3 \mathbf{n}$ be rewritten as $\mathbf{r}_{ij} = u_{ij}^s \mathbf{r}_s + u_{ij}^3 \mathbf{n}$. Dot with $\mathbf{n}$, obtaining $u_{ij}^3 = f_{ij}$; therefore,

$$\mathbf{r}_{ij} = u_{ij}^s \mathbf{r}_s + f_{ij}\mathbf{n} \qquad (1)$$

Dot (*1*) with $\mathbf{r}_k$ and use Problem 10.48:

$$\mathbf{r}_{ij}\mathbf{r}_k = u_{ij}^s \mathbf{r}_s \mathbf{r}_k + 0 \qquad \text{or} \qquad \Gamma_{ijk} = u_{ij}^s g_{sk}$$

Solve for u_{ij}^s:

$$g^{kt}\Gamma_{ijk} = u_{ij}^s g^{kt} g_{sk} \qquad \text{or} \qquad \Gamma_{ij}^t = u_{ij}^s \delta_s^t = u_{ij}^t$$

Substitute back into (*1*):

$$\mathbf{r}_{ij} = \Gamma_{ij}^t \mathbf{r}_t + f_{ij}\mathbf{n}$$

Supplementary Problems

10.30 (*a*) Describe geometrically the curve whose parametric vector equation is

$$\mathbf{r} = (\cos t, \sin t, (1 - t)^{-1}) \qquad (0 \leqq t < 1)$$

What happens as $t \to 1$? (*b*) Use a programmable calculator and Simpson's rule to find the arc length for $0 \leqq t \leqq 1/2$ accurate to 6 places.

10.31 Find the exact length of the space curve $\mathbf{r} = (t^2 + t\sqrt{2}, t^2 - t\sqrt{2}, 2t^3/3)$ $(-1 \leqq t \leqq 1)$.

10.32 (*a*) Using the arc-length parameterization of the right circular helix,

$$\mathbf{r} = \left(a \cos \frac{s}{c}, a \sin \frac{s}{c}, \frac{bs}{c}\right) \qquad (c \equiv \sqrt{a^2 + b^2})$$

find the coordinate equations of the tangent line to the helix at any point $P \equiv \mathbf{r}(s)$. (*b*) Show that the tangent line intersects the xy-plane at a point $Q \equiv \mathbf{r}^*(s)$ such that $PQ = s$. (*c*) By thinking of a string wound along the helix, interpret the result of (*b*).

10.33 Show that for the curve $y = x^5$ in the xy-plane, parameterized as $\mathbf{r} = (t, t^5, 0)$, the vector $\mathbf{T}'/\|\mathbf{T}'\|$ has an essential point of discontinuity at $t = 0$.

10.34 Find the curvature and the torsion of the curve $\mathbf{r} = (t, t^5 + a, t^5 - a)$.

10.35 Prove that a curve is planar if and only if its torsion vanishes.

10.36 Prove that a planar curve with constant nonzero curvature κ is a circle. [*Hint*: $\mathbf{T} = (\cos\theta, \sin\theta, 0)$ and $\mathbf{N} = (-\sin\theta, \cos\theta, 0)$ imply $\kappa = \theta'$ or $\theta = \kappa s + a$; show that the radius is $1/\kappa$.]

10.37 Verify the Serret–Frenet formulas for the circular helix.

10.38 Show that if included in the range of the map $\mathbf{r}(x^1, x^2)$, the vertex of the right circular cone is a singular point.

10.39 Calculate the unit normal for the catenoid as parameterized in Fig. 10-14 and show that the surface is regular.

10.40 Find the length of the curve on the right helicoid (Example 10.4) given by $x^1 = t^2$, $x^2 = \ln t$, with $1 \leqq t \leqq 2$, in the special case when the parameter $a = 1$.

10.41 Find the two possible directions for a curve $\mathscr{C}'$ in the parameter plane (in polar coordinates) whose image on the paraboloid of Fig. 10-17 meets the circle $x^2 + y^2 = 4$, $z = 4$ at $P(0, 2, 4)$ at an angle of $\pi/3$.

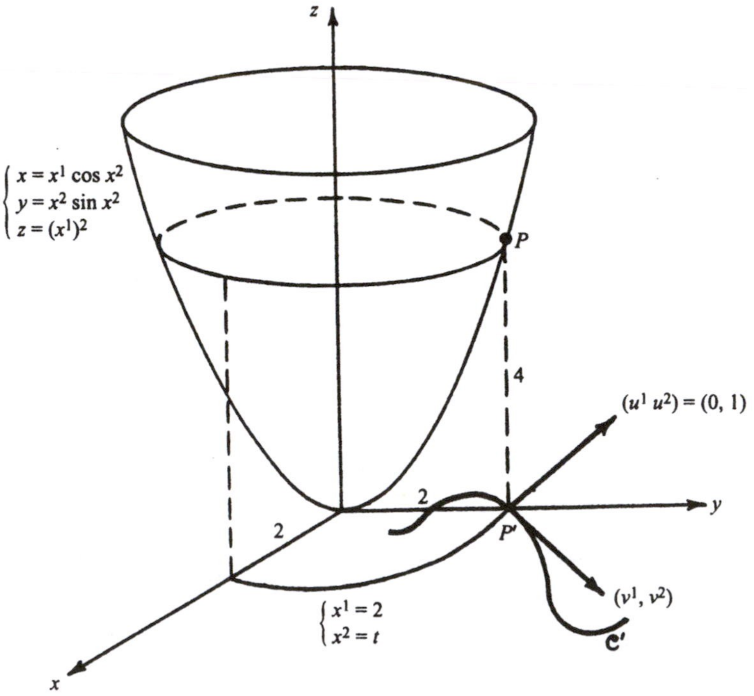

Fig. 10-17

10.42 Use Theorem 10.6 to show that an elliptical helix is not in general a geodesic on an elliptical cylinder.

10.43 Calculate the Christoffel symbols of the second kind for the right helicoid (Example 10.4). Show that circular helics on the surface (x^1 = constant) are *not* geodesics, using (*7.14*).

10.44 Exhibit the SFF for the general surface of revolution (Problem 10.10).

10.45 Establish the formulas below for any surface of revolution, with $G \equiv g'/f'$ (see Problems 10.12 and 10.44):

$$\mathrm{K} = \frac{GG'}{ff'(1+G^2)^2} \qquad \text{and} \qquad \mathrm{H} = \frac{fG' + g'(1+G^2)}{f\,|f'|\,(1+G^2)^{3/2}}$$

Use these formulas to verify that a sphere of radius a has Gaussian curvature $1/a^2$ and mean curvature $2/a$.

10.46 (*a*) Calculate K and H for two different parameterizations of the paraboloid $z = a(x^2 + y^2)$: (i) as the surface of revolution for which $f = x^1$, $g = a(x^1)^2$; (ii) as the surface $\mathbf{r} = (\bar{x}^1, \bar{x}^2, a(\bar{x}^1)^2 + a(\bar{x}^2)^2)$. (*b*) Interpret the results of (*a*).

10.47 Infer from Problem 10.45 that $\mathrm{H} \equiv 0$ for any catenoid. [A surface with $\mathrm{H} = 0$ at all points is called a *minimal surface*. Among minimal surfaces are those that solve "soap-bubble" problems, which naturally involve minimal surface area.]

10.48 Prove that $\Gamma_{ijk} = \mathbf{r}_{ij}\mathbf{r}_k$. [*Hint*: $(\mathbf{r}_i\mathbf{r}_j)_k = \mathbf{r}_{ik}\mathbf{r}_j + \mathbf{r}_i\mathbf{r}_{jk}$.]

10.49 Surfaces for which the Gaussian curvature is a negative constant are very rare. One such surface can be constructed as follows. (*a*) A *tractrix* is the involute of a catenary (see Problem 10.32(*c*)). Write the vector equation for the involute of the catenary $\mathbf{r} = (a \cosh x^1, 0, ax^1)$ (see Fig. 10-14). (*b*) Using Problem 10.45, show that $\mathrm{K} = -1/a^2$ for the *tractroid* generated by revolving the tractrix of (*a*) about the z-axis.

10.50 Prove that the catenoid,

$$\mathrm{I} = (a^2 \cosh^2 x^1)(dx^1)^2 + (a^2 \cosh^2 x^1)(dx^2)^2$$

and the right helicoid,

$$\mathrm{I} = (d\bar{x}^1)^2 + [(\bar{x}^1)^2 + a^2](d\bar{x}^2)^2$$

are locally isometric.

Chapter 11

Tensors in Classical Mechanics

11.1 INTRODUCTION

Classical mechanics originated with the work of Galileo and was developed extensively by Newton (it is often called *Newtonian mechanics*). It deals with the motion of particles in a fixed frame of reference (rectangular coordinate system). The basic premise of Newtonian mechanics is the concept of absolute time measurement between two reference frames at constant velocity relative to each other (called *Galilean frames*). Within those frames, other coordinate systems may be used so long as the metric remains Euclidean. This means that some of the theory of tensors can be brought to bear on this study.

11.2 PARTICLE KINEMATICS IN RECTANGULAR COORDINATES

Let P be a particle whose path in $\mathbf{E}^3$ is given by

$$\mathscr{C} \; : \; \mathbf{x} = (x^i(t)) \tag{11.1}$$

where t represents time. The *velocity vector* of P is defined as

$$\mathbf{v} \equiv \frac{d\mathbf{x}}{dt} \equiv \left(\frac{dx^1}{dt}, \frac{dx^2}{dt}, \frac{dx^3}{dt}\right) \equiv (\dot{x}^1, \dot{x}^2, \dot{x}^3) \tag{11.2}$$

and the (instantaneous) *velocity* or *speed* as the scalar

$$v \equiv \|\mathbf{v}\| \equiv \sqrt{(\dot{x}^1)^2 + (\dot{x}^2)^2 + (\dot{x}^3)^2} \tag{11.3}$$

Further, define the *acceleration vector* as

$$\mathbf{a} \equiv \frac{d\mathbf{v}}{dt} \equiv \left(\frac{d^2x^1}{dt^2}, \frac{d^2x^2}{dt^2}, \frac{d^2x^3}{dt^2}\right) \equiv (\ddot{x}^1, \ddot{x}^2, \ddot{x}^3) \tag{11.4}$$

and the *acceleration* as

$$a = \|\mathbf{a}\| \equiv \sqrt{(\ddot{x}^1)^2 + (\ddot{x}^2)^2 + (\ddot{x}^3)^2} \tag{11.5}$$

If $\mathbf{v} = (v^i)$ and $\mathbf{a} = (a^i)$, the preceding formulas have the component forms

$$v^i = \frac{dx^i}{dt} \qquad v = \sqrt{v^i v^i} \qquad a^i = \frac{d^2x^i}{dt^2} \qquad a = \sqrt{a^i a^i} \tag{11.6}$$

In terms of the geometry of the curve $\mathscr{C}$ we have $\mathbf{v} = v\mathbf{T}$, with $v = ds/dt$. Hence,

$$\mathbf{a} = \frac{d}{dt}(v\mathbf{T}) = \dot{v}\mathbf{T} + v\dot{\mathbf{T}} = \dot{v}\mathbf{T} + v\frac{d\mathbf{T}}{ds}\frac{ds}{dt} = \dot{v}\mathbf{T} + v^2\mathbf{T}'$$

But $\mathbf{T}' = \kappa\mathbf{N}$, and so

$$\mathbf{a} = \dot{v}\mathbf{T} + \kappa v^2\mathbf{N} \tag{11.7}$$

Via the Pythagorean theorem, (*11.7*) implies

$$a = \sqrt{\dot{v}^2 + \kappa^2 v^4} \tag{11.8}$$

EXAMPLE 11.1 Formula (*11.7*) serves to define the *tangential* and *normal accelerations* of P as

$$\dot{v} = \frac{d^2s}{dt^2} \qquad \text{and} \qquad \kappa v^2 = \frac{v^2}{\rho} \quad (\rho \equiv \text{radius of curvature})$$

respectively. For a particle with constant velocity, $\mathbf{a} = \kappa v^2 \mathbf{N}$ and $a = |\kappa| v^2$; i.e., the acceleration is proportional to the absolute curvature and the acceleration vector itself is normal to the direction of the particle.

11.3 PARTICLE KINEMATICS IN CURVILINEAR COORDINATES

There emerges the problem of expressing the preceding formulas in nonrectangular coordinate systems. This is not merely an academic consideration, for there are important situations in which the differential equations of motion can be solved only in polar or spherical coordinates (cf. Example 11.3), not to mention applications in relativistic mechanics.

Let us start with the definitions of the velocity and acceleration vectors in a barred rectangular system:

$$\bar{v}^i = \frac{d\bar{x}^i}{dt} \qquad \text{and} \qquad \bar{a}^i = \frac{d\bar{v}^i}{dt}$$

Because the tangent field of $\mathscr{C}$ is a tensor, the velocity components in an arbitrary coordinate system (x^i) are just $v^i = dx^i/dt$. However, as we found in Chapter 6, the acceleration components must be written as absolute derivatives along $\mathscr{C}$: $a^i = \delta v^i/\delta t$. Hence, in an arbitrary coordinate system (x^i), with (g_{ij}) representing the Euclidean metric, we have:

$$v^i = \frac{dx^i}{dt} \qquad a^i = \frac{dv^i}{dt} + \Gamma^i_{rs} v^r v^s = \frac{d^2x^i}{dt^2} + \Gamma^i_{rs} \frac{dx^r}{dt}\frac{dx^s}{dt} \tag{11.9}$$

and the speed and acceleration scalars are the invariants

$$v = \sqrt{g_{ij} v^i v^j} \qquad a = \sqrt{g_{ij} a^i a^j} \tag{11.10}$$

EXAMPLE 11.2 Formulas (*11.9*) give the *contravariant components* of velocity and acceleration. These are *not* the components used in classical physics and vector analysis. There, the metric for an orthogonal curvilinear coordinate system is written as

$$ds^2 = h_1^2(dx_1)^2 + h_2^2(dx_2)^2 + h_3^2(dx_3)^2$$

and one defines the *physical components* of the velocity vector as

$$v_{(1)} = h_1 \frac{dx_1}{dt} \qquad v_{(2)} = h_2 \frac{dx_2}{dt} \qquad v_{(3)} = h_3 \frac{dx_3}{dt}$$

Thus, the physical components are related to the contravariant components via

$$v_{(\alpha)} = \sqrt{g_{\alpha\alpha}}\, v^\alpha \qquad (\alpha = 1, 2, 3;\ \text{no summation}) \tag{1}$$

Likewise for acceleration and force vectors. (See Problem 11.24.)

Let us illustrate the distinction by calculating the components of acceleration in cylindrical coordinates, $(x^1, x^2, x^3) = (x_1, x_2, x_3) = (r, \theta, z)$, for which $g_{11} = 1$, $g_{22} = (x^1)^2$, $g_{33} = 1$. By Problem 6.26,

$$\Gamma^1_{22} = -x^1 \qquad \Gamma^2_{12} = \Gamma^2_{21} = 1/x^1 \qquad \text{(all others zero)}$$

so that (*11.9*) gives for the contravariant components

$$a^1 = \frac{d^2r}{dt^2} - r\left(\frac{d\theta}{dt}\right)^2 \qquad a^2 = \frac{d^2\theta}{dt^2} + \frac{2}{r}\frac{dr}{dt}\frac{d\theta}{dt} \qquad a^3 = \frac{d^2z}{dt^2} \tag{2}$$

The physical components are then obtained from (*1*) as

$$a_{(r)} = \frac{d^2r}{dt^2} - r\left(\frac{d\theta}{dt}\right)^2 \qquad a_{(\theta)} = r\frac{d^2\theta}{dt^2} + 2\frac{dr}{dt}\frac{d\theta}{dt} \qquad a_{(z)} = \frac{d^2z}{dt^2} \tag{3}$$

Only the θ-component differs between (*2*) and (*3*); but the difference is significant. For instance, the *coriolis acceleration* of a particle is $2\dot{r}\dot{\theta}$, as in (*3*).

11.4 NEWTON'S SECOND LAW IN CURVILINEAR COORDINATES

The *momentum vector* of a particle of mass m is defined as $\mathbf{M} = m\mathbf{v}$. Relative to a rectangular coordinate system (with the property of being an *inertial frame*), Newton's second law of motion effectively defines the *force vector* acting on a particle as $\mathbf{F} = d\mathbf{M}/dt$. Accordingly, in curvilinear coordinates (x^i), the law reads:

$$\mathbf{F} = \frac{\delta \mathbf{M}}{\delta t} = m\,\frac{\delta \mathbf{v}}{\delta t} = m\mathbf{a} \tag{11.11}$$

assuming a constant mass. Therefore, the contravariant components of force are given by

$$F^i = ma^i = m\left(\frac{d^2x^i}{dt^2} + \Gamma^i_{rs}\,\frac{dx^r}{dt}\,\frac{dx^s}{dt}\right) \tag{11.12a}$$

and the covariant components by

$$F_i = g_{ir}F^r = m\left(g_{ir}\,\frac{d^2x^r}{dt^2} + \Gamma_{rsi}\,\frac{dx^r}{dt}\,\frac{dx^s}{dt}\right) \tag{11.12b}$$

By introducing a scalar invariant called the *kinetic energy* of the particle,

$$T \equiv \frac{1}{2}\,mv^2 = \frac{1}{2}\,mg_{ij}v^iv^j$$

one can (see Problem 11.5) put (*11.12b*) into the equivalent *Lagrangian form*

$$F_i = \frac{d}{dt}\left(\frac{\partial T}{\partial v^i}\right) - \frac{\partial T}{\partial x^i} \tag{11.13}$$

The partial derivatives in (*11.13*) are taken with T considered as a function of six independent variables, the x^i (via the g_{ij}) and the v^i.

EXAMPLE 11.3 (*Motion under a Central Force*) (*a*) Obtain the differential equation for the trajectory of a particle acted on by a force that is always directed from (or toward) a fixed point O. (*b*) Solve the equation of (*a*) when the central force is gravitational, thus determining the orbit of a satellite.

(*a*) By Problem 11.18, the motion will be confined to a plane through O. Take O as the origin of polar coordinates $(x^1, x^2) = (r, \theta)$ in the plane; the force field then has the form $\mathbf{F} = (F^1, 0)$. Taking the acceleration components from (*2*) of Example 11.2, we have as the equations of motion:

$$F^1 = ma^1 = m\left[\frac{d^2r}{dt^2} - r\left(\frac{d\theta}{dt}\right)^2\right]$$

$$0 = ma^2 = m\left[\frac{d^2\theta}{dt^2} + \frac{2}{r}\,\frac{dr}{dt}\,\frac{d\theta}{dt}\right] \equiv \frac{m}{r^2}\,\frac{d}{dt}\left(r^2\,\frac{d\theta}{dt}\right)$$

The θ-equation has the first integral

$$r^2\,\frac{d\theta}{dt} = q = \text{const.}$$

(conservation of angular momentum), which can be used to change the parameter of the trajectory from t to θ. Thus, writing $u = 1/r$, we have:

$$\frac{dr}{dt} = \frac{d\theta}{dt}\,\frac{d}{d\theta}\,\frac{1}{u} = (qu^2)\left(-u^{-2}\,\frac{du}{d\theta}\right) = -q\,\frac{du}{d\theta}$$

$$\frac{d^2r}{dt^2} = -q\,\frac{d^2u}{d\theta^2}\,(qu^2) = -q^2u^2\,\frac{d^2u}{d\theta^2}$$

and the r-equation becomes

$$\frac{d^2u}{d\theta^2} + u = g(u, \theta) \tag{1}$$

in which $g(u, \theta) = -F^1(u^{-1}, \theta)/mq^2u^2$.

(*b*) For the gravitational field, $F^1 = -k/r^2 = -ku^2 (k > 0$; attractive force), so that $g(u, \theta) = Q =$ const. using the immediate preceding definition of $g(u, \theta)$. Thus, the solution of (*1*) is $u = P\cos\theta + Q$, or

$$r = \frac{1/Q}{1 + e\cos\theta} \tag{2}$$

which is a conic having eccentricity $e = P/Q$ and focus at O, the classical result.

11.5 DIVERGENCE, LAPLACIAN, CURL

The *divergence* of a contravariant vector $\mathbf{u} = (u^i)$ on $\mathbf{E}^3$ is defined by (*9.11*), with use of Problem 8.14:

$$\operatorname{div}\mathbf{u} = u^i_{,i} = \frac{\partial u^i}{\partial x^i} + \Gamma^i_{ri}u^r = \frac{\partial u^i}{\partial x^i} + u^r \frac{\partial}{\partial x^r}(\ln\sqrt{g}) = \frac{\partial u^i}{\partial x^i} + u^i \frac{1}{\sqrt{g}}\frac{\partial}{\partial x^i}(\sqrt{g})$$

The product rule for partial differentiation yields the compact form

$$\operatorname{div}\mathbf{u} = \frac{1}{\sqrt{g}}\frac{\partial}{\partial x^i}(\sqrt{g}\,u^i) \tag{11.14}$$

Another notation for the divergence is $\nabla \cdot \mathbf{u}$.

The *Laplacian* of a scalar field f is given by $\nabla^2 f \equiv \operatorname{div}(\operatorname{grad} f)$. Since in general coordinates divergence is defined for contravariant tensors only, while $\operatorname{grad} f = (\partial f/\partial x^i)$ is a covariant tensor (Example 3.5), we first raise the subscript and then find the divergence by (*11.14*):

$$\nabla^2 f = \operatorname{div}\left(g^{ij}\frac{\partial f}{\partial x^j}\right) = \frac{1}{\sqrt{g}}\frac{\partial}{\partial x^i}\left(\sqrt{g}\,g^{ij}\frac{\partial f}{\partial x^j}\right) \tag{11.15}$$

The Laplacian figures importantly in electromagnetic theory via the *scalar wave equation*,

$$\frac{\partial^2 f}{\partial t^2} = k^2\,\nabla^2 f \qquad (k = \text{const.} = \text{wave speed}) \tag{11.16a}$$

In cartesian coordinates only, one defines the Laplacian of a vector field as $\nabla^2\mathbf{u} \equiv (\nabla^2 u^i)$, where $\nabla^2 u^i = u^i_{xx} + u^i_{yy} + u^i_{zz}$, and writes the *vector wave equation*,

$$\frac{\partial^2\mathbf{u}}{\partial t^2} = k^2\,\nabla^2\mathbf{u} \tag{11.16b}$$

as an abbreviation for three scalar (component) wave equations.

EXAMPLE 11.4 Write the Laplacian for cylindrical coordinates.
Using $g^{11} = 1$, $g^{22} = 1/(x^1)^2$, $g^{33} = 1$, and $g = (x^1)^2$ in (*11.15*),

$$\begin{aligned}\nabla^2 f &= \frac{1}{x^1}\left[\frac{\partial}{\partial x^1}\left(x^1\frac{\partial f}{\partial x^1}\right) + \frac{\partial}{\partial x^2}\left(\frac{1}{x^1}\frac{\partial f}{\partial x^2}\right) + \frac{\partial}{\partial x^3}\left(x^1\frac{\partial f}{\partial x^3}\right)\right] \\ &= f_{11} + \frac{1}{(x^1)^2}f_{22} + f_{33} + \frac{1}{x^1}f_1\end{aligned} \qquad (x^1 > 0)$$

the last line employing subscript notation for the partial derivatives.

The curl of a vector field $\mathbf{u} = (u^i)$—symbolized $\operatorname{curl}\mathbf{u}$, $\nabla \times \mathbf{u}$, or $\operatorname{rot}\mathbf{u}$—is given *in a rectangular coordinate system* (x^i) by

$$\operatorname{curl}\mathbf{u} \equiv \left(e_{ijk}\frac{\partial u^k}{\partial x^j}\right) \tag{11.17a}$$

where e_{ijk} is the permutation symbol (Chapter 2). The definition may be rewritten as a determinantal operator:

$$\operatorname{curl}\mathbf{u} \equiv \begin{vmatrix} \mathbf{e}_1 & \mathbf{e}_2 & \mathbf{e}_3 \\ \dfrac{\partial}{\partial x^1} & \dfrac{\partial}{\partial x^2} & \dfrac{\partial}{\partial x^3} \\ u^1 & u^2 & u^3 \end{vmatrix} \tag{11.17b}$$

in which $(\mathbf{e}_1, \mathbf{e}_2, \mathbf{e}_3) = (\mathbf{i}, \mathbf{j}, \mathbf{k})$ is the standard orthonormal basis. Unlike the gradient and the divergence, the curl cannot be extended to curvilinear coordinate systems by a tensor formula.

Remark 1: Not everything of significance in mathematical physics is a tensor. Problem 11.11 shows that *(11.17)* defines a direct cartesian tensor, but that is all. This is not to say that the curl operator cannot be formulated and used in curvilinear coordinates (see any text in vector analysis). It is only that the curl in spherical coordinates (say) and the curl in rectangular coordinates are not related *tensorwise*.

Nonrelativistic Maxwell's Equations

Let $\mathbf{E}$ = electric field strength
$\mathbf{D}$ = electric displacement
$\mathbf{H}$ = magnetic field strength
$\mathbf{B}$ = magnetic induction
$\mathbf{J}$ = current density
ρ = charge density
ϵ = dielectric constant
μ = magnetic permeability
c = velocity of light

Then the famous *Maxwell's equations* may be written as follows:

$$\begin{aligned} &\operatorname{curl}\mathbf{E} + \frac{1}{c}\frac{\partial \mathbf{B}}{\partial t} = 0 \qquad && \operatorname{div}\mathbf{B} = 0 \\ &\operatorname{curl}\mathbf{H} - \frac{1}{c}\frac{\partial \mathbf{D}}{\partial t} = \frac{1}{c}\mathbf{J} && \operatorname{div}\mathbf{D} = \rho \end{aligned} \tag{11.18}$$

From standard formulas in electromagnetic theory, $\mathbf{D} = \epsilon\mathbf{E}$, $\mathbf{B} = \mu\mathbf{H}$, and $\mathbf{J} = \rho\mathbf{u}$, where $\mathbf{u}$ denotes the velocity field of the charge distribution; *(11.18)* becomes

$$\begin{aligned} &\operatorname{curl}\mathbf{E} = -\frac{\mu}{c}\frac{\partial \mathbf{H}}{\partial t} \qquad && \operatorname{div}\mathbf{H} = 0 \\ &\operatorname{curl}\mathbf{H} = \frac{\epsilon}{c}\frac{\partial \mathbf{E}}{\partial t} + \frac{\rho}{c}\mathbf{u} && \operatorname{div}\mathbf{E} = \frac{\rho}{\epsilon} \end{aligned}$$

If the charge distribution is in free space ($\epsilon = \epsilon_0$, $\mu = \mu_0$), a proper choice of units brings the equations into the form

$$\begin{aligned} &\operatorname{curl}\mathbf{E} = -\frac{1}{c}\frac{\partial \mathbf{H}}{\partial t} \qquad && \operatorname{div}\mathbf{H} = 0 \\ &\operatorname{curl}\mathbf{H} = \frac{1}{c}\frac{\partial \mathbf{E}}{\partial t} + \frac{\rho}{c}\mathbf{u} && \operatorname{div}\mathbf{E} = \rho \end{aligned} \tag{11.19}$$

Work with Maxwell's equations requires the vector identities listed below (see Problems 11.10 and 11.21).

$$\nabla\cdot(\nabla\times\mathbf{u}) = 0 \qquad \text{(for any } \mathbf{u}\text{)} \tag{11.20}$$

$$\nabla\times(\nabla\times\mathbf{u}) = \nabla(\nabla\cdot\mathbf{u}) - \nabla^2\mathbf{u} \tag{11.21}$$

$$\frac{\partial}{\partial t}(\nabla\cdot\mathbf{u}) = \nabla\cdot\frac{\partial \mathbf{u}}{\partial t} \tag{11.22}$$

$$\frac{\partial}{\partial t}(\nabla\times\mathbf{u}) = \nabla\times\frac{\partial \mathbf{u}}{\partial t} \tag{11.23}$$

Solved Problems

VELOCITY AND ACCELERATION

11.1 Find the velocity and acceleration vectors and the scalars v and a for a particle whose equation of motion (along a twisted cubic) is $\mathbf{x} = (t, t^2, t^3)$ $(-1 \leqq t \leqq 1)$. Determine the extreme values of v and a, and where they are assumed.

$$\mathbf{v} = (1, 2t, 3t^2) \qquad \text{and} \qquad v = \sqrt{1 + 4t^2 + 9t^4}$$
$$\mathbf{a} = (0, 2, 6t) \qquad \text{and} \qquad a = \sqrt{4 + 36t^2}$$

Hence v and a have maxima at $t = \pm 1$, where $v = \sqrt{14}$ and $a = \sqrt{40}$. They have minima at $t = 0$, where $v = 1$ and $a = 2$.

PARTICLE DYNAMICS

11.2 A particle travels at constant speed v on a curve with positive curvature. Show that its acceleration is greatest where the curvature is greatest.

By *(11.8)* with $\dot{v} = 0$, $a = \kappa v^2$ or $a/\kappa = \text{const.}$

11.3 Compute the contravariant acceleration components in a coordinate system (x^i) connected to a rectangular coordinate system $(\bar{x}^i)$ by $\bar{x}^1 = (x^1)^2$, $\bar{x}^2 = x^2$, $\bar{x}^3 = x^3$.

Use *(5.7)*:

$$G = J^T J = \begin{bmatrix} 2x^1 & 0 & 0 \\ 0 & 1 & 0 \\ 0 & 0 & 1 \end{bmatrix} \begin{bmatrix} 2x^1 & 0 & 0 \\ 0 & 1 & 0 \\ 0 & 0 & 1 \end{bmatrix} = \begin{bmatrix} 4(x^1)^1 & 0 & 0 \\ 0 & 1 & 0 \\ 0 & 0 & 1 \end{bmatrix}$$

Hence, the Christoffel symbols are given by

$$\Gamma^1_{11} = \frac{\partial}{\partial x^1}\left[\frac{1}{2}\ln 4(x^1)^2\right] = \frac{1}{x^1} \qquad \text{(all others zero)}$$

and *(11.9)* gives

$$a^1 = \frac{d^2x^1}{dt^2} + \frac{1}{x^1}\left(\frac{dx^1}{dt}\right)^2 \qquad a^2 = \frac{d^2x^2}{dt^2} \qquad a^3 = \frac{d^2x^3}{dt^2}$$

NEWTON'S SECOND LAW

11.4 Show that Newton's second law is consistent with *Newton's first law*: A particle that is not acted upon by an outside force is at rest or is in motion along a straight line at constant velocity. Assume a rectangular coordinate system.

$\mathbf{F} = \mathbf{0}$ implies $d\mathbf{v}/dt = \mathbf{0}$, or $\mathbf{v} = \mathbf{d}$ (constant). Then,

$$\frac{d\mathbf{x}}{dt} = \mathbf{d} \qquad \text{or} \qquad \mathbf{x} = t\mathbf{d} + \mathbf{x}_0$$

which is the parametric equation for a point (if $\mathbf{d} = \mathbf{0}$) or for a straight line (if $\mathbf{d} \neq \mathbf{0}$), along which $v = \|\mathbf{d}\| = \text{const.}$

11.5 Prove the equivalence of *(11.13)* and *(11.12b)*.

For simplicity, take $m = 1$ in *(11.13)*. By the chain rule and the symmetry of (g_{ij}),

$$\frac{d}{dt}\left(\frac{\partial T}{\partial v^i}\right) - \frac{\partial T}{\partial x^i} = \frac{d}{dt}(g_{ir}v^r) - \frac{\partial T}{\partial g_{rs}}\frac{\partial g_{rs}}{\partial x^i} = g_{ir}\frac{dv^r}{dt} - \frac{\partial T}{\partial g_{rs}}\frac{\partial g_{rs}}{\partial x^i} + \frac{dg_{ir}}{dt}v^r$$

$$= g_{ir}\frac{dv^r}{dt} - g_{rsi}\left(\frac{1}{2}v^r v^s\right) + \frac{\partial g_{ir}}{\partial x^s}\frac{dx^s}{dt}v^r = g_{ir}\frac{dv^r}{dt} - \frac{1}{2}g_{rsi}v^r v^s + g_{irs}v^s v^r$$

$$= g_{ir}\frac{dv^r}{dt} - \frac{1}{2}g_{rsi}v^r v^s + \frac{1}{2}g_{sir}v^s v^r + \frac{1}{2}g_{irs}v^s v^r = g_{ir}\frac{dv^r}{dt} + \Gamma_{rsi}v^r v^s$$

The final expression is exactly the right-hand side of (*11.12b*) (for $m = 1$).

11.6 Solve (*1*) of Example 11.3 when the force field is of the form

$$g(u, \theta) = Au + h(\theta)$$

where A is a constant and $h(\theta)$ is periodic of period 2π.

With primes denoting θ-derivatives, we must solve

$$u'' + u = Au + h(\theta) \qquad \text{or} \qquad u'' + (1 - A)u = h(\theta)$$

The general solution to the homogeneous equation is

$$u = \begin{cases} P\cos(\sqrt{1-A}\,\theta + \alpha) & A < 1 \\ \alpha\theta + \beta & A = 1 \\ Q\exp(\sqrt{A-1}\,\theta) + R\exp(-\sqrt{A-1}\,\theta) & A > 1 \end{cases}$$

A particular solution of the nonhomogeneous equation may be obtained in the form $u = u_H w$, where u_H is any particular solution of the homogeneous equation. In fact, substitution in the differential equation yields

$$2u_H' w' + u_H w'' = h \qquad \text{or} \qquad (u_H^2 w')' = u_H h$$

and this last equation can be solved by two quadratures:

$$w'(\theta) = \frac{1}{u_H^2(\theta)}\int_0^\theta u_H(\phi)h(\phi)\,d\phi \qquad \text{and} \qquad w(\theta) = \int_0^\theta \frac{d\psi}{u_H^2(\psi)}\int_0^\psi u_H(\phi)h(\phi)\,d\phi$$

The integrals are easily evaluated when $h(\phi)$ is represented as a Fourier series.

11.7 If $h(\theta) = 0$ in Problem 11.6, identify the orbits corresponding to (*a*) $A = 0$, (*b*) $A = 1$, (*c*) $A = 5/4$.

(*a*) The curve $1/r = P\cos(\theta + \alpha)$, or $r\cos(\theta + \alpha) = 1/P$, is a straight line (Fig. 11-1).

(*b*) The curve $1/r = \alpha\theta + \beta$ is a hyperbolic spiral that degenerates into a circle for $\alpha = 0$.

(*c*) The curve $1/r = Qe^{\theta/2} + Re^{-\theta/2}$ is a complex spiral which, in the case $Q = 0$, $R = 1$, reduces to the simple logarithmic spiral $r = e^{\theta/2}$.

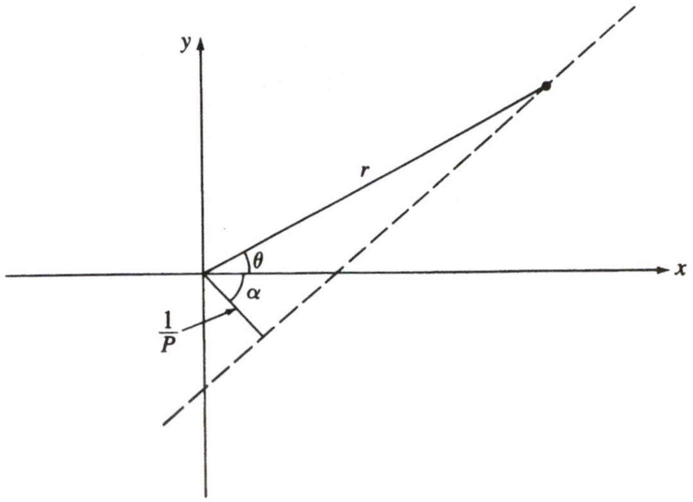

Fig. 11-1

DIFFERENTIAL OPERATORS

11.8 Calculate the Laplacian for spherical coordinates by the tensor formula. (The calculation is very tedious by other methods.)

We have

$$G = \begin{bmatrix} 1 & 0 & 0 \\ 0 & (x^1)^2 & 0 \\ 0 & 0 & (x^1 \sin x^2)^2 \end{bmatrix} \qquad G^{-1} = \begin{bmatrix} 1 & 0 & 0 \\ 0 & (x^1)^{-2} & 0 \\ 0 & 0 & (x^1 \sin x^2)^{-2} \end{bmatrix}$$

and $g = (x^1)^4 \sin^2 x^2$, so that in *(11.15)*,

$$\sqrt{g}\, g^{ij} \frac{\partial f}{\partial x^j} = (x^1)^2(\sin x^2)\left(g^{i1} \frac{\partial f}{\partial x^1} + g^{i2} \frac{\partial f}{\partial x^2} + g^{i3} \frac{\partial f}{\partial x^3}\right)$$

Therefore,

$$\sqrt{g}\, g^{1j} \frac{\partial f}{\partial x^j} = (x^1)^2(\sin x^2) \frac{\partial f}{\partial x^1}$$

$$\sqrt{g}\, g^{2j} \frac{\partial f}{\partial x^j} = (x^1)^2(\sin x^2) \frac{1}{(x^1)^2} \frac{\partial f}{\partial x^2} = (\sin x^2) \frac{\partial f}{\partial x^2}$$

$$\sqrt{g}\, g^{3j} \frac{\partial f}{\partial x^j} = (x^1)^2(\sin x^2) \frac{1}{(x^1 \sin x^2)^2} \frac{\partial f}{\partial x^3} = (\csc x^2) \frac{\partial f}{\partial x^3}$$

and so

$$\frac{\partial}{\partial x^i}\left[\sqrt{g}\, g^{ij} \frac{\partial f}{\partial x^j}\right] = \frac{\partial}{\partial x^1}\left[(x^1)^2(\sin x^2) \frac{\partial f}{\partial x^1}\right] + \frac{\partial}{\partial x^2}\left[(\sin x^2) \frac{\partial f}{\partial x^2}\right] + \frac{\partial}{\partial x^3}\left[(\csc x^2) \frac{\partial f}{\partial x^3}\right]$$

$$= 2x^1(\sin x^2) \frac{\partial f}{\partial x^1} + (x^1)^2(\sin x^2) \frac{\partial^2 f}{(\partial x^1)^2} + (\cos x^2) \frac{\partial f}{\partial x^2} + (\sin x^2) \frac{\partial^2 f}{(\partial x^2)^2} + (\csc x^2) \frac{\partial^2 f}{(\partial x^3)^2}$$

In writing the final steps we convert to $\rho = x^1$, $\varphi = x^2$, and $\theta = x^3$:

$$\nabla^2 f = \frac{1}{\sqrt{g}} \frac{\partial}{\partial x^i}\left[\sqrt{g}\, g^{ij} \frac{\partial f}{\partial x^j}\right]$$

$$= \frac{1}{\rho^2 \sin \varphi}\left[(2\rho \sin \varphi) \frac{\partial f}{\partial \rho} + (\rho^2 \sin \varphi) \frac{\partial^2 f}{\partial \rho^2} + (\cos \varphi) \frac{\partial f}{\partial \varphi} + (\sin \varphi) \frac{\partial^2 f}{\partial \varphi^2} + (\csc \varphi) \frac{\partial^2 f}{\partial \theta^2}\right]$$

$$= \frac{\partial^2 f}{\partial \rho^2} + \frac{1}{\rho^2} \frac{\partial^2 f}{\partial \varphi^2} + \frac{1}{\rho^2 \sin^2 \varphi} \frac{\partial^2 f}{\partial \theta^2} + \frac{2}{\rho} \frac{\partial f}{\partial \rho} + \frac{\cot \varphi}{\rho^2} \frac{\partial f}{\partial \varphi}$$

11.9 Calculate the divergence in spherical coordinates (ρ, φ, θ) of *(a)* a contravariant vector, $\mathbf{u} = (u^i)$; *(b)* a vector specified by its physical components, $\mathbf{u} = u_{(1)}\mathbf{e}_1 + u_{(2)}\mathbf{e}_2 + u_{(3)}\mathbf{e}_3$.

(a) We plug into the formula *(11.14)*:

$$\operatorname{div} \mathbf{u} = \frac{1}{\sqrt{g}} \frac{\partial}{\partial x^i} (\sqrt{g} u^i) = \frac{\partial u^i}{\partial x^i} + u^i \frac{1}{\sqrt{g}} \frac{\partial}{\partial x^i} (\sqrt{g})$$

$$= \frac{\partial u^i}{\partial x^i} + u^i \frac{1}{(x^1)^2 \sin x^2} \frac{\partial}{\partial x^i} [(x^1)^2 \sin x^2]$$

$$= \frac{\partial u^i}{\partial x^i} + u^1\left(\frac{2}{x^1}\right) + u^2\left(\frac{\cos x^2}{\sin x^2}\right) + u^3(0)$$

Thus

$$\operatorname{div} \mathbf{u} = \frac{\partial u^1}{\partial \rho} + \frac{\partial u^2}{\partial \varphi} + \frac{\partial u^3}{\partial \theta} + \frac{2}{\rho} u^1 + (\cot \varphi) u^2$$

(*b*) By Example 11.2, we apply (*11.14*) to the contravariant vector having components

$$u^1 = \frac{u_{(1)}}{1} \qquad u^2 = \frac{u_{(2)}}{x^1} \qquad u^3 = \frac{u_{(3)}}{x^1 \sin x^2}$$

Hence, from (*a*),

$$\begin{aligned}\operatorname{div}\mathbf{u} &= \frac{\partial}{\partial x^1} u_{(1)} + \frac{\partial}{\partial x^2}\left(\frac{u_{(2)}}{x^1}\right) + \frac{\partial}{\partial x^3}\left(\frac{u_{(3)}}{x^1 \sin x^2}\right) + u_{(1)}\left(\frac{2}{x^1}\right) + \frac{u_{(2)}}{x^1}(\cot x^2)\\ &= \frac{\partial u_{(\rho)}}{\partial \rho} + \frac{1}{\rho}\frac{\partial u_{(\varphi)}}{\partial \varphi} + \frac{1}{\rho \sin \varphi}\frac{\partial u_{(\theta)}}{\partial \theta} + \frac{2}{\rho} u_{(\rho)} + \frac{\cot \varphi}{\rho} u_{(\varphi)}\end{aligned}$$

It is in this last form that "the divergence in spherical coordinates" is generally encountered in reference books.

11.10 Establish in rectangular coordinates the identity

$$\nabla \times (\nabla \times \mathbf{u}) = \nabla(\nabla \cdot \mathbf{u}) - \nabla^2 \mathbf{u} \tag{1}$$

("*curl curl* equals *grad div* minus *del-square*").

Both sides of (*1*) are (cartesian) vectors; we shall show that they are componentwise equal.

By (*11.17a*), the ith component of curl $\mathbf{u}$ is $e_{ijk}(\partial u^k/\partial x^j)$. Therefore, the ith component of curl (curl $\mathbf{u}$) is [use (*3.23*)]:

$$\begin{aligned}e_{irs}\frac{\partial}{\partial x^r}\left(e_{sjk}\frac{\partial u^k}{\partial x^j}\right) &= e_{irs}e_{sjk}\frac{\partial^2 u^k}{\partial x^r \partial x^j} = e_{sir}e_{sjk}\frac{\partial^2 u^k}{\partial x^r \partial x^j}\\ &= (\delta_{ij}\delta_{rk} - \delta_{ik}\delta_{rj})\frac{\partial^2 u^k}{\partial x^r \partial x^j} = \frac{\partial^2 u^r}{\partial x^r \partial x^i} - \frac{\partial^2 u^i}{\partial x^r \partial x^r}\\ &\equiv \frac{\partial}{\partial x^i}(\operatorname{div}\mathbf{u}) - \nabla^2 u^i\end{aligned}$$

The first term on the right is recognized as the ith component of grad (div $\mathbf{u}$), and the second term is (by definition) the ith component of the Laplacian of the vector $\mathbf{u}$. QED

11.11 Prove that the array represented in rectangular coordinates (x^i) by

$$\operatorname{curl}\mathbf{u} = \left(e_{ijk}\frac{\partial u^k}{\partial x^j}\right)$$

is a direct cartesian tensor.

It suffices to show that (e_{ijk}) is a direct cartesian tensor, since $(\partial u^i/\partial x^j)$ is known to be a (direct) cartesian tensor. Therefore, given the orthogonal transformation $\bar{x}^i = a^i_j x^j$, with $|a^i_j| = +1$, define the $3^3 = 27$ quantities

$$\tau_{ijk} \equiv e_{rst}\frac{\partial x^r}{\partial \bar{x}^i}\frac{\partial x^s}{\partial \bar{x}^j}\frac{\partial x^t}{\partial \bar{x}^k} = e_{rst}a^i_r a^j_s a^k_t$$

We observe that:

(i) $\tau_{ijk} = 0$ when two subscripts have the same value; e.g.,

$$\tau_{i22} = e_{rst}a^i_r a^2_s a^2_t = -e_{rts}a^i_r a^2_s a^2_t = -e_{rts}a^i_r a^2_t a^2_s = -\tau_{i22}$$

(ii)
$$\tau_{123} = e_{rst}a^1_r a^2_s a^3_t = |a^i_j| = +1$$

(iii) τ_{ijk} changes sign when any two subscripts are interchanged; e.g.

$$\tau_{kji} = e_{rst}a^k_r a^j_s a^i_t = -e_{tsr}a^k_r a^j_s a^i_t = -\tau_{ijk}$$

But these three properties identify τ_{ijk} with $\bar{e}_{ijk}$, and the proof is complete.

11.12 Show that in a vacuum with zero charge ($\rho = 0$), the electric field $\mathbf{E}$ satisfies the vector wave equation

$$\frac{\partial^2 \mathbf{E}}{\partial t^2} = c^2 \nabla^2 \mathbf{E}$$

From Maxwell's equations (*11.19*), along with the identity (*11.23*),

$$\nabla \times (\nabla \times \mathbf{E}) = -\frac{1}{c}\frac{\partial}{\partial t}(\nabla \times \mathbf{H}) = -\frac{1}{c}\left(\frac{1}{c}\frac{\partial^2 \mathbf{E}}{\partial t^2}\right) = -\frac{1}{c^2}\frac{\partial^2 \mathbf{E}}{\partial t^2}$$

But $\nabla \cdot \mathbf{E} = 0$ and Problem 11.10 imply $\nabla \times (\nabla \times \mathbf{E}) = -\nabla^2 \mathbf{E}$, and the wave equation follows.

Supplementary Problems

11.13 Show that if v is constant, a particle describes equal lengths of arc in equal periods of time.

11.14 (*a*) Show that a particle whose path is given by $\mathbf{x} = (\cos t, \sin t, \cot t)$, for $\pi/4 \leqq t < \pi/2$, has velocity decreasing to $\sqrt{2}$ as $t \to \pi/2$. (*b*) What is the behavior of the acceleration as $t \to \pi/2$? (*c*) Find the extreme values of v and a for this particle.

11.15 For what kind of motion, if any, is $a = dv/dt$?

11.16 Develop a formula for $\dot{\mathbf{a}}$ for a particle that has constant speed v.

11.17 Calculate the acceleration components (contravariant) in spherical coordinates (ρ, φ, θ).

11.18 Prove that motion under a central force is planar.

11.19 Calculate the Laplacian for cylindrical coordinates (r, θ, z).

11.20 Show that $\nabla^2 f = g^{ij} f_{,ij}$. [*Hint*: Write (*11.15*) at the origin of Riemannian coordinates.]

11.21 Prove (*11.22*) and (*11.23*).

11.22 Prove that curl (grad f) $= \mathbf{0}$ for any C^2 scalar field f.

11.23 Show that in a charge-free vacuum, $\mathbf{H}$ also satisfies the vector wave equation.

11.24 Show that, relative to an orthogonal curvilinear coordinate system (x^1, x^2, x^3), an arbitrary contravariant vector $\mathbf{v} = (v^i)$ has the representation

$$\mathbf{v} = v_{(1)}\mathbf{e}_1 + v_{(2)}\mathbf{e}_2 + v_{(3)}\mathbf{e}_3$$

where $v_{(\alpha)}$ is the physical component and $\mathbf{e}_\alpha$ is the unit normal to the surface $x^\alpha = \text{const.}$ [*Hint*: Use Problems 5.19 and 5.20].

Chapter 12

Tensors in Special Relativity

12.1 INTRODUCTION

If the motion of a light pulse were an ordinary phenomenon, its velocity c to one observer would appear to a second observer, moving at velocity v relative to the first, to have the value $c - v$. This hypothetical property of light depends on the concept of absolute time measurement for all observers. However, beginning with the landmark Michelson–Morley experiment in 1880, all experimental data force us to abandon this reasonable hypothesis and to accept instead the now undisputed fact that the velocity of light, rather than the measurement of time, is an absolute of nature. Light is observed to have a single velocity, $c = 2.9979 \times 10^8$ m/s, independent of the observer's motion away from or towards its source. This calls for adjustments to the equations of Newtonian mechanics which become major when high-velocity particles are involved.

12.2 EVENT SPACE

It is first necessary to wed the concepts of time and space. Thus, each event (atomic collision, flash of lightning, etc.) is assigned four coordinates (t, x, y, z), where t is the time (in seconds) of the event and (x, y, z) is the location (in meters) of the event in ordinary rectangular coordinates. Such coordinates are called *space-time* coordinates.

Definition 1: An *event space* is an $\mathbf{R}^4$ whose points are *events*, coordinatized by $(x^i) = (x^0, x^1, x^2, x^3)$, where $x^0 = ct$ is the *temporal coordinate*, and $(x^1, x^2, x^3) = (x, y, z)$ the rectangular *positional coordinates*, of an event. Two events $E_1(\mathbf{x}_1)$ and $E_2(\mathbf{x}_2)$ are *identical* if $x_1^i = x_2^i$ for all i; *simultaneous*, if $x_1^0 = x_2^0$; and *copositional*, if $x_1^i = x_2^i$ for $i = 1, 2, 3$. The *spatial distance* between E_1 and E_2 is the number

$$d = \sqrt{(\Delta x^1)^2 + (\Delta x^2)^2 + (\Delta x^3)^2} \tag{12.1}$$

where $\Delta x^i \equiv x_2^i - x_1^i$ for $i = 1, 2, 3$.

Inertial Reference Frames

The general setting for Einstein's Special Theory of Relativity (henceforth abbreviated SR) consists of two or more observers $O, \bar{O}, \bar{\bar{O}}, \ldots$, moving at constant velocities relative to each other, who set up space-time coordinate systems $(x^i), (\bar{x}^i), (\bar{\bar{x}}^i), \ldots$ to record events and make calculations for experiments they conduct. Such coordinate systems in uniform relative motion are called *inertial frames* provided Newton's first law is valid in each system. All the systems are assumed to have a common origin at some instant, which is taken as $t = \bar{t} = \bar{\bar{t}} = \cdots = 0$.

Light Cone

A flash of light at position $(0, 0, 0)$ and time $t = 0$ sends out an expanding spherical wave front, with equation $x^2 + y^2 + z^2 = c^2t^2$, or

$$(x^0)^2 - (x^1)^2 - (x^2)^2 - (x^3)^2 = 0 \tag{12.2}$$

(*12.2*) is the equation of the *light cone* in event space, relative to the inertial frame (x^i). Figure 12-1 shows the projection of the light cone onto the hyperplane $x^3 = 0$. In any other inertial frame, $(\bar{x}^i)$, the equation of the light cone is exactly the same (since all observers measure the velocity of light as c):

$$(\bar{x}^0)^2 - (\bar{x}^1)^2 - (\bar{x}^2)^2 - (\bar{x}^3)^2 = 0$$

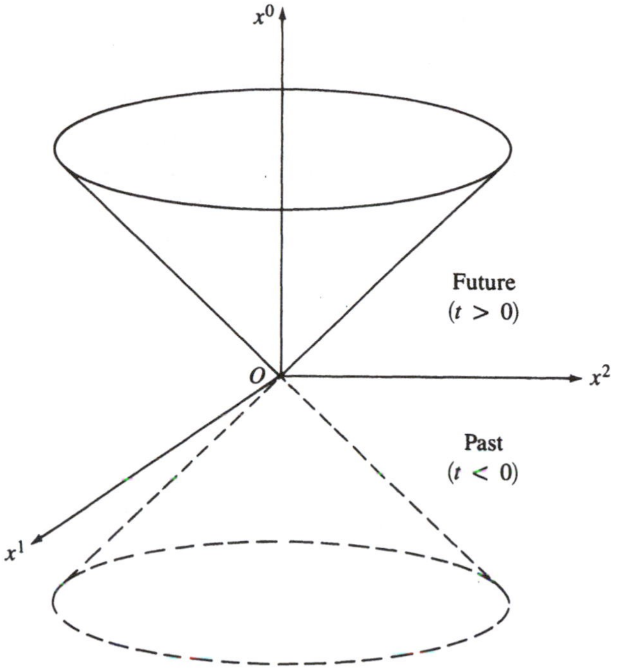

Fig. 12-1

In Example 7.6 (using slightly different notation), we identified the light cone with the null geodesics of $\mathbf{R}^4$ under the metric of SR.

Relativistic Length

For an arbitrary event $E(\mathbf{x})$, the quantity $(x^0)^2 - (x^1)^2 - (x^2)^2 - (x^3)^2$ may be positive, negative, or zero. The *relativistic distance* from $E(\mathbf{x})$ to the origin $E_0(\mathbf{0})$ is the real number $s \geqq 0$ such that

$$\varepsilon s^2 = (x^0)^2 - (x^1)^2 - (x^2)^2 - (x^3)^2 \qquad (\varepsilon = \pm 1)$$

More generally, the *length of interval* or *relativistic distance* between $E_1(\mathbf{x}_1)$ and $E_2(\mathbf{x}_2)$ is the unique real number $\Delta s \geqq 0$ such that

$$\varepsilon(\Delta s)^2 = (\Delta x^0)^2 - (\Delta x^1)^2 - (\Delta x^2)^2 - (\Delta x^3)^2 \qquad (\varepsilon = \pm 1) \tag{12.3}$$

where $\Delta x^i \equiv x_2^i - x_1^i$ for $i = 0, 1, 2, 3$. The chief significance of this length-concept is to be found in

Theorem 12.1: Relativistic distance is an invariant across all inertial frames.

For a proof, see Problem 12.6.

Interval Types

The interval between $E_1(\mathbf{x}_1)$ and $E_2(\mathbf{x}_2)$ is

(1) *spacelike* if $(\Delta x^1)^2 + (\Delta x^2)^2 + (\Delta x^3)^2 > (\Delta x^0)^2$ (or $\varepsilon = -1$; predominance of distance over time);

(2) *lightlike* if $(\Delta x^0)^2 = (\Delta x^1)^2 + (\Delta x^2)^2 + (\Delta x^3)^2$ (equality of time and distance);

(3) *timelike* if $(\Delta x^0)^2 > (\Delta x^1)^2 + (\Delta x^2)^2 + (\Delta x^3)^2$ (or $\varepsilon = +1$; predominance of time over distance).

By Theorem 12.1, the categorization is independent of the particular inertial frame.

12.3 THE LORENTZ GROUP AND THE METRIC OF SR

Imagine two observers, O and $\bar{O}$, in uniform relative motion at speed v. They approach each other during negative time, coincide at zero time, and then recede from each other during positive time (Fig. 12-2(*a*)). Let O and $\bar{O}$ set up independent reference frames (x^i) and $(\bar{x}^i)$ by means of identical but separate clocks, with $t = \bar{t} = 0$ when $O = \bar{O}$, and identical metersticks. Newton's first law will be assumed to hold in both frames, making them inertial frames.

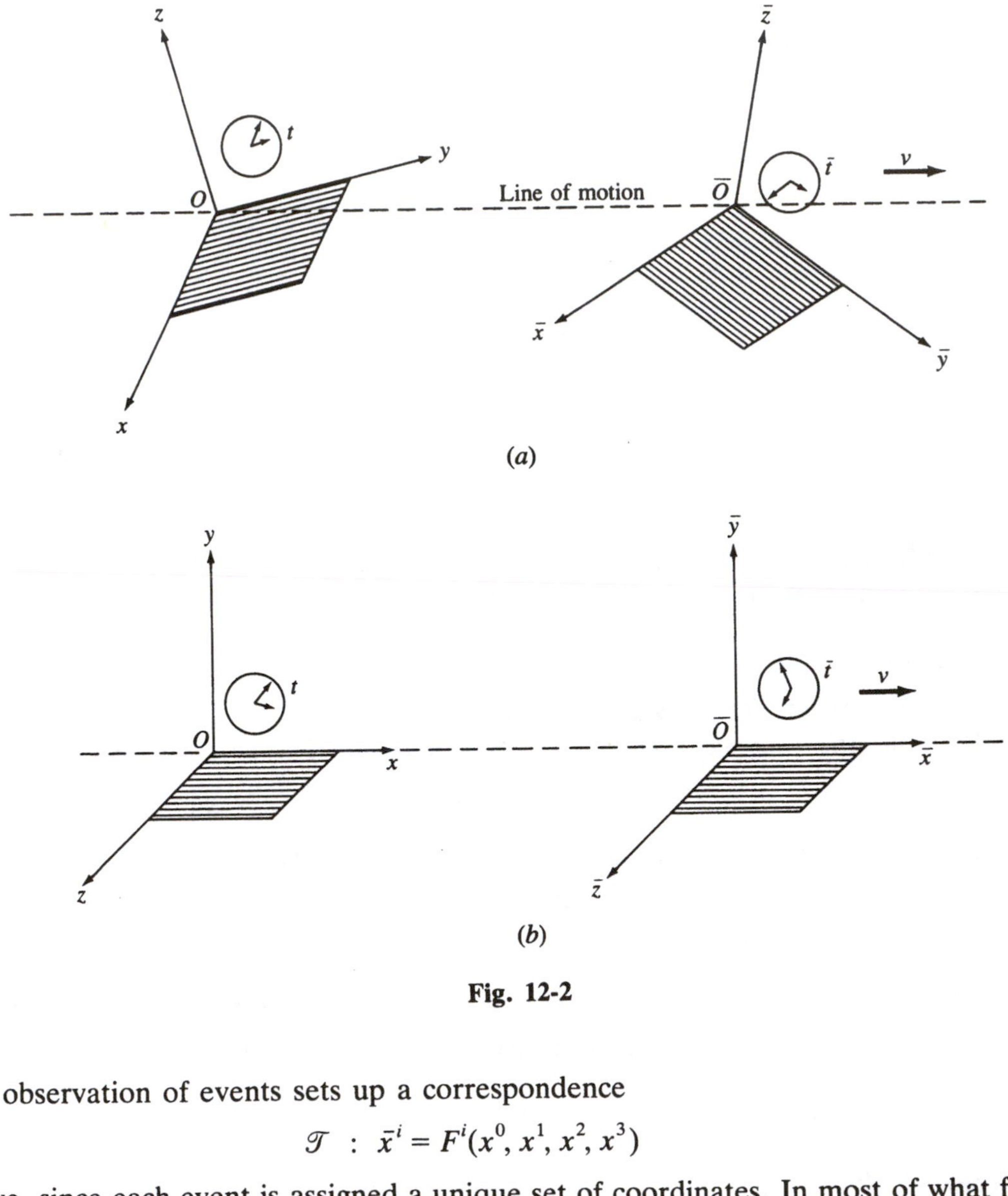

Fig. 12-2

Common observation of events sets up a correspondence

$$\mathscr{T} \; : \; \bar{x}^i = F^i(x^0, x^1, x^2, x^3) \tag{12.4}$$

that is bijective, since each event is assigned a unique set of coordinates. In most of what follows, it will be assumed that O and $\bar{O}$ perform a simplifying maneuver at the instant of coincidence, whereby their x-axes point in the same direction along the line of motion and their y-axes and their z-axes coincide. In the ensuing translation, the y-axes and the z-axes remain parallel (Fig. 12-2(*b*)).

Postulates of SR

(1) *Principle of Relativity:* The laws of physics are the same in all inertial frames.

(2) *Invariance of Uniform Motion:* A particle with constant velocity in one inertial frame will have constant velocity in all inertial frames.

(3) *Invariance of Light Speed:* The speed of light is invariant across all inertial frames.

Postulate 2 requires that the bijective transformation (*12.4*) be such as to map straight lines into straight lines. Consequently, each F^i must be a linear function. Since $F^i(\mathbf{0}) = \mathbf{0}$, constants a^i_j exist such that

$$T \; : \; \bar{x}^i = a^i_j x^j \tag{12.5}$$

Lorentz Matrices and Transformations

The invariance of the equation of the light cone (in consequence of Postulate 3) may be expressed as

$$g_{ij} x^i x^j = 0 = g_{ij} \bar{x}^i \bar{x}^j \tag{12.6}$$

where $g_{00} = 1$, $g_{11} = g_{22} = g_{33} = -1$, and $g_{ij} = 0$ for $i \neq j$. Substitution of (*12.5*) in (*12.6*) yields (Problem 12.4):

$$g_{ij} a^i_r a^j_s = g_{rs} \tag{12.7a}$$

or, in matrix form,

$$A^T G A = G \tag{12.7b}$$

or, written out,

$$\begin{aligned} (a^0_0)^2 - (a^1_0)^2 - (a^2_0)^2 - (a^3_0)^2 &= 1 \\ (a^0_j)^2 - (a^1_j)^2 - (a^2_j)^2 - (a^3_j)^2 &= -1 \qquad (j = 1, 2, 3) \\ a^0_i a^0_j - a^1_i a^1_j - a^2_i a^2_j - a^3_i a^3_j &= 0 \qquad (i \neq j) \end{aligned} \tag{12.7c}$$

It is easy to see (Problem 12.8) that requiring $g_{ij} x^i x^j = 0$ to be invariant is tantamount to requiring the invariance of $g_{ij} x^i x^j = q$ for every value of q. Thus, (*12.7*) is a criterion for the quadratic form $g_{ij} x^i x^j$ to be invariant.

Definition 2: Any 4×4 matrix (or corresponding linear transformation) that preserves the quadratic form $\mathbf{x}^T G \mathbf{x}$ is called *Lorentz*.

In Problem 12.10 it is shown that the set of Lorentz matrices constitutes a group (the *Lorentz group*) under matrix multiplication.

Metric of SR

If the terms $\bar{g}_{ij} \equiv g_{ij}$ are defined for the $(\bar{x}^i)$-system, then (*12.7a*) becomes $g_{rs} = \bar{g}_{ij} a^i_r a^j_s$, which makes (g_{ij}) a covariant tensor of the second order under Lorentz transformations of coordinates. Accordingly, the metric for $\mathbf{R}^4$ is chosen as

$$\varepsilon \, ds^2 = g_{ij} \, dx^i \, dx^j \equiv (dx^0)^2 - (dx^1)^2 - (dx^2)^2 - (dx^3)^2 \tag{12.8}$$

12.4 SIMPLE LORENTZ MATRICES

Let us suppose that O and $\bar{O}$ have performed an alignment of their xyz-axes. Then any right circular cylinder with axis along the line of relative motion must have the same equation in the two systems; i.e., $(x^2)^2 + (x^3)^2$ is invariant. It follows (see Problem 12.11) that the Lorentz transformation for this situation has the form

$$\mathcal{T} \; : \; \begin{cases} \bar{x}^0 = a^0_0 x^0 + a^0_1 x^1 \equiv a x^0 + b x^1 \\ \bar{x}^1 = a^1_0 x^0 + a^1_1 x^1 \equiv d x^0 + e x^1 \\ \bar{x}^2 = x^2 \\ \bar{x}^3 = x^3 \end{cases} \tag{12.9}$$

By (12.7),

$$a^2 - d^2 = 1 \qquad b^2 - e^2 = -1 \qquad ab - de = 0 \tag{12.10}$$

By considering the coordinates which O and $\bar{O}$ would assign to each other's origin (see Problem 12.12), we find that

$$d = -(v/c)a \equiv -\beta a \qquad \text{and} \qquad a = e \tag{12.11}$$

(The notation $\beta = v/c$ is standard in SR.) From (12.10) and the fact that $a > 0$ (since both clocks can be assumed to run in the same sense), it follows that

$$a = (1 - \beta^2)^{-1/2} = e \qquad b = -\beta(1 - \beta^2)^{-1/2} = d \tag{12.12}$$

Therefore, the coordinate transformation takes on the simplified form

$$\mathcal{T}: \begin{cases} \bar{x}^0 = \dfrac{x^0 - \beta x^1}{\sqrt{1-\beta^2}} \\ \bar{x}^1 = \dfrac{-\beta x^0 + x^1}{\sqrt{1-\beta^2}} \\ \bar{x}^2 = x^2 \\ \bar{x}^3 = x^3 \end{cases} \quad \text{or} \quad A = \begin{bmatrix} \dfrac{1}{\sqrt{1-\beta^2}} & \dfrac{-\beta}{\sqrt{1-\beta^2}} & 0 & 0 \\ \dfrac{-\beta}{\sqrt{1-\beta^2}} & \dfrac{1}{\sqrt{1-\beta^2}} & 0 & 0 \\ 0 & 0 & 1 & 0 \\ 0 & 0 & 0 & 1 \end{bmatrix} \tag{12.13}$$

Any 4×4 matrix (linear transformation) of the form

$$A = \begin{bmatrix} a & b & 0 & 0 \\ b & a & 0 & 0 \\ 0 & 0 & 1 & 0 \\ 0 & 0 & 0 & 1 \end{bmatrix}$$

where $a^2 - b^2 = 1$, will be termed *simple Lorentz*. The relative velocity in the physical situation modeled by A is recovered as $\beta = -b/a$.

EXAMPLE 12.1 By Problem 12.9, the inverse of a simple Lorentz matrix is

$$A^{-1} = \begin{bmatrix} a & -b & 0 & 0 \\ -b & a & 0 & 0 \\ 0 & 0 & 1 & 0 \\ 0 & 0 & 0 & 1 \end{bmatrix}$$

which is itself a simple Lorentz matrix, corresponding to a reversal in sign of β. [If the velocity of $\bar{O}$ with respect to O is v, then the velocity of O with respect to $\bar{O}$ is $-v$.]

A Decomposition Theorem

The possibility of simplifying an arbitrary Lorentz matrix by a suitable rotation of axes can be expressed in purely mathematical terms. (See Problems 12.14 and 12.15.)

Theorem 12.2: An arbitrary Lorentz matrix $L = (a^i_j)$ has the representation

$$L = R_1 L^* R_2$$

where L^* is a simple Lorentz matrix with parameters $a = |a^0_0| = \varepsilon a^0_0$ and $b = -\sqrt{(a^0_0)^2 - 1}$, and R_1 and R_2 are orthogonal Lorentz matrices defined by

$$R_1 = L R_2^T (L^*)^{-1} \qquad \text{and} \qquad R_2 = [\mathbf{e}_1 \quad \mathbf{r}' \quad \mathbf{s}' \quad \mathbf{t}']^T \tag{12.14}$$

Here, $\mathbf{e}_1 = (1, 0, 0, 0)$, $\mathbf{r}' = (\varepsilon/b)(0, a^0_1, a^0_2, a^0_3) \equiv (0, \mathbf{r})$, $\mathbf{s}' = (0, \mathbf{s})$, $\mathbf{t}' = (0, \mathbf{t})$, with $\mathbf{s}$ and $\mathbf{t}$ chosen to complete a 3×3 orthogonal matrix $[\mathbf{r} \quad \mathbf{s} \quad \mathbf{t}]$.

Corollary 12.3: If $L = (a^i_j)$ connects two inertial frames, then the relative velocity between the frames is

$$v = c\sqrt{1 - (a^0_0)^{-2}} \tag{12.15}$$

12.5 PHYSICAL IMPLICATIONS OF THE SIMPLE LORENTZ TRANSFORMATION

Length Contraction

For any fixed x^0, (*12.13*) gives

$$\Delta\bar{x}^1 = \frac{1}{\sqrt{1-\beta^2}}\,\Delta x^1 \qquad \text{or} \qquad \Delta x^1 = \sqrt{1-\beta^2}\,\Delta\bar{x}^1 < \Delta\bar{x}^1$$

If the frame $\bar{O}$ is moving at a uniform velocity v relative to O, distances in $\bar{O}$ appear to observer O to be foreshortened in the direction of the motion by the factor $\sqrt{1-\beta^2}$.

Time Dilation

For any fixed $\bar{x}^1$, (*12.13*), inverted, gives

$$\Delta x^0 = \frac{1}{\sqrt{1-\beta^2}}\,\Delta\bar{x}^0 \qquad \text{or} \qquad \Delta t = \frac{1}{\sqrt{1-\beta^2}}\,\Delta\bar{t} > \Delta\bar{t}$$

If the frame $\bar{O}$ is moving at a uniform velocity v relative to O, the clock of observer $\bar{O}$ appears to observer O to run slow by the factor $\sqrt{1-\beta^2}$.

Composition of Velocities

If $\bar{O}$ has velocity v_1 relative to O and $\bar{\bar{O}}$ has velocity v_2 relative to $\bar{O}$, then the Newtonian composition of velocities predicts that $\bar{\bar{O}}$ has velocity $v_3 = v_1 + v_2$ relative to O. Although the error does not show up unless v_1 and v_2 are substantial fractions of the velocity of light, SR shows the Newtonian theory to be incorrect. The correct formula (Problem 12.20) is

$$v_3 = \frac{v_1 + v_2}{1 + v_1 v_2/c^2} \tag{12.16}$$

12.6 RELATIVISTIC KINEMATICS

4-Vectors

We begin with ordinary velocity and acceleration of a particle within a single inertial frame (x^i). By introducing the concept of *proper time*, we shall be able to obtain velocity and acceleration as contravariant vectors with respect to Lorentz transformations (to be called *4-vectors* from now on). In general, (V^i) is a 4-vector if it transforms according to the law $\bar{V}^i = a^i_j V^j$, where (a^i_j) is the Lorentz matrix of (*12.5*). It is customary to use the following notation for 4-vectors:

$$(V^i) \equiv (V^0, \mathbf{V}) \qquad \text{where} \qquad V^0 \equiv V_t \qquad \text{and} \qquad \mathbf{V} \equiv (V^1, V^2, V^3) \equiv (V_x, V_y, V_z)$$

V^0 is referred to as the *time component* of the vector and (V_x, V_y, V_z) are the usual *space components*. All indices are understood to range over the values 0, 1, 2, 3, unless specifically noted otherwise.

Nonrelativistic Velocity and Acceleration

In the inertial frame $(x^i) = (x, y, z)$ let a particle describe the class C^2 curve

$$\mathscr{K} \;:\; (x^i) = (ct, \mathbf{r}(t)) = (ct, x(t), y(t), z(t))$$

Then we have the classical formulas

$$(v_i) = \left(\frac{dx^i}{dt}\right) \equiv (c, \mathbf{v}) \tag{12.17}$$

where $\mathbf{v} = d\mathbf{r}/dt$ and $\hat{v} \equiv \|\mathbf{v}\| = \sqrt{v_x^2 + v_y^2 + v_z^2}$;

$$(a_i) = \left(\frac{d^2x^i}{dt^2}\right) \equiv (0, \mathbf{a}) \tag{12.18}$$

where $\mathbf{a} = d\mathbf{v}/dt$ and $\hat{a} = \|\mathbf{a}\| = \sqrt{a_x^2 + a_y^2 + a_z^2}$.

As defined, neither the velocity nor the acceleration is a tensor under Lorentz transformations. In fact (Problem 12.22), if $(\bar{v}_i)$ and $(\bar{a}_i)$ are the like quantities in $(\bar{x}^i)$, then (*12.13*) yields the relations

$$\bar{v}_0 = c = v_0 \qquad \bar{v}_x = \frac{v_x - v}{1 - v_x v/c^2} \qquad \bar{v}_y = \frac{v_y\sqrt{1-\beta^2}}{1 - v_x v/c^2} \qquad \bar{v}_z = \frac{v_z\sqrt{1-\beta^2}}{1 - v_x v/c^2} \tag{12.19}$$

$$\begin{gathered} \bar{a}_0 = 0 = a_0 \qquad \bar{a}_x = \frac{a_x(1-\beta^2)^{3/2}}{(1 - v_x v/c^2)^3} \qquad \bar{a}_y = \frac{[a_y + (v_y a_x - v_x a_y)(v/c^2)](1-\beta^2)}{(1 - v_x v/c^2)^3} \\ \bar{a}_z = \frac{[a_z + (v_z a_x - v_x a_z)(v/c^2)](1-\beta^2)}{(1 - v_x v/c^2)^3} \end{gathered} \tag{12.20}$$

The inverse relations can be quickly obtained by replacing v by $-v$ and interchanging barred and unbarred terms throughout. For example, the second formula in (*12.19*) inverts to

$$v_x = \frac{\bar{v}_x + v}{1 + \bar{v}_x v/c^2}$$

which is just (*12.16*) as applied to $v_1 = \bar{v}_x$ and $v_2 = v$.

Proper Time; Velocity and Acceleration 4-Vectors

Let us reparameterize the curve $\mathscr{K}$, choosing now the quantity

$$\tau = \frac{s}{c} = \frac{1}{c}\int_{t_0}^{t}\sqrt{\varepsilon g_{ij}\frac{dx^i}{du}\frac{dx^j}{du}}\,du \qquad \text{or} \qquad \frac{d\tau}{dt} = \sqrt{1 - \hat{v}^2/c^2} \tag{12.21}$$

where, as always, $\hat{v} < c$. The new parameter τ (a distance divided by a velocity) is known as the *proper time* for the particle; by Problem 12.23, a clock attached to the particle (and thus accelerating and decelerating along with it) reads τ.

When τ-derivatives replace t-derivatives, velocity and acceleration become tensorial; i.e., the components

$$u^i \equiv \frac{dx^i}{d\tau} \qquad b^i \equiv \frac{du^i}{d\tau} = \frac{d^2x^i}{d\tau^2} \tag{12.22}$$

are taken by (*12.13*) into

$$\bar{u}^0 = \frac{u^0 - \beta u^1}{\sqrt{1-\beta^2}} \qquad \bar{u}^1 = \frac{-\beta u^0 + u^1}{\sqrt{1-\beta^2}} \qquad \bar{u}^2 = u^2 \qquad \bar{u}^3 = u^3 \tag{12.23}$$

$$\bar{b}^0 = \frac{b^0 - \beta b^1}{\sqrt{1-\beta^2}} \qquad \bar{b}^1 = \frac{-\beta b^0 + b^1}{\sqrt{1-\beta^2}} \qquad \bar{b}^2 = b^2 \qquad \bar{b}^3 = b^3 \tag{12.24}$$

The important identities

$$u_i u^i = c^2 \qquad u_i b^i = 0 \tag{12.25}$$

are proved in Problem 12.24, and Problem 12.25 establishes the following connecting formulas between the numerical values of the relativistic and the nonrelativistic components:

$$u^i = \frac{v_i}{\sqrt{1 - \hat{v}^2/c^2}} \qquad b^i = \frac{a_i}{1 - \hat{v}^2/c^2} + \frac{(\mathbf{va})v_i}{c^2(1 - \hat{v}^2/c^2)^2} \tag{12.26}$$

Instantaneous Rest Frame

At time $t = t_1$, the particle moving along $\mathscr{K}$ has instantaneous position $P_1 = \mathbf{r}(t_1)$ and instantaneous speed $\hat{v}(t_1)$. An *instantaneous rest frame* for the particle is an inertial frame that translates at speed $\hat{v}(t_1)$ along the tangent to $\mathscr{K}$ at P_1, in such manner that its origin coincides with P_1 at $t = t_1$. See Fig. 12-3.

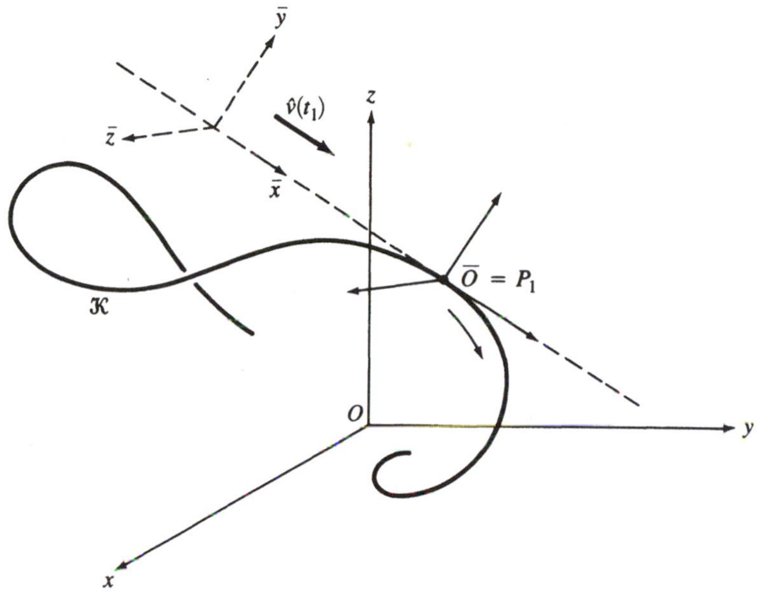

Fig. 12-3

We shall say that a particle's motion (with respect to frame O) is *uniformly accelerated* if its spatial acceleration relative to an instantaneous rest frame $\bar{O}$,

$$\alpha = \hat{\bar{a}} = \sqrt{\bar{a}_x^2 + \bar{a}_y^2 + \bar{a}_z^2}$$

does not vary along the trajectory $\mathscr{K}$.

EXAMPLE 12.2 An electron fired at right angles into a uniform magnetic field undergoes uniformly accelerated motion (in a circle).

12.7 RELATIVISTIC MASS, FORCE, AND ENERGY

The appropriate SR version of Newton's second law depends on the concept of mass to be adopted.

Rest Mass and Relativistic Mass

The *rest mass* of a particle is its mass as measured or as inferred from Newtonian mechanics in any instantaneous rest frame for that particle.

The *relativistic mass* (in O) of a particle with spatial velocity $\mathbf{v}$ (with respect to O) is

$$\hat{m} = \frac{m}{\sqrt{1 - \hat{v}^2/c^2}} \tag{12.27}$$

where m is the rest mass of the particle. As is shown in Problem 12.27, (*12.27*) is a necessary consequence of conservation of momentum.

Relativistic Momentum and Force

The *4-momentum* of SR is defined by

$$(p^i) \equiv (p^0, \mathbf{p}) = (mu^i) = (\hat{m}v_i) \tag{12.28}$$

and the (nontensorial) *Lorentz force* $(F_0, \mathbf{F})$ is defined as the time derivative of the 4-momentum:

$$F_0 \equiv \frac{dp^0}{dt} = \frac{d}{dt}\left(\frac{mc}{\sqrt{1-\hat{v}^2/c^2}}\right) \qquad \mathbf{F} \equiv \frac{d}{dt}\left(\frac{m\mathbf{v}}{\sqrt{1-\hat{v}^2/c^2}}\right) \tag{12.29}$$

Like velocity, force becomes tensorial when proper time is introduced. Thus, the *4-force* (*Minkowski force*) of SR is defined as

$$(K^i) \equiv \left(\frac{dp^i}{d\tau}\right) \tag{12.30}$$

From (*12.21*), we have the connecting formula

$$K^i = \frac{F_i}{\sqrt{1-\hat{v}^2/c^2}} \tag{12.31}$$

The following identities for the Lorentz and Minkowski forces are proved in Problem 12.29:

$$u_i K^i = 0 \qquad K^0 = \frac{1}{c}\,\mathbf{vK} \qquad F_0 = \frac{1}{c}\,\mathbf{vF} \qquad \mathbf{vF} = \frac{d}{dt}\left(\frac{mc^2}{\sqrt{1-\hat{v}^2/c^2}}\right) \tag{12.32}$$

Relativistic Energy

According to the classical *work-energy theorem*, the rate at which work is performed on a particle, $(\mathbf{vF})$, equals the rate of increase of the particle's kinetic energy. Thus, the last identity (*12.32*) suggests the definition for SR

$$\hat{E} = \frac{mc^2}{\sqrt{1-\hat{v}^2/c^2}} \equiv \hat{m}c^2 \tag{12.33}$$

for the energy of a particle moving at speed $\hat{v}$. In the limit as $\hat{v} \to 0$, (*12.33*) becomes

$$E = mc^2 \tag{12.34}$$

This is Einstein's famous formula for the *rest energy* E of a particle with rest mass m.

12.8 MAXWELL'S EQUATIONS IN SR

It is helpful to look briefly at the way the metric for Special Relativity affects the formulas for the divergence and the Laplacian, and to consider a new kind of matrix that will be useful in formulating Maxwell's equations.

Vector Calculus and Lorentz Transformations

For the metric of SR, all Christoffel symbols vanish, so that

$$\operatorname{div}\mathbf{u} = \frac{\partial u^i}{\partial x^i} \tag{12.35}$$

and

$$\begin{aligned}\operatorname{div}(\operatorname{grad} f) \equiv \Box f &= \frac{\partial}{\partial x^i}\left(g^{ij}\frac{\partial f}{\partial x^j}\right) = g^{ij}\frac{\partial^2 f}{\partial x^i \partial x^j}\\ &= \frac{\partial^2 f}{(\partial x^0)^2} - \frac{\partial^2 f}{(\partial x^1)^2} - \frac{\partial^2 f}{(\partial x^2)^2} - \frac{\partial^2 f}{(\partial x^3)^2}\\ &= \frac{1}{c^2}\frac{\partial^2 f}{\partial t^2} - \nabla^2 f\end{aligned} \tag{12.36}$$

Note that in SR the Laplacian operator is notated $\Box$, with ∇^2 reserved for its spatial part. It is verified in Problem 12.31 that $\Box f$ is an invariant under Lorentz transformations, which means that the scalar wave equation has the same form, $\Box f = 0$, in every inertial frame.

If we introduce the differential operator

$$\partial^i \equiv g^{ij} \frac{\partial}{\partial x^j} \tag{12.37}$$

then we can express the *equation of continuity* for a vector $(w_i) \equiv (w_0, \mathbf{w})$ as

$$\partial^i w_i = 0 \tag{12.38}$$

(*12.38*) is equivalent to $\partial w_0/\partial t = c \operatorname{div} \mathbf{w}$.

Maxwell's Equations in Event Space

First, introduce for any two 3-vectors $\mathbf{U} = (U^i)$ and $\mathbf{V} = (V^i)$ the two antisymmetric matrices

$$[f^{ij}]_{44} \equiv \begin{bmatrix} 0 & -V^1 & -V^2 & -V^3 \\ V^1 & 0 & U^3 & -U^2 \\ V^2 & -U^3 & 0 & U^1 \\ V^3 & U^2 & -U^1 & 0 \end{bmatrix} \qquad [\tilde{f}_{ij}]_{44} \equiv \begin{bmatrix} 0 & U^1 & U^2 & U^3 \\ -U^1 & 0 & V^3 & -V^2 \\ -U^2 & -V^3 & 0 & V^1 \\ -U^3 & V^2 & -V^1 & 0 \end{bmatrix} \tag{12.39a}$$

The second matrix may be obtained from the first by making the replacements $\mathbf{V} \to -\mathbf{U}$ and $\mathbf{U} \to \mathbf{V}$; because these replacements constitute an anti-involution—i.e., $\tilde{\tilde{f}}_{ij} = -f^{ij}$—the two matrices are said to be *dual* to each other. In terms of individual components (e_{pqr} denotes the permutation symbol of order 3):

$$\begin{matrix} f^{ij} = -f^{ji} & f^{0q} = -V^q & f^{pq} = e_{pqr} U^r \\ \tilde{f}_{ij} = -\tilde{f}_{ji} & \tilde{f}_{0q} = U^q & \tilde{f}_{pq} = e_{pqr} V^r \end{matrix} \tag{12.39b}$$

in which $i, j \geqq 0$ and $p, q \geqq 1$.

By their concoction, these matrices turn out to be tensors under Lorentz transformations (provided the row-divergences vanish in all inertial frames); a proof is given in Problem 12.32. Moreover, these tensors have the properties (Problem 12.33)

$$\frac{\partial f^{0j}}{\partial x^j} = -\operatorname{div} \mathbf{V} \qquad \frac{\partial \tilde{f}_{0j}}{\partial x^j} = \operatorname{div} \mathbf{U} \tag{12.40}$$

and

$$\left(\frac{\partial f^{1j}}{\partial x^j}, \frac{\partial f^{2j}}{\partial x^j}, \frac{\partial f^{3j}}{\partial x^j} \right) = \operatorname{curl} \mathbf{U} + \frac{1}{c} \frac{\partial \mathbf{V}}{\partial t} \qquad \left(\frac{\partial \tilde{f}_{1j}}{\partial x^j}, \frac{\partial \tilde{f}_{2j}}{\partial x^j}, \frac{\partial \tilde{f}_{3j}}{\partial x^j} \right) = \operatorname{curl} \mathbf{V} - \frac{1}{c} \frac{\partial \mathbf{U}}{\partial t} \tag{12.41}$$

We now show how Maxwell's equations in vacuum, (*11.19*), may be extended to space-time via dual tensors of this sort. The equations are:

$$\operatorname{div} \mathbf{H} = 0 \qquad \operatorname{curl} \mathbf{E} + \frac{1}{c} \frac{\partial \mathbf{H}}{\partial t} = \mathbf{0} \tag{12.42}$$

$$\operatorname{div} \mathbf{E} = \rho \qquad \operatorname{curl} \mathbf{H} - \frac{1}{c} \frac{\partial \mathbf{E}}{\partial t} = \frac{\rho}{c} \mathbf{v} \tag{12.43}$$

in the last of which $\mathbf{v}$ is the classical spatial velocity (*12.17*) of the charge-cloud ρ. Define per (*12.39*) the *tensors*

$$\mathcal{F} = [F^{ij}]_{44} \equiv \begin{bmatrix} 0 & -H_1 & -H_2 & -H_3 \\ H_1 & 0 & E_3 & -E_2 \\ H_2 & -E_3 & 0 & E_1 \\ H_3 & E_2 & -E_1 & 0 \end{bmatrix} \qquad \tilde{\mathcal{F}} = [\tilde{F}^{ij}]_{44} \equiv \begin{bmatrix} 0 & E_1 & E_2 & E_3 \\ -E_1 & 0 & H_3 & -H_2 \\ -E_2 & -H_3 & 0 & H_1 \\ -E_3 & H_2 & -H_1 & 0 \end{bmatrix} \tag{12.44}$$

(with $\mathbf{U} = \mathbf{E}$ and $\mathbf{V} = \mathbf{H}$). In view of the first equations (*12.40*) and (*12.41*), (*12.42*) may be written as

$$\frac{\partial F^{ij}}{\partial x^j} = 0 \tag{12.45a}$$

Similarly, if we make a 4-vector out of $\mathbf{v}$ and ρ by the prescription

$$(s^i) \equiv \left(\rho, \frac{\rho}{c}\,\mathbf{v}\right) \tag{12.46}$$

(see Problem 12.52), then the remaining Maxwell's equations, (*12.43*), are rendered tensorial as

$$\frac{\partial \tilde{F}^{ij}}{\partial x^j} = s^i \tag{12.45b}$$

Equations (*12.45*) are the relativistic Maxwell's equations, valid in every inertial frame. Because $\tilde{\mathbf{F}}$ is antisymmetric, we have from (*12.45b*):

$$\frac{\partial s^i}{\partial x^i} = \frac{\partial^2 \tilde{F}^{ij}}{\partial x^i\,\partial x^j} = 0 \qquad \text{or} \qquad \frac{\partial}{\partial x^i}(g^{ij}s_j) = \left(g^{ji}\frac{\partial}{\partial x^i}\right)s_j \equiv \partial^j s_j = 0$$

so that the covariant vector (s_j) obeys the equation of continuity (*12.38*).

Solved Problems

EVENT SPACE

12.1 Calculate ε and Δs for the event pairs: (*a*) $E_1(5,1,-2,0)$ and $E_2(0,3,1,-3)$, (*b*) $E_1(5,1,3,3)$ and $E_2(2,-1,1,1)$, (*c*) $E_1(7,2,4,4)$ and $E_2(4,1,2,6)$, (*d*) $E_1 \equiv$ flash of light in Chicago at 7 p.m. and $E_2 \equiv$ flash of light in St. Louis (400 miles away) at 7.000 000 61 p.m. (*e*) Determine the interval type in each case.

(*a*) $\varepsilon(\Delta s)^2 = 5^2 - (-2)^2 - (-3)^2 - 3^2 = 25 - 4 - 9 - 9 = 3$, or $\Delta s = \sqrt{3}$ and $\varepsilon = 1$.

(*b*) $\varepsilon(\Delta s)^2 = 9 - 4 - 4 - 4 = -3$, or $\Delta s = \sqrt{3}$ and $\varepsilon = -1$.

(*c*) $\varepsilon(\Delta s)^2 = 9 - 1 - 4 - 4 = 0$, or $\Delta s = 0$ and $\varepsilon = 1$.

(*d*) With $c = 186\,300$ mi/sec, $\varepsilon(\Delta s)^2 = (0.002\,196\,c)^2 - (400)^2 \approx 7375\ \text{mi}^2$, or $\Delta s \approx 85.8$ mi and $\varepsilon = 1$.

(*e*) Timelike, spacelike, lightlike, and timelike, respectively.

12.2 Show that (*a*) simultaneous events have a spacelike interval; (*b*) copositional events have a timelike interval; (*c*) the interval between two light flashes is lightlike if they are simultaneous to an observer who is present at the site of one of the flashes.

(*a*)
$$\varepsilon(\Delta s)^2 = 0^2 - (\Delta x^1)^2 - (\Delta x^2)^2 - (\Delta x^3)^2 < 0$$

(*b*)
$$\varepsilon(\Delta s)^2 = (\Delta x^0)^2 - 0 > 0$$

(*c*) Let the observer measure the proximate flash as $E_1(0,0,0,0)$. The distant flash $E_2(c\,\Delta t, \Delta x^1, \Delta x^2, \Delta x^3)$ will be registered simultaneously, at $x^0 = 0$, if

$$\Delta t = -\frac{\sqrt{(\Delta x^1)^2 + (\Delta x^2)^2 + (\Delta x^3)^2}}{c}$$

But then $\varepsilon(\Delta s)^2 = 0$ and the interval is lightlike. (Note that the (negative) time coordinate of E_2 is *calculated*, not measured.)

THE LORENTZ GROUP

12.3 Prove the following lemma involving the metric of SR, g_{ij}, as given by (*12.6*).

Lemma 12.4: If $C = (c_{ij})$ is a symmetric 4×4 matrix such that $c_{ij}x^i x^j = 0$ for all (x^i) such that $g_{ij}x^i x^j = 0$, there exists a fixed real number λ for which $c_{ij} = \lambda g_{ij}$ $(C = \lambda G)$.

Observe that the vector $(1, \pm 1, 0, 0)$ satisfies $g_{ij}x^i x^j = 0$. Hence, substituting these components into the equation $c_{ij}x^i x^j = 0$ yields

$$c_{00} \pm c_{01} \pm c_{10} + c_{11} = 0 \qquad \text{or} \qquad c_{00} + c_{11} = 0 = c_{01} = c_{10}$$

(by symmetry of C). Similarly, using the vectors $(1, 0, \pm 1, 0)$ and $(1, 0, 0, \pm 1)$, we get

$$c_{00} = -c_{11} = -c_{22} = -c_{33} = \lambda \qquad c_{ij} = 0 \quad (i = 0 \text{ or } j = 0)$$

Finally, employing the vectors $(\sqrt{2}, 1, 1, 0)$, $(\sqrt{2}, 1, 0, 1)$, and $(\sqrt{2}, 0, 1, 1)$, we obtain $c_{12} = c_{13} = c_{23} = 0$.

12.4 Establish the transformation (*12.7*) between inertial frames under the postulates for SR.

From (*12.6*) and (*12.5*),

$$g_{ij}x^i x^j = 0 = g_{ij}\bar{x}^i \bar{x}^j = g_{ij}(a^i_r x^r)(a^j_s x^s) = g_{rs}a^r_i a^s_j x^i x^j$$

that is,

$$g_{rs}a^r_i a^s_j x^i x^j = 0 \qquad \text{whenever} \qquad g_{ij}x^i x^j = 0 \tag{1}$$

Now apply Lemma 12.4 to (*1*), with $g_{rs}a^r_i a^s_j = c_{ij}$, where $C = (c_{ij}) = A^T G A$ is symmetric. We obtain

$$g_{rs}a^r_i a^s_j = \lambda g_{ij} \qquad \text{or} \qquad A^T G A = \lambda G \tag{2}$$

It remains to show that $\lambda = 1$. Since $G^2 = I$, multiplication of (*2*) by the matrix $\lambda^{-1}G$ gives $(G(\lambda^{-1}A^T)G)A = I$, which shows that the inverse of A is

$$B = \frac{1}{\lambda}\, G A^T G = \begin{bmatrix} a^0_0/\lambda & -a^1_0/\lambda & -a^2_0/\lambda & -a^3_0/\lambda \\ -a^0_1/\lambda & a^1_1/\lambda & a^2_1/\lambda & a^3_1/\lambda \\ -a^0_2/\lambda & a^1_2/\lambda & a^2_2/\lambda & a^3_2/\lambda \\ -a^0_3/\lambda & a^1_3/\lambda & a^2_3/\lambda & a^3_3/\lambda \end{bmatrix} \equiv [b^i_j]_{44} \tag{3}$$

In particular, $b^0_0 = a^0_0/\lambda$. Now since observers O and $\bar{O}$ are receding from each other at constant velocity v and are using identical measuring devices, it is clear that each views the other in the same way. It follows that $a^0_0 = b^0_0$ and $\lambda = a^0_0/b^0_0 = 1$ (see Problem 12.5).

12.5 With reference to Problem 12.4, give a "thought-experiment" which leads to the conclusion that $a^0_0 = b^0_0$.

Consider the motion of O in $\bar{O}$'s frame: Transform the point $(ct, 0, 0, 0)$ under $\mathcal{T}$ to get

$$\bar{x}^0 = c\bar{t} = a^0_0 ct \qquad \text{or} \qquad \bar{t} = a^0_0 t$$

Thus, 1 second on O's clock is a^0_0 seconds on $\bar{O}$'s; reciprocally, 1 second on $\bar{O}$'s clock is b^0_0 seconds on O's. Thus, $a^0_0 = b^0_0$.

12.6 Prove Theorem 12.1 from *(12.7)*. [Note that Problem 12.4 did not make use of Theorem 12.1, so the proof will be logically correct.]

By *(12.7)*, (g_{ij}) is a covariant tensor under Lorentz transformations, so that $g_{ij}\,\Delta x^i\,\Delta x^j$ is an invariant (under Lorentz transformations).

12.7 Verify that the following matrix is Lorentz:

$$\begin{bmatrix} \sqrt{3} & \sqrt{2} & 0 & 0 \\ 1 & \dfrac{\sqrt{6}}{2} & \dfrac{1}{2} & \dfrac{1}{2} \\ 1 & \dfrac{\sqrt{6}}{2} & -\dfrac{1}{2} & -\dfrac{1}{2} \\ 0 & 0 & -\dfrac{\sqrt{2}}{2} & \dfrac{\sqrt{2}}{2} \end{bmatrix}$$

We verify directly the conditions *(12.7c)*:

$$(\sqrt{3})^2 - 1^2 - 1^2 - 0^2 = 3 - 2 = 1 \qquad (\sqrt{2})^2 - \left(\frac{\sqrt{6}}{2}\right)^2 - \left(\frac{\sqrt{6}}{2}\right)^2 - 0^2 = 2 - \frac{3}{2} - \frac{3}{2} = -1$$

$$0^2 - \left(\frac{1}{2}\right)^2 - \left(-\frac{1}{2}\right)^2 - \left(\pm\frac{\sqrt{2}}{2}\right)^2 = -1 \qquad (\sqrt{3})(\sqrt{2}) - (1)\left(\frac{\sqrt{6}}{2}\right) - (1)\left(\frac{\sqrt{6}}{2}\right) - 0 = 0$$

$$(\sqrt{3})(0) - (1)\left(\frac{1}{2}\right) - (1)\left(-\frac{1}{2}\right) - (0)\left(\pm\frac{\sqrt{2}}{2}\right) = 0 \quad (\sqrt{2})(0) - \left(\frac{\sqrt{6}}{2}\right)\left(\frac{1}{2}\right) - \left(\frac{\sqrt{6}}{2}\right)\left(-\frac{1}{2}\right) - (0)\left(\pm\frac{\sqrt{2}}{2}\right) = 0$$

$$0 - \left(\frac{1}{2}\right)\left(\frac{1}{2}\right) + \left(\frac{1}{2}\right)\left(-\frac{1}{2}\right) + \left(\frac{\sqrt{2}}{2}\right)\left(\frac{\sqrt{2}}{2}\right) = 0 - \frac{1}{4} - \frac{1}{4} + \frac{1}{2} = 0$$

12.8 Show that a matrix A which preserves $\mathbf{x}^T G \mathbf{x} = 0$ necessarily preserves $\mathbf{x}^T G \mathbf{x} = q$.

This is really Problem 12.6 in another guise. By Problem 12.4, A must satisfy $A^T G A = G$. But then

$$(A\mathbf{x})^T G (A\mathbf{x}) = \mathbf{x}^T (A^T G A)\mathbf{x} = \mathbf{x}^T G \mathbf{x} = q$$

12.9 (*a*) Exhibit the inverse, B, of a given Lorentz matrix, A. (*b*) If we define a matrix A to be *pseudo-orthogonal* when there exists a matrix J whose square is the identity and $A^T J A = J$, show that all Lorentz matrices are pseudo-orthogonal.

(*a*) Set $\lambda = 1$ in *(3)* of Problem 12.4.

(*b*) If A is a Lorentz matrix, then G clearly fills the role of J in the definition of pseudo-orthogonal matrix.

12.10 Prove that the Lorentz matrices compose a group under matrix multiplication.

We are required to show that (*a*) the product of two Lorentz matrices is Lorentz, (*b*) the inverse of a Lorentz matrix is Lorentz.

(*a*)
$$(PQ)^T G (PQ) = Q^T (P^T G P) Q = Q^T G Q = G$$

(*b*) Using Problem 12.4 with $\lambda = 1$, $B = A^{-1} = G A^T G$, and

$$B^T G B = (G A^T G)^T G B = G A G^2 B = G A B = G$$

SIMPLE LORENTZ MATRICES

12.11 Derive the simple form (*12.9*) of the transformation equations for SR by considering how observers O and $\bar{O}$ will view events occurring on a circular cylinder about their common x-axis.

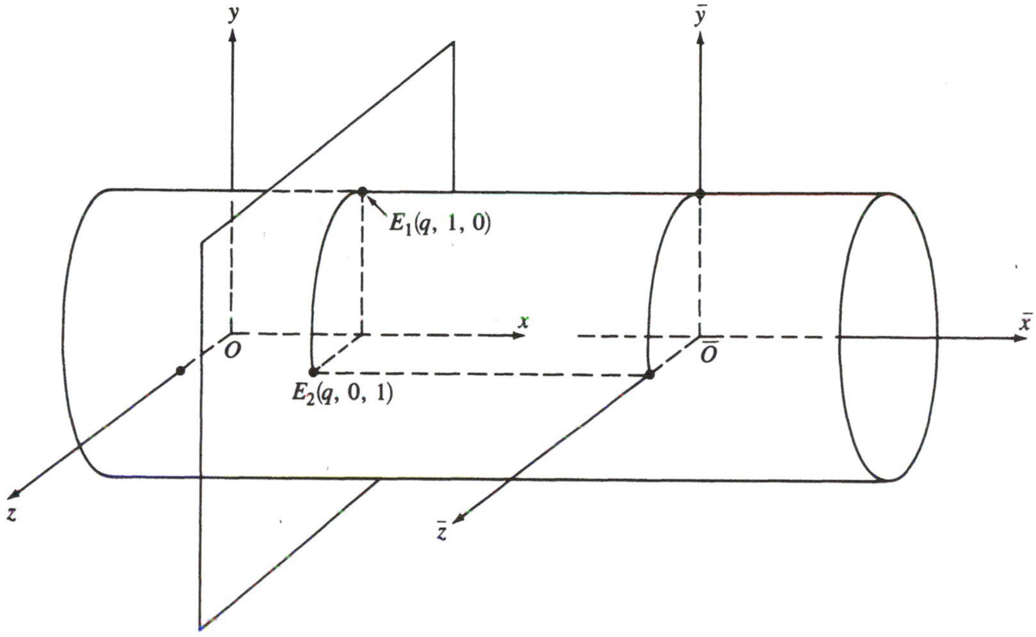

Fig. 12-4

At any time t, let E_1 and E_2 be two events taking place at the points of space $(q, 1, 0)$ and $(q, 0, 1)$, respectively, which lie on a unit cylinder about O's x-axis (Fig. 12-4). Thus, with $p = ct$, we have space-time coordinates $E_1(p, q, 1, 0)$ and $E_2(p, q, 0, 1)$. Since the axes of O are not turning with respect to $\bar{O}$'s, these two events will be viewed by observer $\bar{O}$ as $E_1(\bar{p}, \bar{q}, 1, 0)$ and $E_2(\bar{p}^*, \bar{q}^*, 0, 1)$, respectively. The transformation equations (*12.5*) give:

$$\text{(I)}\ \begin{cases} \bar{p} = a^0_0 p + a^0_1 q + a^0_2 \\ \bar{q} = a^1_0 p + a^1_1 q + a^1_2 \\ 1 = a^2_0 p + a^2_1 q + a^2_2 \\ 0 = a^3_0 p + a^3_1 q + a^3_2 \end{cases} \qquad \text{(II)}\ \begin{cases} \bar{p}^* = a^0_0 p + a^0_1 q + a^0_3 \\ \bar{q}^* = a^1_0 p + a^1_1 q + a^1_3 \\ 0 = a^2_0 p + a^2_1 q + a^2_3 \\ 1 = a^3_0 p + a^3_2 q + a^3_3 \end{cases}$$

Observing just the last equation of (I) and the third equation of (II), we may, since p and q are arbitrary, take $p = q = 0$, then $p = 1$, $q = 0$, and $p = 0$, $q = 1$. It follows that all six of the coefficients vanish: $a^2_0 = a^2_1 = a^2_3 = a^3_0 = a^3_1 = a^3_2 = 0$. Using the third equation of (I) and the last equation of (II), we find that $a^2_2 = a^3_3 = 1$. It follows that the last two equations of $\mathcal{T}$ reduce to $\bar{x}^2 = x^2$ and $\bar{x}^3 = x^3$. Now to concentrate on the first two: If $p = q = 0$, then event E_1 is $(0, 0, 1, 0)$—occurring when $t = \bar{t} = 0$ at $x^1 = 0$, the instant when $\bar{x}^1 = 0$. That is, $\bar{p} = \bar{q} = 0$, with the result $a^0_2 = a^1_2 = 0$. Similarly, using E_2, $p = q = 0$ implies $\bar{p}^* = \bar{q}^* = 0$ and $a^0_3 = a^1_3 = 0$.

12.12 Consider event E_1, a lightning flash at the point $(v, 0, 0)$ at time $t = 1$ s in O's frame, and event E_2, a lightning flash at $(-v, 0, 0)$ at time $\bar{t} = 1$ s in $\bar{O}$'s frame. By determining the corresponding events in the opposing frames of reference, deduce (*12.11*).

Since at $t = 1$ observer $\bar{O}$ has reached the point $(v, 0, 0)$, the lightning strikes $\bar{O}$'s origin at time $\bar{t}$. Hence, E_1 has coordinates $(c, v, 0, 0)$ in O and $(c\bar{t}, 0, 0, 0)$ in $\bar{O}$. Substituting these into $\mathcal{T}$ we obtain

$$c\bar{t} = ac + bv \qquad 0 = dc + ev$$

The second equation gives $d = -\beta e$.

Since O has progressed backwards to the point $(-v, 0, 0)$ in $\bar{O}$ at the time $\bar{t} = 1$ at which E_2 occurs, this event has coordinates $(ct, 0, 0, 0)$ in O and $(c, -v, 0, 0)$ in $\bar{O}$. Substituting these into $\mathcal{T}$ yields

$$c = act + b(0) \qquad -v = dct + e(0)$$

which upon division give $d = -\beta a$. Hence, $a = e$.

12.13 Show that a 4×4 matrix is both Lorentz and orthogonal if and only if it has the form

$$R = \begin{bmatrix} \pm 1 & 0 & 0 & 0 \\ 0 & r_1 & s_1 & t_1 \\ 0 & r_2 & s_2 & t_2 \\ 0 & r_3 & s_3 & t_3 \end{bmatrix} \tag{1}$$

where the 3×3 matrix $[\mathbf{r} \quad \mathbf{s} \quad \mathbf{t}]$ is orthogonal.

A Lorentz matrix $A = (a^i_j)$ is also orthogonal if and only if its inverse B, as obtained in Problem 12.4 (with $\lambda = 1$), is equal to A^T and is itself orthogonal. This observation immediately yields the form (*1*).

12.14 Prove Theorem 12.2.

Since $\|\mathbf{r}\|^2 = b^{-2}[(a^0_1)^2 + (a^0_2)^2 + (a^0_3)^2] = b^{-2}[(a^0_0)^2 - 1] = 1$ (using Problem 12.36), the matrix $[\mathbf{r} \quad \mathbf{s} \quad \mathbf{t}]$ is orthogonal and R^T_2 has the form of the matrix in Problem 12.13, making it Lorentz and orthogonal. It follows that R_2 is orthogonal (and Lorentz), with $R_2^{-1} = R_2^T$; hence, $L = R_1 L^* R_2$.

Now, as the product of Lorentz matrices, R_1 is Lorentz; to show it is orthogonal, consider $LR_2^T(L^*)^{-1}$, which may be written as

$$\begin{bmatrix} a_0 & b_0 & c_0 & d_0 \\ a_1 & b_1 & c_1 & d_1 \\ a_2 & b_2 & c_2 & d_2 \\ a_3 & b_3 & c_3 & d_3 \end{bmatrix} \begin{bmatrix} 1 & 0 & 0 & 0 \\ 0 & r_1 & s_1 & t_1 \\ 0 & r_2 & s_2 & t_2 \\ 0 & r_3 & s_3 & t_3 \end{bmatrix} \begin{bmatrix} a & -b & 0 & 0 \\ -b & a & 0 & 0 \\ 0 & 0 & 1 & 0 \\ 0 & 0 & 0 & 1 \end{bmatrix}$$

$$= \begin{bmatrix} a_0 & b_0r_1 + c_0r_2 + d_0r_3 & b_0s_1 + c_0s_2 + d_0s_3 & b_0t_1 + c_0t_2 + d_0t_3 \\ & & \cdots & \\ & & \cdots & \\ & & \cdots & \end{bmatrix} \begin{bmatrix} a & -b & 0 & 0 \\ -b & a & 0 & 0 \\ 0 & 0 & 1 & 0 \\ 0 & 0 & 0 & 1 \end{bmatrix}$$

[The omitted rows have the form $(a_i, b_ir_1 + c_ir_2 + d_ir_3, b_is_1 + c_is_2 + d_is_3, b_it_1 + c_it_2 + d_it_3)$, with $i = 1, 2, 3$.] We first concentrate on proving that the top row and first column of this product are $(\pm 1, 0, 0, 0)$. The 00-element of the product is

$$a_0a + (b_0r_1 + c_0r_2 + d_0r_3)(-b) = \varepsilon a_0^2 + \frac{\varepsilon}{b}(b_0^2 + c_0^2 + d_0^2)(-b) = \varepsilon(a_0^2 - b_0^2 - c_0^2 - d_0^2) = \varepsilon$$

again using the fact that the transpose of a Lorentz matrix is Lorentz. The next element in the top row of the product is

$$-a_0b + (b_0r_1 + c_0r_2 + d_0r_3)a = -a_0b + \frac{b}{\varepsilon}(r^2)\varepsilon a_0 = -a_0b + ba_0 = 0$$

For the third and fourth elements,

$$b_0s_1 + c_0s_2 + d_0s_3 = \frac{b}{\varepsilon}\,\mathbf{rs} = 0 \qquad \text{and} \qquad b_0t_1 + c_0t_2 + d_0t_3 = \frac{b}{\varepsilon}\,\mathbf{rt} = 0$$

Now for the first column of the product; its elements, beginning with the second, are (for $i = 1, 2, 3$)

$$a_i a + (b_i r_1 + c_i r_2 + d_i r_3)(-b) = \varepsilon a_i a_0 - \varepsilon(b_i b_0 + c_i c_0 + d_i d_0) = 0$$

Hence, the product matrix becomes

$$R_1 = \begin{bmatrix} \varepsilon & 0 & 0 & 0 \\ 0 & & & \\ 0 & & R & \\ 0 & & & \end{bmatrix}$$

and the 3×3 matrix R must be orthogonal, since R_1 is Lorentz.

12.15 Apply Theorem 12.2 to the Lorentz matrix of Problem 12.7, and demonstrate the physical significance of this matrix by computing the velocity v between the two observers involved.

We proceed to calculate a, b, and the vectors $\mathbf{r}$, $\mathbf{s}$, and $\mathbf{t}$:

$$a = \sqrt{3} \qquad \varepsilon = 1 \qquad b = -\sqrt{3-1} = -\sqrt{2} \qquad \mathbf{r} = -\frac{1}{\sqrt{2}}(\sqrt{2}, 0, 0) = (-1, 0, 0)$$

Hence, we may take $\mathbf{s} = (0, 1, 0)$ and $\mathbf{t} = (0, 0, 1)$, yielding

$$R_2 = \begin{bmatrix} 1 & 0 & 0 & 0 \\ 0 & -1 & 0 & 0 \\ 0 & 0 & 1 & 0 \\ 0 & 0 & 0 & 1 \end{bmatrix}$$

and

$$R_1 = \begin{bmatrix} \sqrt{3} & \sqrt{2} & 0 & 0 \\ 1 & \sqrt{6}/2 & 1/2 & 1/2 \\ 1 & \sqrt{6}/2 & -1/2 & -1/2 \\ 0 & 0 & -\sqrt{2}/2 & \sqrt{2}/2 \end{bmatrix} \begin{bmatrix} 1 & 0 & 0 & 0 \\ 0 & -1 & 0 & 0 \\ 0 & 0 & 1 & 0 \\ 0 & 0 & 0 & 1 \end{bmatrix} \begin{bmatrix} \sqrt{3} & \sqrt{2} & 0 & 0 \\ \sqrt{2} & \sqrt{3} & 0 & 0 \\ 0 & 0 & 1 & 0 \\ 0 & 0 & 0 & 1 \end{bmatrix}$$

$$= \begin{bmatrix} \sqrt{3} & -\sqrt{2} & 0 & 0 \\ 1 & -\sqrt{6}/2 & 1/2 & 1/2 \\ 1 & -\sqrt{6}/2 & -1/2 & -1/2 \\ 0 & 0 & -\sqrt{2}/2 & \sqrt{2}/2 \end{bmatrix} \begin{bmatrix} \sqrt{3} & \sqrt{2} & 0 & 0 \\ \sqrt{2} & \sqrt{3} & 0 & 0 \\ 0 & 0 & 1 & 0 \\ 0 & 0 & 0 & 1 \end{bmatrix} = \begin{bmatrix} 1 & 0 & 0 & 0 \\ 0 & -\sqrt{2}/2 & 1/2 & 1/2 \\ 0 & -\sqrt{2}/2 & -1/2 & -1/2 \\ 0 & 0 & -\sqrt{2}/2 & \sqrt{2}/2 \end{bmatrix}$$

By Corollary 12.3,

$$v = c\sqrt{1 - (\sqrt{3})^{-2}} = \sqrt{\frac{2}{3}}\, c$$

LENGTH CONTRACTION, TIME DILATION

12.16 A pole-vaulter runs at the rate $(\sqrt{3}/2)c$ (in m/s) and carries a pole that is 20 m long in his reference frame [the *rest length* of the pole is 20 m]. He approaches a barn that is open at both ends and is 10 m long, as measured by a ground observer. To the ground observer, will the pole fit inside the barn? What is the pole-vaulter's conclusion?

To the ground observer, the pole undergoes length contraction with the factor $\sqrt{1-\beta^2}$, where $\beta = \sqrt{3}/2$. Hence, the length of the pole in the frame of the ground observer is

$$20\sqrt{1 - (\sqrt{3}/2)^2} = 10 \text{ m}$$

and so, for her, the pole exactly fits inside the barn (instantaneously). To the runner, however, the barn is $10(1/2) = 5$ m long, so that the 20-m pole does not fit.

This example shows that order relations are not preserved under the Lorentz transformation.

12.17 (*the Twin Paradox*) One of a pair of twins embarks on a journey into outer space, taking one year (earth time) to accelerate to $(3/4)c$, then spends the next 20 years cruising to reach a galaxy 15 light-years away. An additional year is spend in decelerating in order to explore one of its solar systems. After one year of exploration ($\beta = 0$), the twin returns to earth by the same schedule—one year of acceleration, 20 years of cruising, and one year of deceleration. Estimate the difference in the ages of the twins after the journey has ended.

In order to apply SR, replace the four periods of acceleration or deceleration by four periods of uniform motion at speed $(3/8)c$ (the time-average speed under constant acceleration). These account for 4 years by the earth clock; but to the space twin, who measures proper (shortest) time intervals, the time lapse is ($\beta = 3/8$)

$$4\sqrt{1-(3/8)^2} \approx 3.71 \text{ years}$$

Similarly, the 40 earth-years of cruising at $\beta = 3/4$ corresponds to a proper-time interval of

$$40\sqrt{1-(3/4)^2} \approx 26.46 \text{ years}$$

Thus, the space twin has aged $3.71 + 26.46 + 1 \approx 31$ years while the earth twin has aged $4 + 40 + 1 = 45$ years.

The space twin returns biologically younger by some 14 years. While the accelerations and decelerations between the two twins were reciprocal, the *forces* in the situation acted on the space twin alone.

12.18 Prove the basic integrity of (*12.16*) by solving algebraically for v_2 as a function of v_1 and v_3 to verify that v_2 follows the correct format for composition of velocities.

Solving,

$$v_2 = \frac{-v_1 + v_3}{1 - v_1 v_3/c^2}$$

which is precisely (*12.16*) under the substitution $(v_1, v_2, v_3) \to (-v_1, v_3, v_2)$.

12.19 A light source at O sends a spherical wavefront (Fig. 12-5(*a*)) advancing in all directions at velocity c; it reaches the ends of a diameter AB centered at O simultaneously, as determined by O. But as far as $\bar{O}$ is concerned, the spherical wave, centered at $\bar{O}$, moves with him (invariance of the light cone) and therefore reaches point B before it reaches point A. Calculate the time difference on $\bar{O}$'s clock for these two events (light reaching B and light reaching A) if $\beta = 1/2$ and if $AB = 6$ m.

Since $AB = 6$ m and O is the midpoint of segment AB, O assigns spatial coordinates $B(3,0,0)$ and $A(-3,0,0)$ to the endpoints. It takes $3/c$ seconds for light to reach A and B, so O calculates the time coordinate as $x^0 = c(3/c) = 3$ m. The space-time coordinates of the two events are thus

$$E_B(3,3,0,0) \qquad \text{and} \qquad E_A(3,-3,0,0)$$

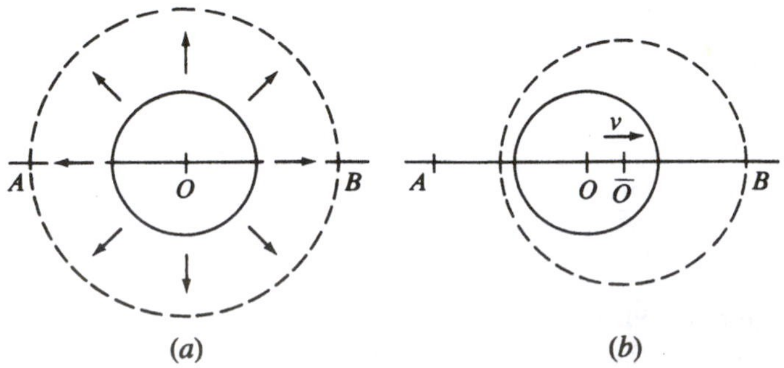

Fig. 12-5

Substitute these values and $\beta = 1/2$ into the first equation (12.13) to obtain $\bar{t}_B = \sqrt{3}/c$, $\bar{t}_A = 3\sqrt{3}/c$. Hence, $\Delta\bar{t} = 2\sqrt{3}/c$ (in s), while $\Delta t = 0$.

It is seen that simultaneity is not an invariant of Lorentz transformations.

12.20 Derive the composition of velocities formula, (12.16).

According to Section 12.4, we must have $v_i = -b_i c/a_i$ for $i = 1, 2, 3$. Composing the simple Lorentz transformations, we have

$$\begin{bmatrix} a_1 & b_1 & 0 & 0 \\ b_1 & a_1 & 0 & 0 \\ 0 & 0 & 1 & 0 \\ 0 & 0 & 0 & 1 \end{bmatrix}\begin{bmatrix} a_2 & b_2 & 0 & 0 \\ b_2 & a_2 & 0 & 0 \\ 0 & 0 & 1 & 0 \\ 0 & 0 & 0 & 1 \end{bmatrix} = \begin{bmatrix} a_1a_2 + b_1b_2 & a_1b_2 + a_2b_1 & 0 & 0 \\ a_2b_1 + a_1b_2 & b_1b_2 + a_1a_2 & 0 & 0 \\ 0 & 0 & 1 & 0 \\ 0 & 0 & 0 & 1 \end{bmatrix}$$

whence $a_3 = a_1a_2 + b_1b_2$, $b_3 = a_1b_2 + a_2b_1$, and

$$v_3 = -\frac{(a_1b_2 + a_2b_1)c}{a_1a_2 + b_1b_2} = \frac{-\dfrac{a_1b_2c}{a_1a_2} - \dfrac{a_2b_1c}{a_1a_2}}{\dfrac{a_1a_2}{a_1a_2} + \dfrac{b_1b_2}{a_1a_2}} = \frac{-\dfrac{b_2c}{a_2} - \dfrac{b_1c}{a_1}}{1 + \dfrac{b_1b_2}{a_1a_2}} = \frac{v_2 + v_1}{1 + v_1v_2/c^2}$$

12.21 A physicist wants to compose two equal velocities $v = v_1 = v_2$ to produce a resultant velocity that is 90% of the velocity of light. What velocity must he use?

From (12.16),

$$0.90c = \frac{2v}{1 + v^2/c^2} \qquad \text{or} \qquad 0.90 = \frac{2\beta}{1 + \beta^2}$$

Solving the quadratic, $\beta \approx 0.627$ and $v \approx 0.627c$ (as compared to the Newtonian value $0.45c$).

VELOCITY AND ACCELERATION IN RELATIVITY

12.22 Establish (12.19) and (12.20), the Lorentz transformations of velocity and acceleration, that define how $\bar{O}$ tracks the motion of a particle in O's frame.

To simplify notation, let $\gamma \equiv (1 - \beta^2)^{-1/2}$. Then $\mathcal{T}$ is

$$c\bar{t} = \gamma(ct - \beta x) \qquad \bar{x} = \gamma(-\beta ct + x) \qquad \bar{y} = y \qquad \bar{z} = z$$

Differentiate the first equation with respect to $\bar{t}$ and use the chain rule:

$$c = \gamma(c - \beta v_x)\frac{dt}{d\bar{t}} \qquad \text{or} \qquad \frac{dt}{d\bar{t}} = \frac{1}{\gamma(1 - vv_x/c^2)}$$

Now differentiate the last three equations:

$$\bar{v}_x = \gamma(-\beta c + v_x)\frac{dt}{d\bar{t}} = \frac{\gamma(-v + v_x)}{\gamma(1 - vv_x/c^2)} = \frac{v_x - v}{1 - v_xv/c^2}$$

$$\bar{v}_y = v_y\frac{dt}{d\bar{t}} = \frac{v_y}{\gamma(1 - vv_x/c^2)} = \frac{v_y\sqrt{1 - \beta^2}}{1 - v_xv/c^2}$$

$$\bar{v}_z = v_z\frac{dt}{d\bar{t}} = \frac{v_z\sqrt{1 - \beta^2}}{1 - v_xv/c^2}$$

By differentiation of the velocity components just found,

$$\bar{a}_x = \frac{d\bar{v}_x}{dt}\frac{dt}{d\bar{t}} = \frac{(a_x - 0)(1 - v_xv/c^2) - (v_x - v)(0 - a_xv/c^2)}{(1 - v_xv/c^2)^2}\,\frac{1}{\gamma(1 - v_xv/c^2)}$$

$$= \frac{a_x - a_xv_xv/c^2 + v_xa_xv/c^2 - a_xv^2/c^2}{\gamma(1 - v_xv/c^2)^3} = \frac{a_x(1 - \beta^2)^{3/2}}{(1 - v_xv/c^2)^3}$$

$$\bar{a}_y = \frac{a_y(1 - v_x v/c^2) - v_y(0 - a_x v/c^2)}{(1 - v_x v/c^2)^2} \, \frac{1 - \beta^2}{1 - v v_x/c^2} = \frac{a_y + (a_x v_y - v_x a_y)(v/c^2)}{(1 - v_x v/c^2)^3}(1 - \beta^2)$$

The formula for $\bar{a}_z$ is derived as that for $\bar{a}_y$, with z replacing y throughout.

12.23 Show that if the curve of motion in O's frame is the path of $\bar{O}$ itself, the clock in $\bar{O}$'s frame (the clock moving with the particle) measures proper time.

By (3) of Problem 12.4,

$$x^0 = a_0^0\bar{x}^0 - a_0^1\bar{x}^1 - a_0^2\bar{x}^2 - a_0^3\bar{x}^3$$
$$x^i = -a_i^0\bar{x}^0 + a_i^1\bar{x}^1 + a_i^2\bar{x}^2 + a_i^3\bar{x}^3 \qquad (i = 1, 2, 3)$$

Now the motion of $\bar{O}$ relative to itself is obviously $\bar{x}^1 = \bar{x}^2 = \bar{x}^3 = 0$. Hence,

$$x^0 = a_0^0 cu \qquad x^1 = -a_1^0 cu \qquad x^2 = -a_2^0 cu \qquad x^3 = -a_3^0 cu$$

give the trajectory of $\bar{O}$ in O's frame, with parameter $u = \bar{t}$. Therefore, the tangent field to the trajectory is

$$\left(\frac{dx^i}{du}\right) = (a_0^0 c, -a_1^0 c, -a_2^0 c, -a_3^0 c)$$

so that the proper time parameter for this curve is defined as

$$\tau = \frac{1}{c}\int_0^{\bar{t}} \sqrt{|(a_0^0 c)^2 - (a_1^0 c)^2 - (a_2^0 c)^2 - (a_3^0 c)^2|}\, du = \sqrt{|(a_0^0)^2 - (a_1^0)^2 - (a_2^0)^2 - (a_3^0)^2|}\int_0^{\bar{t}} du$$

Because the inverse transformation is Lorentz, the factor in front of the integral sign equals 1, and so $\tau = \bar{t}$.

12.24 Derive the identities (*12.25*).

By (*12.21*),

$$u_i u^i = g_{ij}u^i u^j \equiv (u^0)^2 - (u^1)^2 - (u^2)^2 - (u^3)^2$$
$$= [(v_t)^2 - (v_x)^2 - (v_y)^2 - (v_z)^2]\left(\frac{dt}{d\tau}\right)^2 = [c^2 - \hat{v}^2]\,\frac{1}{1 - \hat{v}^2/c^2} = c^2$$

and from this,

$$0 = \frac{d}{d\tau}(c^2) = \frac{d}{d\tau}(u_i u^i) = 2u_i b^i$$

12.25 Establish the formulas in (*12.26*).

$$u^i = \frac{dx^i}{d\tau} = \frac{dx^i}{dt}\frac{dt}{d\tau} = \frac{v_i}{\sqrt{1 - \hat{v}^2/c^2}}$$

$$b^i = \frac{du^i}{d\tau} = \left[\frac{d}{dt}\left(\frac{v_i}{\sqrt{1 - \hat{v}^2/c^2}}\right)\right]\frac{dt}{d\tau}$$

$$= \frac{a_i(1 - \hat{v}^2/c^2)^{1/2} - v_i(1/2)(1 - \hat{v}^2/c^2)^{-1/2}(-2a_x v_x - 2a_y v_y - 2a_z v_z)/c^2}{1 - \hat{v}^2/c^2}\,\frac{dt}{d\tau}$$

$$= \frac{a_i(1 - \hat{v}^2/c^2) + v_i(a_x v_x + a_y v_y + a_z v_z)/c^2}{(1 - \hat{v}^2/c^2)^2} = \frac{a_i}{1 - \hat{v}^2/c^2} + \frac{(\mathbf{va})v_i}{c^2(1 - \hat{v}^2/c^2)^2}$$

12.26 Derive the equation for uniformly accelerated motion along the x-axis of an inertial frame:

$$x^2 - c^2t^2 = \frac{c^4}{\alpha^2}$$

Let $\bar{O}$ be an instantaneous rest frame at some point t_1, and let O be a given (stationary) frame in which the motion curve is traced. Since the motion is along the x-axis of O,

$$v_y = v_z = a_y = a_z = 0 \qquad \text{and} \qquad \bar{v}_y = \bar{v}_z = \bar{a}_y = \bar{a}_z = 0$$

whereby $\bar{a}_x = \alpha = \text{const.}$ (assuming $\bar{a}_x > 0$). At $t = t_1$, $v = v_x$ (the constant velocity of $\bar{O}$ is by definition equal to the instantaneous velocity of the particle); thus, from (12.20),

$$\alpha = \frac{a_x(1 - v^2/c^2)^{3/2}}{(1 - v_x v/c^2)^3} = \frac{a_x(1 - v_x^2/c^2)^{3/2}}{(1 - v_x^2/c^2)^3} = \frac{a_x}{(1 - v_x^2/c^2)^{3/2}} \tag{1}$$

Since t_1 is arbitrary, (1) must hold for all t. Writing $\dot{x}$, $\ddot{x}$ for the derivatives of $x(t)$, we have from (1):

$$c^3 \ddot{x} = \alpha(c^2 - \dot{x}^2)^{3/2} \tag{2}$$

Make the substitution $y = \dot{x}$ and (2) becomes

$$c^3 \frac{dy}{dt} = \alpha(c^2 - y^2)^{3/2} \qquad \text{or} \qquad \int \frac{c^3\, dy}{(c^2 - y^2)^{3/2}} = \int \alpha\, dt \tag{3}$$

Standard techniques of integration yield the first integral

$$\frac{cy}{\sqrt{c^2 - y^2}} = \alpha t \tag{4}$$

(where we have taken the initial velocity to be zero). Solving (4) for y (assumed positive for positive t) and then integrating the equation $\dot{x} = y(t)$, we obtain

$$x = c\sqrt{c^2 + \alpha^2 t^2}/\alpha \qquad \text{or} \qquad x^2 - c^2 t^2 = c^4/\alpha^2$$

(where we also take the initial position as zero). This is the desired equation, which represents a hyperbola in the xt plane. By contrast, the Newtonian equation is the parabola $x = \frac{1}{2}\alpha t^2$.

RELATIVISTIC MASS, FORCE, AND ENERGY

12.27 Show that the observed mass of a particle with rest mass m, moving at velocity v, is $\hat{m} = m(1 - v^2/c^2)^{-1/2}$, by considering the following experiment. Let each observer O and $\bar{O}$ carry a ball with rest mass m near his origin and so situated as to collide obliquely at $t = \bar{t} = 0$ (when their origins coincide). See Fig. 12-6. Suppose this collision imparts reciprocal velocities of ε in the negative y-direction and positive $\bar{y}$-direction. Calculate the momentum of the system before and after collision (which is preserved), and what each observer sees based on the equations of SR; then take the limit as $\varepsilon \to 0$.

The velocity vectors $\mathbf{v}_1$ and $\mathbf{v}_2$ of balls B_1 and B_2 before collision are, as seen by O, $(0, 0, 0) = \mathbf{0}$ and $(v, 0, 0) = v\mathbf{i}$. Observer $\bar{O}$ calculates these vectors as $\bar{\mathbf{v}}_1 = (-v, 0, 0)$ and $\bar{\mathbf{v}}_2 = (0, 0, 0)$ (either by

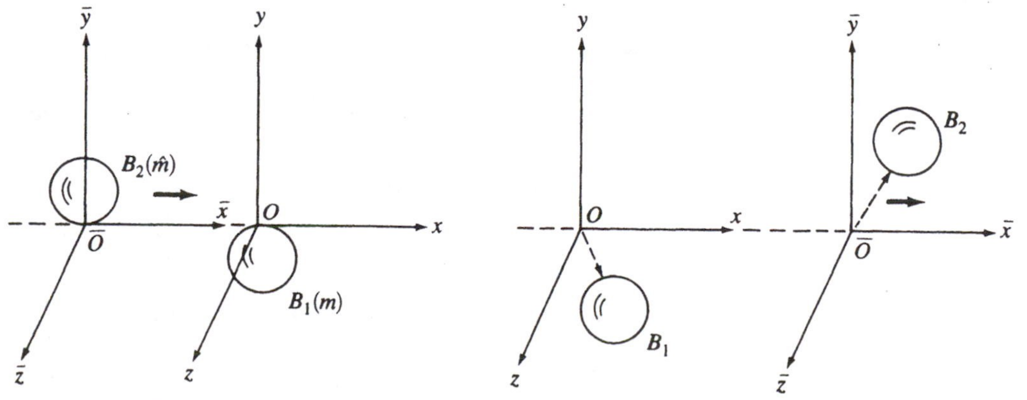

(a) Before impact (b) After impact

Fig. 12-6

reciprocity or by use of (*12.19*). After collision, observer O calculates the velocity of B_1 as $\mathbf{v}_1 = (\varepsilon, -\varepsilon, 0) = \varepsilon\mathbf{i} - \varepsilon\mathbf{j}$, assuming B_1 has the proper alignment with B_2. Reciprocally, observer $\bar{O}$ calculates the velocity of B_2 as $\bar{\mathbf{v}}_2 = (\varepsilon, \varepsilon, 0)$. To find $\mathbf{v}_2$, use the inverse of (*12.19*), with $\bar{v}_x = \varepsilon$ and $\bar{v}_y = \varepsilon$:

$$\mathbf{v}_2 = v_x\mathbf{i} + v_y\mathbf{j} = \left(\frac{\varepsilon + v}{1 + \varepsilon v/c^2}\right)\mathbf{i} + \left(\frac{\varepsilon\sqrt{1-\beta^2}}{1 + \varepsilon v/c^2}\right)\mathbf{j}$$

Thus, observer O calculates the net momentum vector of the system as follows, using the rest mass m of B_1 for m_1 and the "perceived mass" $\hat{m}$ of B_2 for m_2:

before impact $\quad m_1\mathbf{v}_1 + m_2\mathbf{v}_2 = m_1(\mathbf{0}) + m_2(v\mathbf{i}) = \hat{m}v\mathbf{i}$

after impact $\quad m_1\mathbf{v}_1 + m_2\mathbf{v}_2 = m(\varepsilon\mathbf{i} - \varepsilon\mathbf{j}) + \hat{m}\left[\left(\frac{\varepsilon + v}{1 + \varepsilon v/c^2}\right)\mathbf{i} + \left(\frac{\varepsilon\sqrt{1-\beta^2}}{1 + \varepsilon v/c^2}\right)\mathbf{j}\right]$

$$= \left(m\varepsilon + \hat{m}\,\frac{\varepsilon + v}{1 + \varepsilon v/c^2}\right)\mathbf{i} + \left(m\varepsilon - \hat{m}\,\frac{\varepsilon\sqrt{1-\beta^2}}{1 + \varepsilon v/c^2}\right)\mathbf{j}$$

Since O is using the universal laws of physics as they apply to his frame (Postulate 1 of SR), the two momentum vectors above must be the same. Hence,

$$\hat{m}v = m\varepsilon + \hat{m}\,\frac{\varepsilon + v}{1 + \varepsilon v/c^2} \qquad \text{and} \qquad 0 = -m + \hat{m}\,\frac{\sqrt{1-\beta^2}}{1 + \varepsilon v/c^2}$$

(after division by δ). Now take the limit as $\varepsilon \to 0$:

$$\hat{m}v = \hat{m}v \qquad \text{and} \qquad 0 = -m + \hat{m}\sqrt{1-\beta^2}$$

The right-hand equation is the connection between m and $\hat{m}$.

12.28 Show that the Minkowski force is a 4-vector.

We must show that $\bar{K}^i = a^i_j K^j$, if $\bar{x}^i = a^i_j x^j$, where (a^i_j) is any Lorentz matrix. Since τ is invariant and $a^i_j = \text{const.}$ we may differentiate the coordinate transformation with respect to τ across the equal sign:

$$\frac{d}{d\tau}(\bar{x}^i) = \frac{d}{d\tau}(a^i_j x^j) \qquad \text{or} \qquad \bar{u}^i = a^i_j u^j$$

(proving that (u^i) is a 4-vector). Multiply both sides by m and differentiate again, using the fact that the rest mass of a particle is invariant:

$$\bar{K}^i = \frac{d}{d\tau}(\bar{m}\bar{u}^i) = \frac{d}{d\tau}(m\bar{u}^i) = \frac{d}{d\tau}(a^i_j m u^j) = a^i_j\,\frac{d}{d\tau}(mu^j) = a^i_j K^j$$

12.29 Establish (*12.32*).

Definition (*12.30*), $K^i = d(mu^i)/d\tau = mb^i$, along with the second identity (*12.25*), gives at once $u_i K^i = 0$.

From $u_i K^i = g_{ij}u^i K^j = u^0 K^0 - u^q K^q = 0$ and the first formula (*12.26*), we have

$$\frac{v_0 K^0}{\sqrt{1 - \hat{v}^2/c^2}} - \frac{v_q K^q}{\sqrt{1 - \hat{v}^2/c^2}} = 0 \qquad \text{or} \qquad cK^0 = \mathbf{vK}$$

By (*12.31*) and $cK^0 = \mathbf{vK}$,

$$\frac{1}{c}\,\mathbf{vF} = \frac{1}{c}\,\mathbf{vK}\sqrt{1 - \hat{v}^2/c^2} = K^0\sqrt{1 - \hat{v}^2/c^2} = F_0$$

Using the first definition (*12.29*),

$$\mathbf{vF} = cF_0 = \frac{d}{dt}\left(\frac{mc^2}{\sqrt{1 - \hat{v}^2/c^2}}\right)$$

12.30 Show that as $\hat{v} \to 0$, $\hat{E} = mc^2 + \frac{1}{2}m\hat{v}^2 + O(\hat{v}^4/c^2)$. Interpret this result.

The expression for relativistic energy, $\hat{E} = mc^2(1 - \hat{v}^2/c^2)^{-1/2}$ may be expanded by the binomial theorem:

$$(1+x)^\alpha = 1 + \alpha x + \frac{\alpha(\alpha-1)}{2!}x^2 + \cdots \qquad (-1 < x < 1)$$

The result is

$$\hat{E} = mc^2 + \frac{1}{2}m\hat{v}^2 + \frac{3m\hat{v}^4}{8c^2} + \cdots$$

Thus, at low speeds, the total energy of particle is very nearly the sum of its rest energy (which includes all sorts of potential energy) and its classical kinetic energy.

MAXWELL'S EQUATIONS IN SR

12.31 Prove that $\bar{\Box}\bar{f} = \Box f$.

As the g_{ij} are constants, $\Box f \equiv g^{ij}f_{,ij} = \text{invariant}$.

12.32 Prove that if (F^{ij}) is any matrix of functions of the 3-vectors **U** and **V** such that $\partial F^{ij}/\partial x^j = 0$ $(i = 0, 1, 2, 3)$ for all inertial frames and $F^{ij}(\mathbf{0}, \mathbf{0}) = 0$ for all i, j, where $\mathbf{0} = (0, 0, 0)$, then (F^{ij}) is a second-order contravariant tensor under Lorentz transformations.

Let (u_i) be any constant, covariant vector under Lorentz transformations [hence, $(\bar{u}_i) = (b_i^k u_k)$ is also constant]. Define

$$S^i \equiv u_k F^{ki} \qquad \bar{S}^i \equiv \bar{u}_k \bar{F}^{ki}$$

By the given conditions $\partial \bar{F}^{ij}/\partial \bar{x}^j = 0$,

$$\frac{\partial \bar{S}^i}{\partial \bar{x}^i} = \bar{u}_k \frac{\partial \bar{F}^{ki}}{\partial \bar{x}^i} = 0 = \frac{\partial S^i}{\partial x^i}$$

Suppose that at some point (x_0^i), $\bar{S}^i = h^i(S^0, S^1, S^2, S^3)$; then,

$$\frac{\partial \bar{S}^i}{\partial \bar{x}^i} = 0 = \frac{\partial h^i}{\partial S^j}\frac{\partial S^j}{\partial x^k}\frac{\partial x^k}{\partial \bar{x}^i} \qquad \text{or} \qquad \left(b_i^k \frac{\partial h^i}{\partial S^j}\right)\frac{\partial S^j}{\partial x^k} = 0$$

for an arbitrary matrix $(\partial S^j/\partial x^k)$ having $\partial S^i/\partial x^i = 0$. By a well-known lemma (Problem 12.57), there exists a real number $\lambda = \lambda(S^0, S^1, S^2, S^3)$ such that

$$b_i^k \frac{\partial h^i}{\partial S^j} = \lambda \delta_j^k \tag{1}$$

Now differentiate both sides of (*1*) with respect to S^l:

$$b_i^k \frac{\partial^2 h^i}{\partial S^j \partial S^l} = \frac{\partial \lambda}{\partial S^l}\delta_j^k \tag{2}$$

which is symmetrical in j and l; therefore,

$$\frac{\partial \lambda}{\partial S^l}\delta_j^k = \frac{\partial \lambda}{\partial S^j}\delta_l^k \tag{3}$$

for all j, k, l. Let $k = l \neq j$ in (*3*):

$$\frac{\partial \lambda}{\partial S^k}\cdot 0 = \frac{\partial \lambda}{\partial S^j}\cdot 1 \qquad \text{or} \qquad \frac{\partial \lambda}{\partial S^j} = 0$$

Hence λ is constant with respect to the S^i and (*1*) inverts to give

$$\frac{\partial h^i}{\partial S^j} = \lambda a_j^i \tag{4}$$

Integrating (*4*),

$$h^i \equiv \bar{S}^i = \lambda a^i_j S^j + T^i \tag{5}$$

For the special assignment $\mathbf{U} = \mathbf{V} = \mathbf{0}$, we have (since $\bar{\mathbf{0}} = \mathbf{0}$):

$$\begin{aligned} S^i &= (u_1)(0) + (u_2)(0) + (u_3)(0) + (u_4)(0) = 0 \\ \bar{S}^i &= (\bar{u}_1)(0) + (\bar{u}_2)(0) + (\bar{u}_3)(0) + (\bar{u}_4)(0) = 0 \end{aligned} \tag{6}$$

Together, (*5*) and (*6*) imply $T^i = 0 \quad (i = 0, 1, 2, 3)$; consequently,

$$\bar{S}^i = \lambda a^i_j S^j \tag{7}$$

Similarly, there exists a real number μ such that

$$S^i = \mu b^i_j \bar{S}^j \tag{8}$$

It follows that $\bar{S}^i = \lambda a^i_j \mu b^j_k \bar{S}^k = \lambda\mu \bar{S}^i$, or $\lambda\mu = 1$. But we can exploit the reciprocal relationship between observers O and $\bar{O}$, as in Problem 12.4, to show that $\lambda = \mu$. Therefore, $\lambda = \mu = 1$ and (*7*) or (*8*) becomes the transformation law of a (contravariant) 4-vector. Finally, we conclude from the Quotient Theorem that if $F^{ki}u_k \equiv S^i$ is a tensor for an arbitrary covariant vector (u_i), (F^{ij}) is a second-order contravariant tensor.

12.33 Prove the relations (*12.40*) and (*12.41*).

By (*12.39*) and the constancy of the g_{ij},

$$\frac{\partial f^{0j}}{\partial x^j} = \frac{\partial f^{00}}{\partial x^0} + \frac{\partial f^{0q}}{\partial x^q} = \frac{\partial}{\partial x^q}(-V^q) = -\frac{\partial V^q}{\partial x^q} = -\operatorname{div} \mathbf{V}$$

and, for $p = 1, 2, 3$,

$$\frac{\partial f^{pj}}{\partial x^j} = \frac{\partial f^{p0}}{\partial x^0} + \frac{\partial f^{pq}}{\partial x^q} = -\frac{\partial f^{0p}}{\partial x^0} + \frac{\partial}{\partial x^q}(\varepsilon_{pqr}U^r) = \frac{\partial V^p}{\partial x^0} - \varepsilon_{prq}\frac{\partial U^r}{\partial x^q} = \left(\frac{1}{c}\frac{\partial \mathbf{V}}{\partial t} + \operatorname{curl} \mathbf{U}\right)_p$$

The other two formulas are derived from these by replacing $\mathbf{U}, \mathbf{V}$ by $\mathbf{V}, -\mathbf{U}$.

Supplementary Problems

12.34 Suppose two events consist of light signals, and an observer sends one of the signals himself. Classify the space-time interval between the events if the observer sees the distant light signal (*a*) before he sends his own signal, (*b*) after he sends his own signal, (*c*) at the same time he sends his own signal.

12.35 Assuming that any velocity less than c is attainable, suppose that a concert in Los Angeles begins at 8:0508 p.m. and one in New York City, 3000 miles away (consider this the accurate distance), begins at 8:0506 p.m. could a person physically attend both events (opening measures only)? Is the pair of events timelike or spacelike?

12.36 Show that the transpose of a Lorentz matrix is Lorentz.

12.37 Verify the expressions (*12.12*).

12.38 An event occurs at $\bar{O}$'s origin at some time $\bar{t}$. (*a*) How does O view this event? (*b*) What is the significance of $a^0_0 > 0$?

12.39 Write out the simple Lorentz transformation connecting inertial frames O and $\bar{O}$ that move apart at 80% of the velocity of light.

12.40 (*a*) Confirm that a photon (a particle with the velocity of light in some inertial frame) will be viewed as having the velocity of light in all other inertial frames. (*b*) What must be the rest mass of such a particle?

12.41 Show that the following matrix is Lorentz, and use Theorem 12.2 to find the matrices L^*, R_1, and R_2, and the velocity v between the two observers.

$$L = \begin{bmatrix} 5/4 & 1/2 & 1/4 & -1/2 \\ -3/4 & -5/6 & -5/12 & 5/6 \\ 0 & 2/3 & 2/15 & 11/15 \\ 0 & -1/3 & 14/15 & 2/15 \end{bmatrix}$$

12.42 Verify that the following matrix is Lorentz and calculate the velocity between the two observers without finding the simple Lorentz matrix L^*.

$$L = \begin{bmatrix} 3/\sqrt{3} & 1/\sqrt{3} & 2/\sqrt{3} & -1/\sqrt{3} \\ 1 & 1 & 1 & 0 \\ 1 & 0 & 1 & -1 \\ 0 & 1/\sqrt{3} & -1/\sqrt{3} & -1/\sqrt{3} \end{bmatrix}$$

12.43 Show that by definition the proper-time parameter τ is an invariant with respect to all Lorentz transformations.

12.44 Verify the formula for the composition of velocities by (i) multiplying the two simple Lorentz matrices below; (ii) calculating from (*12.15*) the velocities belonging to the two matrices and to their product; (iii) showing that the three velocities obey (*12.16*).

$$L_1 = \begin{bmatrix} 13/12 & 5/12 & 0 & 0 \\ 5/12 & 13/12 & 0 & 0 \\ 0 & 0 & 1 & 0 \\ 0 & 0 & 0 & 1 \end{bmatrix} \qquad L_2 = \begin{bmatrix} 17/8 & -15/8 & 0 & 0 \\ -15/8 & 17/8 & 0 & 0 \\ 0 & 0 & 1 & 0 \\ 0 & 0 & 0 & 1 \end{bmatrix}$$

12.45 An electron gun shoots particles in opposite directions at one-half the velocity of light. At what relative velocity are the particles receding from each other?

12.46 Show that the composition of two velocities less than c is also less than c.

12.47 How slow would your watch run relative to a stationary clock if you were moving at 2/3 the velocity of light?

12.48 At the age of 20, an astronaut left her twin brother on earth to go exploring in outer space. The first two years the spaceship gradually accelerated to a cruising speed 95 percent of the velocity of light. Traveling at that speed for 25 years, it reached a distant galaxy (23.75 light years away) and then decelerated for two years. Two years were spent exploring the galaxy before the journey back home, which followed the schedule of the trip outward. How old is the astronaut when she rejoins her 80-year-old brother? (Use an average rate for clock-retardation during the 8 years in acceleration/deceleration.)

12.49 How fast would a pole-vaulter have to run for his 20-foot pole to fit (instantaneously) inside a barn, in the judgment of a ground observer for whom the barn is 19 feet 11 inches long?

12.50 An alternate definition of uniformly accelerated motion is motion under a *constant Lorentz force*. Verify that the two definitions are equivalent for one-dimensional motion.

12.51 Show that $g^{rs}a^i_r a^j_s = g^{ij}$.

12.52 Prove that the array (*12.46*) is a 4-vector.

12.53 Show that the matrices $\tilde{\mathcal{F}}$ and $\mathcal{F}$ of (*12.44*) are connected via $\tilde{F}^{ij} = \frac{1}{2}e_{ijkl}g_{kr}g_{ls}F^{rs}$. [*Hint*: First evaluate the matrix product $GFG \equiv P$.]

12.54 Define *Faraday's two-form* by

$$\Phi \equiv G\tilde{\mathcal{F}}G = \begin{bmatrix} 0 & -E_1 & -E_2 & -E_3 \\ E_1 & 0 & H_3 & -H_2 \\ E_2 & -H_3 & 0 & H_1 \\ E_3 & H_2 & -H_1 & 0 \end{bmatrix} = [\Phi_{ij}]_{44}$$

or, inversely, $\tilde{\mathcal{F}} = G\Phi G$. Show that (*a*) $\mathcal{F}$ is related to Φ through $F^{ij} = -\frac{1}{2}e_{ijkl}\Phi_{kl}$; (*b*) Maxwell's equations can be written in terms of the single matrix Φ as

$$\frac{\partial \Phi_{ij}}{\partial x^k} + \frac{\partial \Phi_{ki}}{\partial x^j} + \frac{\partial \Phi_{jk}}{\partial x^i} = 0 \qquad g_{ik}g_{jl}\frac{\partial \Phi_{kl}}{\partial x^j} = s^i$$

12.55 The energy flux in an electromagnetic field is specified by the *Poynting vector*, $\mathbf{p} = \mathbf{E} \times \mathbf{H}$. By direct matrix multiplication or otherwise, derive the formula

$$\frac{1}{2}(\tilde{\mathcal{F}}\mathcal{F} - \mathcal{F}\tilde{\mathcal{F}}) = \begin{bmatrix} 0 & 0 & 0 & 0 \\ * & 0 & p_3 & -p_2 \\ * & * & 0 & p_1 \\ * & * & * & 0 \end{bmatrix} \quad \text{(antisymmetric matrix)}$$

12.56 Verify that for simple Lorentz matrices A (hence, no rotation of axes allowed): (*a*) $\tilde{\mathcal{F}}(\mathbf{U}, \mathbf{V}) = G\mathcal{F}(\mathbf{V}, \mathbf{U})G$; (*b*) $\bar{\mathcal{F}}(\bar{\mathbf{V}}, \bar{\mathbf{U}}) = B^T\mathcal{F}(\mathbf{V}, \mathbf{U})B$, where $B = A^{-1}$; and (*c*) $\bar{\tilde{\mathcal{F}}}(\bar{\mathbf{U}}, \bar{\mathbf{V}}) = A\tilde{\mathcal{F}}(\mathbf{U}, \mathbf{V})A^T$ (thereby proving that (F^{ij}) is a contravariant tensor under simple Lorentz transformations).

12.57 Prove that if $A \equiv [A_{ij}]_{nn}$ satisfies $A_{ij}B_{ij} = 0$ for every $B \equiv [B_{ij}]_{nn}$ that has zero trace ($B_{ii} = 0$), then $A = \lambda I$, for some real λ. [*Hint*: First take B as having all elements zero, except for one off-diagonal element. Then choose $B_{\alpha\alpha} = -B_{\beta\beta} = 1$ ($\alpha \neq \beta$; no summation), with all other B_{ij} zero.]

Chapter 13

Tensor Fields on Manifolds

13.1 INTRODUCTION

The modern, noncoordinate approach to tensors will be introduced as an important alternative to the coordinate-component approach employed exclusively in the previous chapters. This will entail somewhat more sophisticated mathematics.

13.2 ABSTRACT VECTOR SPACES AND THE GROUP CONCEPT

Linear algebra provides a means of systematically studying the algebraic interplay between real numbers (*scalars*) and a wide variety of different types of objects (*vectors*). Vectors can be matrices, n-tuples of real numbers, functions, differential operators, etc. In this chapter, we shall adopt the convention of using uppercase boldface characters for sets (of points, of real numbers, of elements of a group, etc.), and lowercase boldface for vectors (as in the preceding chapters). However, the latter will be gradually phased out in favor of light uppercase characters, not only for easier reading, but also in conformity with notation used in many standard textbooks.

The concept of vector spaces requires a careful distinction between the scalars $a, b, c, \ldots$, and the objects of study (vectors), $\mathbf{u}, \mathbf{v}, \mathbf{w}, \ldots$. We shall always identify the scalars with the field of real numbers, although any field could serve for the construction of an abstract vector space.

Algebraic Properties of a Vector Space

In terms of two binary operations, the axioms for a vector space are as follows.

Addition Axioms

1. $\mathbf{u} + \mathbf{v}$ is always a vector
2. $\mathbf{u} + \mathbf{v} = \mathbf{v} + \mathbf{u}$
3. $(\mathbf{u} + \mathbf{v}) + \mathbf{w} = \mathbf{u} + (\mathbf{v} + \mathbf{w})$
4. There is a vector $\mathbf{0}$ such that $\mathbf{u} + \mathbf{0} = \mathbf{u}$.
5. For each $\mathbf{u}$ there is a vector $-\mathbf{u}$ such that $\mathbf{u} + (-\mathbf{u}) = \mathbf{0}$.

Scalar Multiplication Axioms

6. $a \cdot \mathbf{u} \equiv a\mathbf{u}$ is always a vector
7. $a(\mathbf{u} + \mathbf{v}) = a\mathbf{u} + a\mathbf{v}$
8. $(a + b)\mathbf{u} = a\mathbf{u} + b\mathbf{u}$
9. $(ab)\mathbf{u} = a(b\mathbf{u})$
10. $1\mathbf{u} = \mathbf{u}$

EXAMPLE 13.1 We give notation for four familiar vector spaces.

(*a*) $\mathbf{R}^n \equiv$ the n-tuples of reals under componentwise addition and scalar multiplication.

(*b*) $\mathbf{P}^n \equiv$ the polynomials (in a variable t) of degree n or less. If $p(t) \equiv a_i t^i$, $q(t) \equiv b_i t^i$, let $p(t) + q(t) = (a_i + b_i)t^i$ and $r \cdot p(t) = (ra_i)t^i$.

(*c*) $C^k(\mathbf{R}) \equiv$ the continuously k-times differentiable functions (of t), $f : \mathbf{R} \to \mathbf{R}$ (mapping the reals into the reals). To define $+$ and $\cdot$, write $f(t) + g(t) = (f + g)(t)$ and $r \cdot f(t) = (rf)(t)$.

(*d*) $\mathbf{M}^n(\mathbf{R}) \equiv$ the $n \times n$ matrices over $\mathbf{R}$. If $A = (a_{ij})$ and $B = (b_{ij})$, addition and scalar multiplication are defined by $A + B = (a_{ij} + b_{ij})$ and $rA = (ra_{ij})$.

Algebraic Properties of a Group

Axioms 1–5 make a vector space an abelian (commutative) group under addition. In the general definition of a group, the binary operation is designated as "multiplication" and the commutative requirement is dropped.

Multiplication Axioms

1. uv belongs to the group.
2. $(uv)w = u(vw)$.
3. There is an identity element e such that $eu = ue = u$.
4. For each u there is an inverse element u^{-1} such that $uu^{-1} = u^{-1}u = e$.

EXAMPLE 13.2 Some frequently encountered groups follow.

(*a*) The reals **R** over ordinary addition; the reals over ordinary multiplication if 0 is removed from the set.

(*b*) The cube roots of unity, $\mathbf{C}^3 = \{1, \omega, \omega^2\}$, over ordinary multiplication of complex numbers, where $\omega = \frac{1}{2}(-1 + i\sqrt{3})$. Groups of this type are called *cyclic* and are generally denoted by $\mathbf{C}^k$ (the cyclic group of order k). $\mathbf{C}^k$ is necessarily abelian.

(*c*) The 4-group $\{e, u, s, b\}$, under the rules $u^2 = s^2 = b^2 = e$, $b = us$, and the associative law of multiplication. The 4-group is abelian, but it is not equivalent to the cyclic group on four elements, $\mathbf{C}^4$.

(*d*) $\mathbf{M}^n(\mathbf{R})$, under matrix addition.

(*e*) $\mathbf{GL}(n, \mathbf{R}) \equiv$ the real, nonsingular $n \times n$ matrices under matrix multiplication; this is the *general linear group* (nonabelian). $\mathbf{GL}(n, \mathbf{R})$ contains many very important smaller groups (called *subgroups*). Some of these are: $\mathbf{SL}(n, \mathbf{R}) \equiv$ the real $n \times n$ matrices with determinant $+1$; $\mathbf{SO}(n) \equiv$ the $n \times n$ orthogonal matrices; and $\mathbf{L}(n) \equiv$ the $n \times n$ Lorentz matrices [see the definition of $\mathbf{L}(4)$ in Section 12.3].

(*f*) $\mathbf{GL}(n, \mathbf{C}) \equiv$ the complex, nonsingular $n \times n$ matrices under matrix multiplication. An important subgroup is the *unitary group*, $\mathbf{U}(n)$, consisting of all $n \times n$ Hermitian matrices (such that $A = \bar{A}^T$, where the bar denotes complex conjugation).

13.3 IMPORTANT CONCEPTS FOR VECTOR SPACES

Basis

A *basis* for a vector space is a maximal, linearly independent set of vectors $\mathbf{b}_1, \mathbf{b}_2, \ldots$. If this set is finite, possessing n elements, the vector space is *finite-dimensional*, of dimension n. Otherwise, the space is said to be *infinite-dimensional*.

EXAMPLE 13.3 (1) It is obvious that a basis for $\mathbf{R}^n$ is the set of vectors

$$\mathbf{e}_1 = (1, 0, 0, \ldots, 0), \quad \mathbf{e}_2 = (0, 1, 0, \ldots, 0), \quad \ldots, \quad \mathbf{e}_n = (0, 0, \ldots, 0, 1)$$

called the *standard* basis. (2) $\mathbf{P}^n$ is finite-dimensional, of dimension $n + 1$; one basis is $\{t^i\}$, $0 \leqq i \leqq n$. (3) The vector space of all polynomials is infinite-dimensional, as is the vector space $C^k(\mathbf{R})$. See Problem 13.4.

Isomorphisms, Linear Mappings

Two mathematical systems of the same type (such as two vector spaces or two groups) are called *isomorphic* if they are structurally identical and differ only in nomenclature. In the case of two vector spaces, an *isomorphism* is a one-to-one (bijective) linear mapping φ from one space to the other, where the term *linear* refers to the properties (for all vectors **u**, **v**, and scalars a):

$$\varphi(\mathbf{u} + \mathbf{v}) = \varphi(\mathbf{u}) + \varphi(\mathbf{v}) \qquad \text{and} \qquad \varphi(a\mathbf{u}) = a\varphi(\mathbf{u}) \tag{13.1}$$

For groups, an isomorphism would be a bijection ψ with the property $\psi(uv) = \psi(u)\psi(v)$, for all elements of the group. A more general mapping that is important for groups is a *homomorphism*, which merely requires that $\psi(uv) = \psi(u)\psi(v)$ for all u and v, without necessarily requiring one-to-oneness.

Product of Vector Spaces

If $\mathbf{U}$ and $\mathbf{V}$ are any two vector spaces, the ordinary cartesian product $\mathbf{U} \times \mathbf{V}$, the set of ordered pairs $(\mathbf{u}, \mathbf{v})$ with $\mathbf{u}$ in $\mathbf{U}$ and $\mathbf{v}$ in $\mathbf{V}$, can be made into a vector space by defining addition and scalar multiplication of pairs via

$$(\mathbf{p}, \mathbf{q}) + (\mathbf{r}, \mathbf{s}) = (\mathbf{p} + \mathbf{r}, \mathbf{q} + \mathbf{s}) \qquad \text{and} \qquad a(\mathbf{p}, \mathbf{q}) = (a\mathbf{p}, a\mathbf{q})$$

Such a product space is denoted $\mathbf{U} \otimes \mathbf{V}$; if $\mathbf{U} = \mathbf{V}$, write $\mathbf{U} \otimes \mathbf{V}$ as $\mathbf{U}^2$. More generally, the product of any number of vector spaces $\mathbf{V}_1, \mathbf{V}_2, \ldots, \mathbf{V}_k$ may be easily defined as above; this product is denoted $\mathbf{V}_1 \otimes \mathbf{V}_2 \otimes \mathbf{V}_3 \otimes \cdots \otimes \mathbf{V}_k$. If $\mathbf{V}_1 = \mathbf{V}_2 = \cdots = \mathbf{V}_k = \mathbf{V}$, the product space is written $\mathbf{V}^k$. (This notation is also often used for the *tensor product* of two vector spaces, a concept which will not be treated here.)

13.4 THE ALGEBRAIC DUAL OF A VECTOR SPACE

If a vector space $\mathbf{V}$ be mapped linearly into the reals $\mathbf{R}$, satisfying *(13.1)*, the mapping is called a *linear functional*, or *one-form*. As in Example 13.1(*c*), we can make the set of all linear functionals on $\mathbf{V}$ into a vector space itself, with the zero functional as that mapping which takes every vector in $\mathbf{V}$ into 0 in $\mathbf{R}$.

Definition 1: The *algebraic dual* of a vector space $\mathbf{V}$ is the set $\mathbf{V}^*$ of all linear functionals made into a vector space under ordinary pointwise addition and scalar multiplication:

$$(f + g)(\mathbf{v}) = f(\mathbf{v}) + g(\mathbf{v}) \qquad (\lambda f)(\mathbf{v}) = \lambda f(\mathbf{v})$$

Since any linear functional on $\mathbf{R}^n$ can be expressed as a linear function of the coordinates,

$$\mathbf{v} = v^1\mathbf{e}_1 + v^2\mathbf{e}_2 + \cdots + v^n\mathbf{e}_n \quad \rightarrow \quad f(\mathbf{v}) = a_1v^1 + a_2v^2 + \cdots + a_nv^n$$

where $a_i = f(\mathbf{e}_i)$ for each i, the functional is completely determined by the n-tuple $(a_1, a_2, \ldots, a_n)$.

Differential Notation: One-Forms

Thus, different functionals correspond to different n-tuples, as

$$f \leftrightarrow (a_1, a_2, \ldots, a_n) \qquad g \leftrightarrow (b_1, b_2, \ldots, b_n) \qquad \ldots$$

and it has become customary to represent linear functionals by the compact notation of *one-forms*:

$$\boldsymbol{\omega} = a_1\,dx^1 + a_2\,dx^2 + \cdots + a_n\,dx^n \qquad \boldsymbol{\sigma} = b_1\,dx^1 + b^2\,dx^2 + \cdots + b_n\,dx^n \qquad \ldots$$

But why dx^i for the coordinates? The motivation comes from differential geometry. Recall that any class C^1 multivariate function $F(x^1, x^2, \ldots, x^n)$ on $\mathbf{R}^n$ has the gradient $\nabla F = (\partial F/\partial x^i)$ and the directional derivative (in the direction $(dx^1, dx^2, \ldots, dx^n)$)

$$dF = \frac{\partial F}{\partial x^1}\,dx^1 + \frac{\partial F}{\partial x^2}\,dx^2 + \cdots + \frac{\partial F}{\partial x^n}\,dx^n$$

which, at a specific point in space, is a one-form that defines a linear functional on $\mathbf{R}^n$ (i.e., the set of all directions). Recall too that, just as in ordinary one-dimensional calculus,

$$dx = \Delta x \equiv \text{an unspecified real number}$$

not necessarily small.

EXAMPLE 13.4 (*a*) In $\mathbf{R}^3$, find the image of $\mathbf{v} = (1, 3, 5)$ under the one-forms (linear functionals)

$$\boldsymbol{\omega} = 4\,dx^1 - dx^2 \qquad \boldsymbol{\sigma} = 2\,dx^1 + 3\,dx^2 - dx^3 \qquad \boldsymbol{\omega} + \boldsymbol{\sigma} = 6\,dx^1 + 2\,dx^2 - dx^3$$

(*b*) What is the relationship among $\boldsymbol{\omega}(\mathbf{v})$, $\boldsymbol{\sigma}(\mathbf{v})$, and $(\boldsymbol{\omega} + \boldsymbol{\sigma})(\mathbf{v})$?

(*a*)
$$\omega(\mathbf{v}) = 4\cdot 1 - 1\cdot 3 + 0\cdot 5 = 4 - 3 + 0 = 1$$
$$\sigma(\mathbf{v}) = 2\cdot 1 + 3\cdot 3 - 1\cdot 5 = 2 + 9 - 5 = 6$$
$$(\omega + \sigma)(\mathbf{v}) = 6\cdot 1 + 2\cdot 3 - 1\cdot 5 = 6 + 6 - 5 = 7$$

(*b*)
$$\omega(\mathbf{v}) + \sigma(\mathbf{v}) = 1 + 6 = 7 = (\omega + \sigma)(\mathbf{v})$$

For vector spaces different from $\mathbf{R}^n$ we agree to use the procedure of Example 13.4 on the components of vectors relative to an arbitrary basis. That is, to evaluate the image of $\mathbf{v} = v^1\mathbf{b}_1 + v^2\mathbf{b}_2 + \cdots + v^n\mathbf{b}_n \equiv v^i\mathbf{b}_i$ under the one-form $\omega = a_i\, dx^i$, simply write

$$\omega(\mathbf{v}) = \omega(v^j\mathbf{b}_j) \equiv (a_i\, dx^i)(v^j\mathbf{b}_j) = a_i v^i \tag{13.2}$$

A dual reading of (*13.2*) gives a better understanding of the relationship between **V** and **V*** (between vectors and one-forms). If we regard the a_i as fixed (tantamount to fixing a basis in **V**) while the v^i vary—the "normal" situation—then a linear map from **V** to **R** is uniquely defined. If, on the other hand, the vector components v^i are held fixed and the coefficients a_i are allowed to vary (this amounts to fixing a basis in **V***), a linear map from **V*** to **R** is defined (the latter map is actually an element of the space **V****). The expression $a_i v^i$ is *bilinear* in the two vector variables **v** and ω.

Theorem 13.1: If **V** is a finite-dimensional vector space, then **V*** is finite-dimensional, of the same dimension, and is isomorphic to **V**.

A proof is given in Problem 13.6.

Dual Basis

A basis $\mathbf{b}_1, \mathbf{b}_2, \ldots, \mathbf{b}_n$ for **V** determines one for the dual space **V*** in a very natural way. Each **v** in **V** has a representation $\mathbf{v} = v^j\mathbf{b}_j$ and thus defines a linear functional

$$\varphi(\mathbf{v}) = v^1\, dx^1 + v^2\, dx^2 + \cdots + v^n\, dx^n \tag{13.3}$$

Then the n linear functionals (vectors in **V***) defined by

$$\varphi(\mathbf{b}_i) \equiv \boldsymbol{\beta}^i \qquad (i = 1, 2, \ldots, n) \tag{13.4a}$$

form a basis for **V*** (see Problem 13.6); we say that the basis $\{\boldsymbol{\beta}^i\}$ in **V*** is the *dual* of the basis $\{\mathbf{b}_i\}$ in **V**. The evaluation rule (*13.2*) provides a simpler characterization of the dual basis:

$$\boldsymbol{\beta}^i(\mathbf{v}) = (\varphi(\mathbf{b}_i))(v^j\mathbf{b}_j) = (0\cdot dx^1 + 0\cdot dx^2 + \cdots + 1\cdot dx^i + \cdots + 0\cdot dx^n)(v^j\mathbf{b}_j) = v^i \tag{13.4b}$$

Thus, $\boldsymbol{\beta}^i = dx^i$ is the linear functional that picks out the ith component relative to $\{\mathbf{b}_k\}$ of any vector in **V**. A special application of (*13.4b*) gives

$$\boldsymbol{\beta}^i(\mathbf{b}_j) = \delta^i_j \tag{13.5}$$

for all i, j.

EXAMPLE 13.5 The standard basis $\mathbf{e} = \{\mathbf{e}_1, \mathbf{e}_2, \ldots, \mathbf{e}_n\}$ for $\mathbf{R}^n$ generates the standard basis for $(\mathbf{R}^n)^*$, given in terms of one-forms as

$$\boldsymbol{\beta}^1(\mathbf{e}) = dx^1 \qquad \boldsymbol{\beta}^2(\mathbf{e}) = dx^2 \qquad \ldots \qquad \boldsymbol{\beta}^n(\mathbf{e}) = dx^n$$

Suppose, then, that $\mathbf{R}^3$ is referred to the (nonstandard) basis

$$\mathbf{b}_1 = (1,\ 1,\ 0) \quad \mathbf{b}_2 = (1,\ 0 1) \quad \mathbf{b}_3 = (0,\ 1,\ 1)$$

This may be written in terms of the standard basis $\{\mathbf{e}_i\}$ through a formal matrix multiplication:

$$\begin{bmatrix} \mathbf{b}_1 \\ \mathbf{b}_2 \\ \mathbf{b}_3 \end{bmatrix} = \begin{bmatrix} 1 & 1 & 0 \\ 1 & 0 & 1 \\ 0 & 1 & 1 \end{bmatrix} \begin{bmatrix} \mathbf{e}_1 \\ \mathbf{e}_2 \\ \mathbf{e}_3 \end{bmatrix}$$

Express the dual basis $\{\boldsymbol{\beta}^i\}$ for $(\mathbf{R}^3)^*$ in terms of its standard basis (dx^i) as a similar matrix product.

Let $\boldsymbol{\beta}^i = a_1^i\, dx^1 + a_2^i\, dx^2 + a_3^i\, dx^3$; we must solve for the components a_j^i. For $i = 1$, we have from (*13.5*):

$$\boldsymbol{\beta}^1(\mathbf{b}_1) = \boldsymbol{\beta}^1(1,1,0) = a_1^1 \cdot 1 + a_2^1 \cdot 1 + a_3^1 \cdot 0 = a_1^1 + a_2^1 \equiv x + y = 1$$
$$\boldsymbol{\beta}^1(\mathbf{b}_2) = \boldsymbol{\beta}^1(1,0,1) = x \cdot 1 + y \cdot 0 + z \cdot 1 = x + z = 0$$
$$\boldsymbol{\beta}^1(\mathbf{b}_3) = \boldsymbol{\beta}^1(0,1,1) = x \cdot 0 + y \cdot 1 + z \cdot 1 = y + z = 0$$

(where $x \equiv a_1^1$, $y \equiv a_2^1$, $z \equiv a_3^1$). Solving, $x = \frac{1}{2} = y$, $z = -\frac{1}{2}$. A similar analysis may be used to determine the a_j^2 and a_j^3. The final result is

$$\begin{aligned} \boldsymbol{\beta}^1 &= \tfrac{1}{2}dx^1 + \tfrac{1}{2}dx^2 - \tfrac{1}{2}dx^3 \\ \boldsymbol{\beta}^2 &= \tfrac{1}{2}dx^1 - \tfrac{1}{2}dx^2 + \tfrac{1}{2}dx^3 \\ \boldsymbol{\beta}^3 &= -\tfrac{1}{2}dx^1 + \tfrac{1}{2}dx^2 + \tfrac{1}{2}dx^3 \end{aligned} \qquad \text{or} \qquad \begin{bmatrix} \boldsymbol{\beta}^1 \\ \boldsymbol{\beta}^2 \\ \boldsymbol{\beta}^3 \end{bmatrix} = \begin{bmatrix} \frac{1}{2} & \frac{1}{2} & -\frac{1}{2} \\ \frac{1}{2} & -\frac{1}{2} & \frac{1}{2} \\ -\frac{1}{2} & \frac{1}{2} & \frac{1}{2} \end{bmatrix} \begin{bmatrix} dx^1 \\ dx^2 \\ dx^3 \end{bmatrix}$$

Observe that the two basis-connecting matrices are formal inverses of each other.

Change of Basis in V and V*

The result of Example 13.5 can be generalized. Let $\{\mathbf{b}_i\}$ and $\{\bar{\mathbf{b}}_i\}$ be two bases for **V**, and let $\{\boldsymbol{\beta}^i\}$ and $\{\bar{\boldsymbol{\beta}}^i\}$ be the respective dual bases for **V***. Then

$$\bar{\mathbf{b}}_i = A_i^j \mathbf{b}_j \quad \rightarrow \quad \bar{\boldsymbol{\beta}}^i = \bar{A}_j^i \boldsymbol{\beta}^j \qquad \text{with} \qquad (\bar{A}_j^i) = (A_j^i)^{-1} \tag{13.6}$$

the arrow denoting implication. (See Problem 13.7.)

13.5 TENSORS ON VECTOR SPACES

The concept of a *multilinear functional* is needed: If $f(\mathbf{v}^1, \mathbf{v}^2, \ldots, \mathbf{v}^m)$ represents a mapping of m vector variables into the reals such that the restricted mapping obtained by holding all but one of the variables fixed is a linear functional, then f is said to be *multilinear* in all its variables.

Definition 2: A *type-*$\binom{p}{q}$ *tensor* is any multilinear functional $T : (\mathbf{V}^*)^p \otimes \mathbf{V}^q \rightarrow \mathbf{R}$ mapping p one-forms and q vectors into the reals; the real image is denoted

$$T(\boldsymbol{\omega}^1, \ldots, \boldsymbol{\omega}^p; \mathbf{v}^1, \ldots, \mathbf{v}^q)$$

EXAMPLE 13.6 Let T represent a linear functional in what follows. A type-$\binom{1}{0}$ tensor takes on real values $T(\boldsymbol{\omega})$ for all one-forms $\boldsymbol{\omega}$ as argument. As we shall see later, such a tensor can be identified with a contravariant vector. A type-$\binom{0}{1}$ tensor takes on real values $T(\mathbf{v})$ for all vectors $\mathbf{v}$ as argument; it can be shown to correspond to a covariant vector. A type-$\binom{1}{1}$ tensor takes on real values $U(\boldsymbol{\omega}; \mathbf{v})$ for all ordered pairs in $\mathbf{V}^* \otimes \mathbf{V}$ as argument, with U a bilinear functional.

EXAMPLE 13.7 For n-dimensional vectors, the ordinary scalar product $\mathbf{u} \cdot \mathbf{v} \equiv \mathbf{uv}$ defines a type-$\binom{0}{2}$ tensor, in the form $G(\mathbf{u}, \mathbf{v}) = \mathbf{uv}$, since the elementary properties of the scalar product make G a bilinear mapping of a vector pair into the reals. More generally, an inner product defined arbitrarily by the quadratic form

$$G(\mathbf{u}, \mathbf{v}) = \mathbf{u}^T E \mathbf{v}$$

where E is an $n \times n$ matrix, defines a type-$\binom{0}{2}$ tensor.

Definition 3: A type-$\binom{0}{2}$ tensor $G(\mathbf{u}, \mathbf{v})$ is (i) *symmetric* if, for every $\mathbf{u}$ and $\mathbf{v}$,

$$G(\mathbf{u}, \mathbf{v}) = G(\mathbf{v}, \mathbf{u})$$

(ii) *nonsingular* if

$$[G(\mathbf{u}, \mathbf{v}) = 0, \text{ identically in } \mathbf{u}] \quad \rightarrow \quad \mathbf{v} = \mathbf{0}$$

and (iii) *positive definite* if, for any nonzero vector $\mathbf{u}$,

$$G(\mathbf{u}, \mathbf{u}) > 0$$

A type-$\binom{0}{2}$ tensor that is symmetric and nonsingular is called a *metric tensor*. (A positive definite tensor is necessarily nonsingular.)

EXAMPLE 13.8 Let $C = [C^i_j]_{nn}$ be a square matrix and let (a_i) and (v^i) be the respective components of $\boldsymbol{\omega}$ and $\mathbf{v}$ relative to the standard basis in $\mathbf{R}^n$ and its dual. Then the matrix product

$$T(\boldsymbol{\omega}; \mathbf{v}) = \boldsymbol{\omega} C \mathbf{v} \equiv a_i C^i_j v^j \qquad \text{(a bilinear form)}$$

defines a type-$\binom{1}{1}$ tensor over the vector space $\mathbf{R}^n$.

Tensor Components

In the three types of tensors considered in Examples 13.6–13.8, we may define tensor components in the following manner, which may be generalized to arbitrary tensors in an obvious way. Let $\mathbf{b}_1, \ldots, \mathbf{b}_n$ be a basis for $\mathbf{V}$, and $\boldsymbol{\beta}^1, \ldots, \boldsymbol{\beta}^n$ its dual in $\mathbf{V}^*$. Then, for each i, write

$$\begin{aligned} &\textit{type } \tbinom{1}{0} \qquad T^i = T(\boldsymbol{\beta}^i) \\ &\textit{type } \tbinom{0}{1} \qquad T_i = T(\mathbf{b}_i) \\ &\textit{type } \tbinom{1}{1} \qquad T^i_j = T(\boldsymbol{\beta}^i; \mathbf{b}_j) \end{aligned}$$

EXAMPLE 13.9 Find the components, relative to the standard basis for $\mathbf{V} = \mathbf{R}^n$, of a type-$\binom{1}{1}$ tensor on $\mathbf{V}$ constructed by the recipe of Example 13.8.

By construction, $T(\boldsymbol{\omega}; \mathbf{v}) = a_i C^i_j v^j$, for all $\boldsymbol{\omega}$ and $\mathbf{v}$. Substituting $\boldsymbol{\omega} = \boldsymbol{\beta}^p = dx^p$ and $\mathbf{v} = \mathbf{b} = \mathbf{e}_q$, we find

$$T^p_q \equiv T(dx^p; \mathbf{e}_q) = \delta^p_i C^i_j \delta^j_q = C^p_q$$

Thus, the components of T are independent of those of the arguments, $\boldsymbol{\omega}$ and $\mathbf{v}$, and depend only on the components of the matrix C.

Effect of Change of Basis on Tensor Components

Under a change of basis, *(13.6)*,

$$\begin{aligned} &\textit{type } \tbinom{1}{0} \qquad \bar{T}^i = T(\bar{\boldsymbol{\beta}}^i) = T(\bar{A}^i_r \boldsymbol{\beta}^r) = \bar{A}^i_r T(\boldsymbol{\beta}^r) = T^r \bar{A}^i_r \\ &\textit{type } \tbinom{0}{1} \qquad \bar{T}_i = T(\bar{\mathbf{b}}_i) = T(A^r_i \mathbf{b}_r) = A^r_i T(\mathbf{b}_r) = T_r A^r_i \\ &\textit{type } \tbinom{2}{1} \qquad \bar{T}^{ij}_k = T(\bar{\boldsymbol{\beta}}^i, \bar{\boldsymbol{\beta}}^j; \bar{\mathbf{b}}_k) = T(\bar{A}^i_r \boldsymbol{\beta}^r, \bar{A}^j_s \boldsymbol{\beta}^s; A^t_k \mathbf{b}_t) = \bar{A}^i_r \bar{A}^j_s A^t_k T(\boldsymbol{\beta}^r, \boldsymbol{\beta}^s; \mathbf{b}_t) = T^{rs}_t \bar{A}^i_r \bar{A}^j_s A^t_k \\ &\cdots\cdots \qquad \cdots\cdots\cdots\cdots\cdots\cdots\cdots\cdots\cdots\cdots\cdots\cdots \end{aligned}$$

EXAMPLE 13.10 If $\mathbf{V}$ is Euclidean $\mathbf{R}^n$, the change of basis $\bar{\mathbf{b}}_i = A^j_i \mathbf{b}_j$ induces the change of coordinates $x^i = A^i_j \bar{x}^j$, for which

$$\bar{J} \equiv \left(\frac{\partial x^i}{\partial \bar{x}^j}\right) = A \qquad \text{and so} \qquad J = \bar{J}^{-1} = \bar{A}$$

The above transformation formulas then reduce to the classical laws for affine tensors—compare *(3.21)*.

13.6 THEORY OF MANIFOLDS

A manifold is the natural extension of a surface to higher dimensions, and also to spaces more general than $\mathbf{R}^n$. It is helpful at first to think of a manifold as just a hypersurface in $\mathbf{R}^n$.

By the term *neighborhood of a point* we shall understand either the set of all points in $\mathbf{R}^n$ within some fixed distance from the given point, or any set containing such points. A neighborhood of p will be denoted $\mathbf{U}_p$. If the concept used for distance in $\mathbf{R}^m$ is Euclidean, then every neighborhood $\mathbf{U}_p$ contains a solid, spherical ball (or "hyperball," if $n > 3$), having some positive radius and centered at p. A neighborhood of a point *p in a set* is the intersection of a neighborhood $\mathbf{U}_p$ and the set. A set is *open* if each of its points has a neighborhood completely composed of points of the set. An open neighborhood is simply a neighborhood that is also an open set (in the case of a solid-ball neighborhood, the outer boundary of the ball would have to be removed in order to make it an open neighborhood).

Descriptive Definition of a Manifold

A *manifold* is a set which has the property that each point can serve as the origin of local coordinates that are valid in an open neighborhood of the point, which neighborhood is an exact "copy" of an open neighborhood of a point in $\mathbf{R}^n$. Though such a definition allows the manifold to lie in a metric space, topological space, Banach space, or other abstract mathematical system, it is best that we begin with manifolds in a simpler space, like $\mathbf{R}^m$. Accordingly:

Definition 4: A *manifold* is any set $\mathbf{M}$ in $\mathbf{R}^m$ which has the property that for each point p in the manifold there exists an open neighborhood $\mathbf{U}_p$ in $\mathbf{M}$ and a mapping φ_p which carries $\mathbf{U}_p$ into a neighborhood in $\mathbf{R}^n$. The mapping is required to be a *homeomorphism*; i.e.

(1) φ_p is continuous.
(2) φ_p is bijective from $\mathbf{U}_p$ onto its range, $\varphi_p(\mathbf{U}_p)$.
(3) φ_p^{-1} is continuous.

See Fig. 13-1.

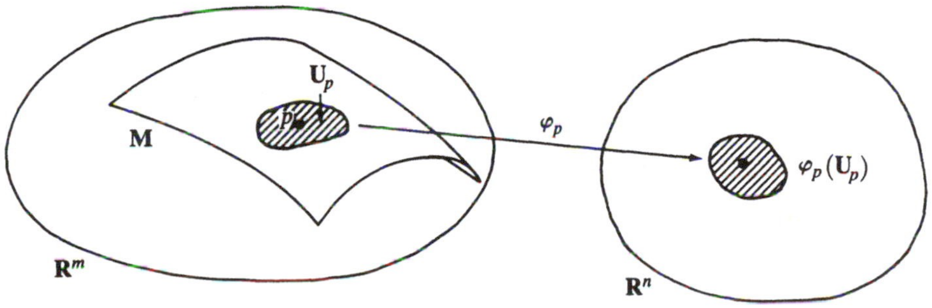

Fig. 13-1

Coordinate Patches, Atlas

The neighborhoods $\mathbf{U}_p$ for p in $\mathbf{M}$ provide a means for locally ascribing coordinates to $\mathbf{M}$ which have the correct dimension (e.g., a plane lying in 3-space is actually 2-dimensional and, as a manifold, has a coordinatization by pairs of reals instead of triples, as in Section 10.4). For any point p in $\mathbf{M}$, the pair $(\mathbf{U}_p, \varphi_p)$ is called a *coordinate patch* (also *chart*, or *local coordinatization*) for $\mathbf{M}$, while any collection of such pairs for which the neighborhoods $\mathbf{U}_p$ together cover $\mathbf{M}$ is called an *atlas* for $\mathbf{M}$. Since the coordinate patches make $\mathbf{M}$ n-dimensional at each point, $\mathbf{M}$ is sometimes referred to as an *n-manifold*.

Often a finite number of charts is sufficient for an atlas (Example 13.11). It can be proved that if a manifold in $\mathbf{R}^m$ is closed, and bounded in terms of the distance in $\mathbf{R}^m$, a finite number of charts will always be sufficient.

EXAMPLE 13.11

(*a*) The *2-sphere*, denoted $\mathbf{S}^2$, is the ordinary sphere in 3-dimensional space (y^i), centered at $(0, 0, 0)$ having radius a. It may be coordinatized by an atlas of only two charts, as follows. [Note that the usual spherical coordinates (φ, θ) fail to give a one-one mapping at the poles, where θ is indeterminate.]

$$y^1 = \frac{2a^2x^1}{(x^1)^2 + (x^2)^2 + a^2}$$

$$y^2 = \frac{2a^2x^2}{(x^1)^2 + (x^2)^2 + a^2}$$

$$y^3 = \varepsilon a \frac{(x^1)^2 + (x^2)^2 - a^2}{(x^1)^2 + (x^2)^2 + a^2} \qquad (\varepsilon = \pm 1)$$

As illustrated in Fig. 13-2, the chart corresponding to $\varepsilon = +1$ has $\mathbf{U}_p$ centered on the south pole (whose coordinates are $x^1 = x^2 = 0$) and including every point of the sphere except the north pole. The other chart ($\varepsilon = -1$) is the mirror image of the first chart in the equatorial plane. For a derivation of this atlas, see Problem 13.15.

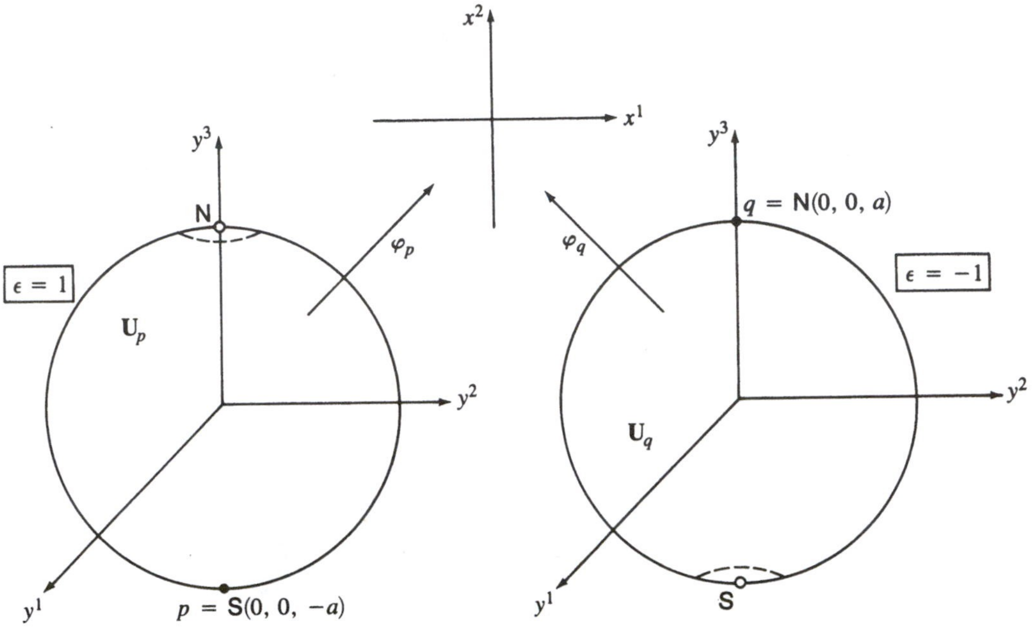

Fig. 13-2

(*b*) The *n-sphere* $\mathbf{S}^n$ in $\mathbf{R}^{n+1}$ may be defined as the set of points (y^i) in $\mathbf{R}^{n+1}$ such that

$$(y^1)^2 + (y^2)^2 + (y^3)^2 + \cdots + (y^{n+1})^2 = a^2$$

(centered at $(0, 0, 0, \ldots, 0)$, radius a). A coordinate patch for a neighborhood of $(0, 0, \ldots, a)$ is:

$$y^1 = x^1 \qquad y^2 = x^2 \qquad \cdots \qquad y^n = x^n \qquad y^{n+1} = \sqrt{a^2 - (x^1)^2 - (x^2)^2 - \cdots - (x^n)^2}$$

where the mapping is into the n-dimensional neighborhood $(x^1)^2 + \cdots + (x^n) < a^2$ (the interior of $\mathbf{S}^{n-1}$). Establishing the analogous patches around the other "diametrically opposite" endpoints, we obtain an atlas of $2n + 2$ charts. (A smaller atlas requires a more clever approach.)

Differentiable Manifolds

Inevitably, there will exist pairs $(\mathbf{U}_p, \varphi_p)$ and $(\mathbf{U}_q, \varphi_q)$ whose neighborhoods overlap in $\mathbf{M}$ (Fig. 13-3); so the common region $\mathbf{U}_p \cap \mathbf{U}_q \equiv \mathbf{W}$, called an *overlapping set*, generates a map φ between the images of $\mathbf{W}$ under φ_p and φ_q. Explicitly (trace the circuit in Fig. 13-3), $\varphi = \varphi_q \circ \varphi_p^{-1}$.

It is clear that φ and φ^{-1} are both continuous. If φ and φ^{-1} are of class C^k (have continuous partial derivatives of order k at each point) then the overlapping set $\mathbf{W}$ is said to be of class C^k.

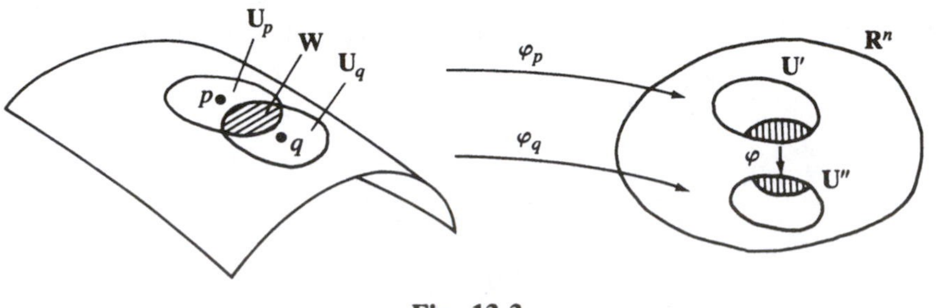

Fig. 13-3

Definition 5: A *differentiable* manifold is a manifold which possesses an atlas such that all overlapping sets are of class C^1. A C^k (C^∞ or C^ω) manifold has an atlas whose overlapping sets are of class C^k (C^∞ or C^ω).

Remark 1: Recall the distinction between *infinitely differentiable* (C^∞) and *analytic* (C^ω).

One way to ensure that a manifold be C^k in the present context is to demand that each φ_p and φ_p^{-1} be class C^k. As a matter of convenience, we assume from now on that all manifolds are C^∞ manifolds.

EXAMPLE 13.12 In the case of the spherical manifolds of Example 13.11, the mapping functions φ_p^{-1} are either rational with nonvanishing denominators or square roots of positive polynomials. These are certainly C^∞ manifolds (in fact, C^ω).

To bring the notation closer to that of differential geometry (cf. Section 10.4), we now redesignate the maps φ_p^{-1} linking **M** with coordinates (x^i): let

$$\varphi_p^{-1}(x^1, x^2, \ldots, x^n) \equiv \mathbf{r}(x^1, x^2, \ldots, x^n) \equiv (y^j(x^1, x^2, \ldots, x^n))$$

for $1 \leqq j \leqq m$. (See Fig. 13-4.)

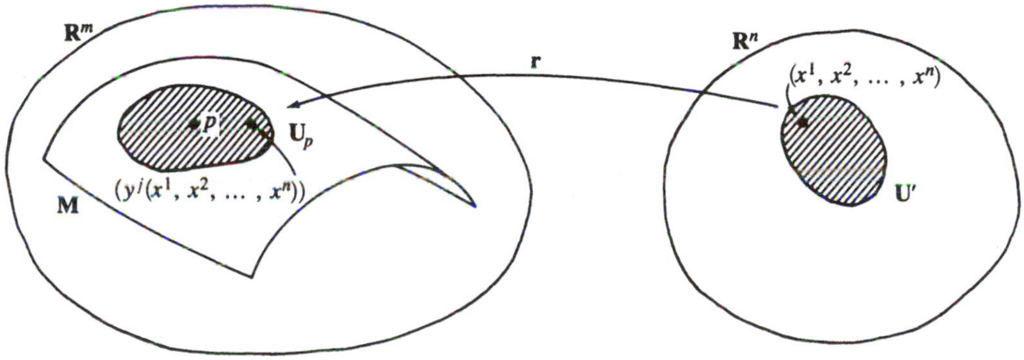

Fig. 13-4

13.7 TANGENT SPACE; VECTOR FIELDS ON MANIFOLDS

Intuitively expressed, a vector field V on a manifold **M** is simply a tangent vector to **M** which varies in some continuous (and differentiable) manner from point to point (Fig. 13-5). More precisely, it is a rule that gives a tangent vector at every point of **M**. One way to obtain a vector field (if we are in $\mathbf{R}^3$) is to take the variable normal vector **n** and cross it with some fixed vector **a**; thus $V = \mathbf{n} \times \mathbf{a}$ is a differentiable vector field. But this definition takes us outside the manifold (is *extrinsic*). We seek a way to remain on the manifold itself (which is immediately applicable to abstract manifolds not imbeddable in a familiar space); such endeavors are called *intrinsic* methods.

The clue is to consider some curve on **M** and to define V as the tangent field of that curve. If the curve is defined through a coordinate patch by

$$\mathbf{c} = \mathbf{r}(x^1(t), x^2(t), \ldots, x^n(t)) = (y^j(x^1, x^2, \ldots, x^n)) \qquad (1 \leqq j \leqq m)$$

then the chain rule gives

$$\frac{d\mathbf{c}}{dt} = \frac{d\mathbf{r}}{dt} = \frac{\partial \mathbf{r}}{\partial x^i}\frac{dx^i}{dt} = V \qquad \text{or} \qquad V = V^i\mathbf{r}_i$$

where the vectors $\mathbf{r}_i \equiv \partial\mathbf{r}/\partial x^i$ are themselves tangent to **M**.

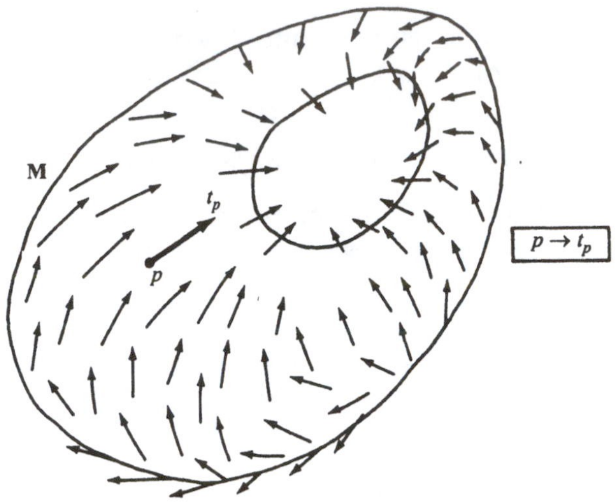

Fig. 13-5

Definition 6: For any (differentible) manifold **M** having coordinate patch $\mathbf{U}_p$ at any point p, the span of the vectors $\mathbf{r}_1, \mathbf{r}_2, \ldots, \mathbf{r}_n$ [evaluated at $\varphi_p(p)$] is the *tangent space at* p, denoted $T_p(\mathbf{M})$. The union of all tangent spaces $T_p(\mathbf{M})$, for all p in **M**, is called the *tangent bundle* of **M**, denoted $T(\mathbf{M})$.

Although each $T_p(\mathbf{M})$ is a vector space, this does not guarantee that $T(\mathbf{M})$ is a vector space. For example, the sum of a vector in $T_p(\mathbf{M})$ and a vector in $T_q(\mathbf{M})$ will not generally be tangent to **M**.

Definition 7: A *vector field V* on a manifold **M** is any C^∞ function that maps **M** to its tangent bundle $T(\mathbf{M})$. That is, for each point p in **M**, the image $\mathbf{V}(p) = \mathbf{V}_p$ is a vector belonging to the tangent space $T_p(\mathbf{M})$ at p. Explicitly, for certain scalar functions V^i,

$$V = V^i \mathbf{r}_i = \left(V^i \frac{\partial y^j}{\partial x^i} \right) \qquad (j = 1, 2, \ldots, m) \tag{13.7a}$$

A fundamental theorem regarding vector fields on manifolds may be proved from the basic theory of systems of ordinary differential equations.

Theorem 13.2: Every vector field on a manifold **M** possesses a system of *flow curves* or *integral curves* on **M**, defined as curves for which the tangent vector at each point coincides with the given vector field at that point.

Notation

To de-emphasize the particular choice of coordinatization map $\varphi_p : \mathbf{U}_p \to \mathbf{R}^n$, it is customary to omit the vector **r** from the above description of the basis for $T_p(\mathbf{M})$, and to write

$$\frac{\partial}{\partial x^1}, \frac{\partial}{\partial x^2}, \ldots, \frac{\partial}{\partial x^n} \quad \text{in place of} \quad \frac{\partial \mathbf{r}}{\partial x^1}, \frac{\partial \mathbf{r}}{\partial x^2}, \ldots, \frac{\partial \mathbf{r}}{\partial x^n}$$

or, even more cursorily, $\partial_1, \partial_2, \ldots, \partial_n$. Many textbooks use this last notation exclusively, and write (*13.7a*) as

$$V = V^i \partial_i \tag{13.7b}$$

Such shorthand is particularly convenient when the coordinate maps φ_p^{-1} for a particular manifold are unspecified (for example, when the manifold is defined by an equation $F(y^1, y^2, \ldots, y^m) = 0$ for some real-valued function F). In this situation, since $\mathbf{r}_1, \mathbf{r}_2, \ldots, \mathbf{r}_n$ are not explicitly defined, we use the notation $E_1, E_2, \ldots, E_n$ to denote the *coordinate frame* on **M**, whose restriction to $T_p(\mathbf{M})$, for each p in **M**, is a basis for $T_p(\mathbf{M})$. We make the identifications $E_i \equiv \partial_i \quad (i = 1, 2, \ldots, n)$, giving

$$V = V^i E_i \tag{13.7c}$$

Extrinsic Representation of Vector Fields

It is possible to represent a vector field V on a manifold without reference to coordinate patches (which often have the disadvantage of being complicated or difficult to construct); one can stay entirely in the (y^i) system, which we assume to be rectangular. Suppose **M** is given by a single equation $F(y^1, y^2, \ldots, y^m) = 0$, for some C^k function F. A one-form $\sigma = \omega_i\, dy^i$, where $\omega_i = \omega_i(y^1, y^2, \ldots, y^m)$, is said to be *restricted to* **M** if the point (y^i) is required to lie on **M**; that is, $F(y^1, y^2, \ldots, y^m) = 0$. As is well known from multidimensional calculus, the gradient $\nabla F = (\partial F/\partial y^i)$ is normal to **M**, so that if we further require that the restriction of σ map ∇F into zero,

$$\omega_i \frac{\partial F}{\partial y^i} = 0$$

then $V = \sigma$ is a vector field on **M** (having components dy^i).

EXAMPLE 13.13 Consider the paraboloid **P** in $\mathbf{R}^3$ given by

$$F(y^1, y^2, y^3) = (y^1)^2 + (y^2)^2 - y^3 = 0$$

Show that the restriction of $\sigma = y^1y^2\, dy^1 + (y^2)^2\, dy^2 + 2y^2y^3\, dy^3$ to **P** is a vector field on **P**.

We must show that the scalar product of $(y^1y^2, (y^2)^2, 2y^2y^3)$ and $\nabla F = (2y^1, 2y^2, -1)$ is zero:

$$(y^1y^2)(2y^1) + (y^2)^2(2y^2) + (2y^2y^3)(-1) = [(y^1)^2 + (y^2)^2 - y^3]2y^2 = 0$$

13.8 TENSOR FIELDS ON MANIFOLDS

Dual Tangent Bundle

At each p in **M**, let $T_p^*(\mathbf{M})$ denote the dual of the vector space $T_p(\mathbf{M})$, and denote by $T^*(\mathbf{M})$ the union of all spaces $T_p^*(\mathbf{M})$. The set $T^*(\mathbf{M})$, called the *dual tangent bundle* of **M**, is not necessarily a vector space (just as $T(\mathbf{M})$ was not).

We need to make explicit certain elements of $T^*(\mathbf{M})$.

Differentials on M

The differential of a function $f : \mathbf{R}^n \to \mathbf{R}$ is rigorously defined as a two-vector function $df : (\mathbf{R}^n)^2 \to \mathbf{R}$ which maps each pair $(\mathbf{x}, \mathbf{v})$—where $\mathbf{x}$ is a point in $\mathbf{R}^n$ and $\mathbf{v} = (dx^1, dx^2, \ldots, dx^n)$ is a direction in $\mathbf{R}^n$—to the real number

$$df(\mathbf{x}, \mathbf{v}) \equiv \frac{\partial f}{\partial x^1}\, dx^1 + \frac{\partial f}{\partial x^2}\, dx^2 + \cdots + \frac{\partial f}{\partial x^n}\, dx^n = f_i\, dx^i \tag{13.8}$$

Here, the f_i are evaluated at $\mathbf{x}$. If f is any real-valued C^k function on **M**, then the differential of

$$f(x^1, x^2, \ldots, x^n) \equiv f(y^1(x^1, \ldots, x^n), y^2(x^1, \ldots, x^n), \ldots, y^m(x^1, \ldots, x^n))$$

called a *differential field* on **M**, is

$$df = \frac{\partial}{\partial x^i}\,(f(\mathbf{r}(x)))\, dx^i = \frac{\partial f}{\partial y^k}\frac{\partial y^k}{\partial x^i}\, dx^i = (\nabla f \cdot \mathbf{r}_i)\, dx^i$$

—a one-form. Thus, df may be thought of as a mapping from **M** to $T^*(\mathbf{M})$: we agree that the evaluation of df at p is the one-form $(\nabla f(p) \cdot \mathbf{r}_i(p))\, dx^i$ in $T_p^*(\mathbf{M})$.

Let us now compare the two kinds of fields on **M**.

	Mapping	Restricted to $\mathbf{U}_p$
vector field	$V : \mathbf{M} \to T(\mathbf{M})$	$V = V^iE_i$
differential field	$df : \mathbf{M} \to T^*(\mathbf{M})$	$\omega = (\nabla f \cdot E_i)\, dx^i$

Definition 8: A *tensor field of type* $\binom{r}{s}$ on a manifold **M** is a mapping $T : [T^*(\mathbf{M})]^r \otimes [T(\mathbf{M})]^s \to C^k(\mathbf{R}^m)$ taking r differential fields and s vector fields on **M** to real-valued C^k-functions f on $\mathbf{R}^m$. It is assumed that the evaluation of T at a point p on **M** is given by

$$T_p(\omega^1, \ldots, \omega^r; V_1, \ldots, V_s) = T(\omega^1_p, \ldots, \omega^r_p; V_{1p}, \ldots, V_{sp}) \equiv f(p)$$

and that each map T_p is multilinear.

EXAMPLE 13.14 (*a*) At each fixed p on **M**, the mapping T_p is a *tensor* [on the vector space $[T^*_p(\mathbf{M})]^r \otimes [T_p(\mathbf{M})]^s$, of type $\binom{r}{s}$], per Definition 2. (*b*) Any vector field V on **M** can be interpreted as a type-$\binom{1}{0}$ tensor field via a mapping $T(\omega) = \omega(V)$; compare Problem 13.20.

Solved Problems

ABSTRACT VECTOR SPACES AND THE GROUP CONCEPT

13.1 (*a*) Show that the set of polynomials

$$p_1(t) = 1 + t \qquad p_2(t) = t + t^2 \qquad p_3(t) = t^2 + t^3 \qquad p_4(t) = t^3 - 1$$

is a basis for the vector space $\mathbf{P}^3$ (polynomials of degree $\leqq 3$). (*b*) Find the components of the polynomial $p(t) = t^3$ relative to this basis.

(*a*) Since the dimension of $\mathbf{P}^3$ is 4 and there are 4 vectors, it suffices to show they are linearly independent. Suppose that, for all t,

$$\lambda^1(1+t) + \lambda^2(t+t^2) + \lambda^3(t^2+t^3) + \lambda^4(t^3-1) = 0$$

or

$$(\lambda^1 - \lambda^4)\cdot 1 + (\lambda^1 + \lambda^2)t + (\lambda^2 + \lambda^3)t^2 + (\lambda^3 + \lambda^4)t^3 = 0$$

Since this is an identity, we must have

$$0 = \lambda^1 - \lambda^4 = \lambda^1 + \lambda^2 = \lambda^2 + \lambda^3 = \lambda^3 + \lambda^4$$

Thus $\lambda^1 = \lambda^4$, $\lambda^1 = -\lambda^2 = \lambda^3$; so the last equation gives $\lambda^1 + \lambda^1 = 0$ or $\lambda^1 = 0$, and all λ^i vanish, thus proving linear independence.

(*b*) To find the linear combination yielding $p(t) = t^3$, write

$$\lambda^1(1+t) + \lambda^2(t+t^2) + \lambda^3(t^2+t^3) + \lambda^4(t^3-1) = t^3$$

i.e.

$$(\lambda^1 - \lambda^4)\cdot 1 + (\lambda^1 + \lambda^2)t + (\lambda^2 + \lambda^3)t^2 + (\lambda^3 + \lambda^4 - 1)t^3 = 0$$

or

$$\lambda^1 = \lambda^4 \qquad \lambda^1 = -\lambda^2 = \lambda^3 \qquad \lambda^1 + \lambda^1 - 1 = 0$$

Hence, $\lambda^1 = -\lambda^2 = \lambda^3 = \lambda^4 = 1/2$.

13.2 (*a*) Model the 4-group by manipulating an ordinary $8\frac{1}{2}$ by 11 sheet of paper, in the following way: Let s be the operation of turning the sheet over *sideways* (as in a book) and setting it on its original location; u, the operation of turning the sheet *upside down* (end-for-end); b, *both* operations (s followed by u, resulting in a 180° rotation of the page, face up); and e, *doing nothing* (identity). Interpret the group operation (multiplication) as *one operation followed by another* (thus, for example, by the above definitions, $b = su$, reading from left to right). (*b*) Show that the 4-group cannot be isomorphic to the cyclic group on four elements, $\mathbf{C}^4$.

(*a*) This is one of those problems in mathematics that is best handled without formulas or equations. By simple observation, the operation us also results in a 180° rotation; hence, $b = su = us$. It is also clear that if we apply s twice, or u twice, the sheet is left in its original state; $s^2 = u^2 = e$. Next, observe that the associative law is valid, so long as we keep the *order* of the operations intact. It follows that

$$b^2 = (su)(us) = s(u)^2s = s^2 = e$$

When we multiply all the group elements by b, we obtain:

$$be = b \qquad bb = e \qquad bu = (su)u = su^2 = s \qquad bs = (su)s = (us)s = u$$

Hence, the multiplication table for this group may be displayed and may be seen to coincide with that for the 4-group:

·	e	s	u	b
e	e	s	u	b
s	s	e	b	u
u	u	b	e	s
b	b	u	s	e

(*b*) For the cyclic group, $\{e, z, z^2, z^3\}$, with $z^4 = e$ for some z, we could not have $z^2 = e$, which is the characteristic property of all elements of the 4-group.

13.3 The simple Lorentz group can be studied by compressing the 4×4 matrices down to 2×2 matrices:

$$\begin{bmatrix} a & b & 0 & 0 \\ b & a & 0 & 0 \\ 0 & 0 & 1 & 0 \\ 0 & 0 & 0 & 1 \end{bmatrix} \rightarrow \begin{bmatrix} a & b \\ b & a \end{bmatrix} \qquad (a^2 - b^2 = 1)$$

Show explicitly that all real 2×2 matrices of the above form constitute an abelian group (the group $\mathbf{L}(2)$) under matrix multiplication, and that $\mathbf{L}(2)$ is a subgroup of the following two larger groups:

$$\mathbf{GL}(2, \mathbf{R}) \ : \ \text{matrices of the form} \begin{bmatrix} a & b \\ c & d \end{bmatrix}, \quad ad \neq bc$$

$$\mathbf{SU}(2) \ : \ \text{matrices of the form} \begin{bmatrix} a & b \\ c & d \end{bmatrix}, \quad ad - bc = 1$$

Since for a matrix in $\mathbf{L}(2)$,

$$ad - bc = a^2 - b^2 = 1$$

all such matrices belong to $\mathbf{SU}(2)$, which, in turn, is a subgroup of $\mathbf{GL}(2, \mathbf{R})$. Now verify the group properties:

(1) uv belongs to the group for all u, v.

If
$$A = \begin{bmatrix} a & b \\ b & a \end{bmatrix} \qquad B = \begin{bmatrix} c & d \\ d & c \end{bmatrix}$$

then
$$AB = \begin{bmatrix} a & b \\ b & a \end{bmatrix}\begin{bmatrix} c & d \\ d & c \end{bmatrix} = \begin{bmatrix} ac + bd & ad + bc \\ bc + ad & bd + ac \end{bmatrix} \equiv \begin{bmatrix} x & y \\ y & x \end{bmatrix}$$

and
$$x^2 - y^2 = \det AB = (\det A)(\det B) = (1)(1) = 1$$

(2) $(uv)w = u(vw)$. Yes: matrix multiplication is associative.

(3) For some e and all u, $eu = ue = u$. Yes: the identity matrix has $1^2 - 0^2 = 1$, so is a member of $\mathbf{L}(2)$.

(4) Given u, $u^{-1}u = uu^{-1} = e$ for some u^{-1}.

$$\begin{bmatrix} a & b \\ b & a \end{bmatrix}^{-1} = \frac{1}{a^2 - b^2}\begin{bmatrix} a & -b \\ -b & a \end{bmatrix} = \begin{bmatrix} a & -b \\ -b & a \end{bmatrix}$$

which is in $\mathbf{L}(2)$.

(5) $uv = vu$ (abelian group).

$$BA = \begin{bmatrix} c & d \\ d & c \end{bmatrix}\begin{bmatrix} a & b \\ b & a \end{bmatrix} = \begin{bmatrix} ca+db & cb+da \\ da+cb & db+ca \end{bmatrix} = \begin{bmatrix} x & y \\ y & x \end{bmatrix} = AB$$

VECTOR SPACE CONCEPTS

13.4 (*a*) Show that the space **P** of all real-valued polynomials in a real variable x is infinite-dimensional. (*b*) Conclude that $C^k(\mathbf{R})$ is infinite-dimensional.

(*a*) Suppose that **P** had the finite basis $\{p_1, p_2, \ldots, p_n\}$. Then, for any real polynomial $p(x)$, there exist constants $a_1, \ldots, a_n$ such that

$$a_1p_1(x) + a_2p_2(x) + \ldots + a_np_n(x) = p(x) \tag{1}$$

Write (*1*) for the $n+1$ values $x_1 < x_2 < \cdots < x_{n+1}$ as a matrix equation:

$$a_1\begin{bmatrix} p_1(x_1) \\ p_1(x_2) \\ \cdots \\ p_1(x_{n+1}) \end{bmatrix} + a_2\begin{bmatrix} p_2(x_1) \\ p_2(x_2) \\ \cdots \\ p_2(x_{n+1}) \end{bmatrix} + \cdots + a_n\begin{bmatrix} p_n(x_1) \\ p_n(x_2) \\ \cdots \\ p_n(x_{n+1}) \end{bmatrix} = \begin{bmatrix} p(x_1) \\ p(x_2) \\ \cdots \\ p(x_{n+1}) \end{bmatrix} \tag{2}$$

The column vectors on the left are elements of $\mathbf{R}^{n+1}$, and as there are n of them, they do not span $\mathbf{R}^{n+1}$ (see Problem 13.5). To finish the proof, we have only to choose a vector on the right of (*2*) that is not in the span of those on the left—say, $(z_1, z_2, \ldots, z_{n+1})$—and then to exhibit a polynomial p that takes on those values at $x_1, x_2, \ldots, x_{n+1}$. The polynomial provided by *Lagrange's interpolation formula* does the job.

(*b*) For any k, the vector space $C^k(\mathbf{R})$ contains the infinite-dimensional subspace **P**; thus, it too is infinite-dimensional.

13.5 The set **S** of all linear combinations of a fixed set of n vectors, $\{\mathbf{b}_1, \mathbf{b}_2, \ldots, \mathbf{b}_n\}$, is called the *span* of the given vectors; it is obviously a vector space. Prove that this space has dimension $m \leqq n$, with equality if and only if the given vectors are linearly independent.

First we show that any $n+1$ vectors in **S** are linearly dependent. Suppose, on the contrary, that $\{\mathbf{u}_1, \mathbf{u}_2, \ldots, \mathbf{u}_{n+1}\}$ are linearly independent. Then, because the sequence of vectors

$$\mathbf{u}_1 \ \mathbf{b}_1 \ \mathbf{b}_2 \ \ldots \ \mathbf{b}_n$$

is necessarily dependent, the well-known *exchange lemma* tells us that the sequence

$$\mathbf{u}_1 \ \mathbf{b}_1 \ \ldots \ \mathbf{b}_{j-1} \ \mathbf{b}_{j+1} \ \ldots \ \mathbf{b}_n$$

also spans S for some j. Repeating the argument $n-1$ more times, we arrive at the result that the vectors

$$\mathbf{u}_n \ \mathbf{u}_{n-1} \ \ldots \ \mathbf{u}_2 \ \mathbf{u}_1$$

span **S**, making $\mathbf{u}_{n+1}$ dependent on them—a contradiction.

If, therefore, $\{\mathbf{b}_i\}$ is linearly independent, it constitutes a basis for **S**, and $m = n$. On the other hand, if only $m < n$ of the $\mathbf{b}_i$ are linearly independent, the above argument shows that any basis consists of exactly m vectors.

DUAL SPACE

13.6 Prove Theorem 13.1.

It is almost trivial that any two vector spaces of dimension n are isomorphic [if $\{\mathbf{b}_i^{(1)}\}$ and $\{\mathbf{b}_i^{(2)}\}$ are bases, set up the correspondence $v^i\mathbf{b}_i^{(1)} \leftrightarrow v^i\mathbf{b}_i^{(2)}$]. Thus it is necessary to prove only that $\mathbf{V}^*$ is n-dimensional if **V** is; in other words, to prove that the set of vectors $\{\boldsymbol{\beta}^i\}$ defined by (*13.4*) (i) is linearly independent and (ii) has $\mathbf{V}^*$ as its span. (Problem 13.5 will then immediately yield Theorem 13.1.)

Proof of (i): By (*13.5*), for $j = 1, 2, \ldots, n$,

$$\lambda_i \boldsymbol{\beta}^i(\mathbf{v}) = 0 \quad \rightarrow \quad \lambda_i \boldsymbol{\beta}^i(\mathbf{b}_j) = 0 \quad \rightarrow \quad \lambda_i \delta^i_j = 0 \quad \rightarrow \quad \lambda_j = 0$$

Proof of (ii): If $\boldsymbol{\beta}(\mathbf{v})$ is an arbitrary element of $\mathbf{V}^*$, then, by (*13.4b*),

$$\boldsymbol{\beta}(\mathbf{v}) = \boldsymbol{\beta}(v^i \mathbf{b}_i) = \boldsymbol{\beta}(\mathbf{b}_i)\, v^i = \boldsymbol{\beta}(\mathbf{b}_i)\, \boldsymbol{\beta}^i(\mathbf{v})$$

that is, $\boldsymbol{\beta}$ is a linear combination of the $\boldsymbol{\beta}^i$.

13.7 Prove the inverse relation between the matrices A and $\bar{A}$ of (*13.6*).

By definition $\bar{\mathbf{b}}_i = A^j_i \mathbf{b}_j$ and $\bar{\boldsymbol{\beta}}^j = \bar{A}^j_k \boldsymbol{\beta}^k$, so look at (*13.5*): $\bar{\boldsymbol{\beta}}^j(\bar{\mathbf{b}}_i) = \delta^j_i$. By the algebra of mappings and the fact that each $\bar{\boldsymbol{\beta}}^j$ and $\boldsymbol{\beta}^k$ is linear, we have

$$\begin{aligned}\delta^j_i = \bar{\boldsymbol{\beta}}^j(\bar{\mathbf{b}}_i) &= (\bar{A}^j_k \boldsymbol{\beta}^k)(\bar{\mathbf{b}}_i) = \bar{A}^j_k \boldsymbol{\beta}^k(\bar{\mathbf{b}}_i) = \bar{A}^j_k \boldsymbol{\beta}^k(A^r_i \mathbf{b}_r) \\ &= \bar{A}^j_k A^r_i \boldsymbol{\beta}^k(\mathbf{b}_r) = \bar{A}^j_k A^r_i \delta^k_r = \bar{A}^j_k A^k_i\end{aligned}$$

that is, $\bar{A}A = I$.

TENSORS ON VECTOR SPACES

13.8 Which of the following represent linear mappings of $(\mathbf{R}^3)^*$ (taking the one-forms on $\mathbf{R}^3$ into the reals), and so constitute (contravariant) tensors of type $\binom{1}{0}$?

(*a*) $T(a_1\,dx^1 + a_2\,dx^2 + a_3\,dx^3) = a_1 a_2 a_3$ (*b*) $T(a_i\,dx^i) = a_1 - a_3$

(*c*) $T(a_i\,dx^i) = 1$ (*d*) $T(a_i\,dx^i) = 0$

(*b*) and (*d*)—the only *linear* mappings.

13.9 Associated with a particular basis $\{\mathbf{b}_i\}$ of a vector space of dimension n, we are given some set of numbers $\{C^{ij}_k;\ \ i, j, k = 1, \ldots, n\}$. Then we define another set of numbers (and assume a similar definition for all changes of bases), $\{\bar{C}^{ij}_k;\ \ i, j, k = 1, \ldots, n\}$, such that

$$\bar{C}^{ij}_k = \bar{A}^i_r \bar{A}^j_s A^t_k C^{rs}_t$$

and call these numbers the components of the "tensor" C on the new basis $\{\bar{\mathbf{b}}_i\}$. Show that this "tensor" is indeed a tensor per Definition 2.

We have only to define the functional

$$T(\boldsymbol{\omega}_1, \boldsymbol{\omega}_2; \mathbf{v}) = T(a_i \boldsymbol{\beta}^i, b_j \boldsymbol{\beta}^j; v^k \mathbf{b}_k) = a_i b_j C^{ij}_k v^k$$

which, by inspection, is a type $\binom{2}{1}$ tensor. We have:

$$\begin{aligned}T^{ij}_k &= T(\boldsymbol{\beta}^i, \boldsymbol{\beta}^j; \mathbf{b}_k) = T(\delta^i_r \boldsymbol{\beta}^r, \delta^j_s \boldsymbol{\beta}^s; \delta^t_k \mathbf{b}_t) = \delta^i_r \delta^j_s C^{rs}_t \delta^t_k = C^{ij}_k \\ \bar{T}^{ij}_k &= T(\bar{\boldsymbol{\beta}}^i, \bar{\boldsymbol{\beta}}^j; \bar{\mathbf{b}}_k) = T(\bar{A}^i_r \boldsymbol{\beta}^r, \bar{A}^j_s \boldsymbol{\beta}^s; A^t_k \mathbf{b}_t) = \bar{A}^i_r \bar{A}^j_s A^t_k T(\beta^r, \beta^s; b_t) = \bar{A}^i_r \bar{A}^j_s A^t_k C^{rs}_t \equiv \bar{C}^{ij}_k\end{aligned}$$

which show that T and C coincide in all coordinate systems.

13.10 In terms of the components g_{ij} of a metric tensor $G(\mathbf{u}, \mathbf{v})$, show that:

(*a*) G is *symmetric* if and only if $g_{ij} = g_{ji}$ for all i, j.

(*b*) G is *nonsingular* if and only if $|g_{ij}| \neq 0$.

(*c*) G is *positive definite* if, for all vectors $(u^i) \neq \mathbf{0}$, $g_{ij} u^i u^j \neq 0$ and $g_{11} > 0$.

By Section 13.5, $g_{ij} = G(\mathbf{b}_i, \mathbf{b}_j)$ where $\{\mathbf{b}_i\}$ is some basis for $\mathbf{V}$. Then, if $\mathbf{u} = u^i \mathbf{b}_i$ and $\mathbf{v} = v^i \mathbf{b}_i$ are any two vectors in $\mathbf{V}$,

$$G(\mathbf{u}, \mathbf{v}) = u^i v^j G(\mathbf{b}_i, \mathbf{b}_j) = g_{ij} u^i v^j$$

(*a*) $G(\mathbf{u}, \mathbf{v}) = G(\mathbf{v}, \mathbf{u})$, for all $\mathbf{u}, \mathbf{v}$, if and only if

$$g_{ij}u^i v^j = g_{ij}v^i u^j = g_{ji}u^i v^j \qquad \text{or} \qquad (g_{ij} - g_{ji})u^i v^j = 0$$

for all real u^i, v^j, which is true if and only if $g_{ij} = g_{ji}$.

(*b*) In matrix form, the nonsingularity criterion reads:

$$[u^T G v = 0, \text{ for all } u] \quad \rightarrow \quad v = 0$$

But $u^T G v$ vanishes for all u if and only if Gv is the zero vector. Hence the criterion takes the form

$$Gv = 0 \quad \rightarrow \quad v = 0$$

which defines G as a nonsingular matrix (a matrix with nonvanishing determinant).

(*c*) For each fixed $\mathbf{u}$ and a scalar parameter λ, we have

$$g_{ij}(u^i + \lambda \mathbf{b}_1^i)(u^j + \lambda \mathbf{b}_1^j) = g_{ij}(u^i + \lambda\delta_1^i)(u^j + \lambda\delta_1^j) = G(\mathbf{u}, \mathbf{u}) + b\lambda + g_{11}\lambda^2 \equiv P(\lambda)$$

where $b \equiv (g_{1j} + g_{j1})u^j$. If $\mathbf{u}$ is not in the span of $\mathbf{b}_1$, the quadratic form is, by hypothesis, nonzero. Hence, the discriminant of $P(\lambda)$ is negative:

$$b^2 - 4g_{11}G(\mathbf{u}, \mathbf{u}) < 0 \qquad \text{or} \qquad G(\mathbf{u}, \mathbf{u}) > \frac{b^2}{4g_{11}} \geqq 0$$

It only remains to note that if $\mathbf{u} = \kappa \mathbf{b}_1$ $(\kappa \neq 0)$, then $G(\mathbf{u}, \mathbf{u}) = \kappa^2 g_{11}$, which is again positive.

13.11 Show that positive-definiteness of a type-$\binom{0}{2}$ tensor G implies its nonsingularity.

If $G(\mathbf{u}, \mathbf{v}) = 0$ for all $\mathbf{u}$ and some $\mathbf{v}$, then $G(\mathbf{v}, \mathbf{v}) = 0$; and so, by positive-definiteness, $\mathbf{v} = 0$.

13.12 A covariant tensor $A(\mathbf{u}, \mathbf{v})$ is *antisymmetric* if and only if $A(\mathbf{u}, \mathbf{v}) = -A(\mathbf{v}, \mathbf{u})$, for all $\mathbf{u}, \mathbf{v}$. Show that a criterion for antisymmetry is:

$$A(\mathbf{u}, \mathbf{u}) = 0 \qquad (\text{all } \mathbf{u})$$

By bilinearity,

$$A(\mathbf{u} + \mathbf{v}, \mathbf{u} + \mathbf{v}) = A(\mathbf{u}, \mathbf{u}) + A(\mathbf{u}, \mathbf{v}) + A(\mathbf{v}, \mathbf{u}) + A(\mathbf{v}, \mathbf{v})$$

Thus, if $A(\mathbf{u}, \mathbf{u}) = 0$ for all $\mathbf{u}$,

$$0 = 0 + A(\mathbf{u}, \mathbf{v}) + A(\mathbf{v}, \mathbf{u}) + 0 \qquad \text{or} \qquad A(\mathbf{u}, \mathbf{v}) = -A(\mathbf{v}, \mathbf{u})$$

Conversely, suppose $A(\mathbf{u}, \mathbf{v}) = -A(\mathbf{v}, \mathbf{u})$, for all $\mathbf{u}$ and $\mathbf{v}$. Then, with $\mathbf{u} = \mathbf{v}$, we have $A(\mathbf{u}, \mathbf{u}) = -A(\mathbf{u}, \mathbf{u})$, or $A(\mathbf{u}, \mathbf{u}) = 0$.

MANIFOLDS

13.13 (*a*) Show that the 1-sphere $\mathbf{S}^1$ (a circle in $\mathbf{R}^2$) can be made into a C^∞ 1-manifold by constructing an atlas with two charts. (*b*) Show that a one-chart atlas does not exist [thus, a circle is not homeomorphic to a line or interval].

(*a*) The standard parameterization of the circle,

$$\varphi^{-1} : \begin{cases} y^1 = a\cos\theta \\ y^2 = a\sin\theta \end{cases} \qquad (0 \leq \theta < 2\pi)$$

is insufficient, since the inverse map φ is discontinuous at point p (Fig. 13-6). But if we define

$$\varphi_p^{-1} : \begin{cases} y^1 = a\cos x^1 \\ y^2 = a\sin x^1 \end{cases} \quad (-\pi < x^1 < \pi) \qquad \varphi_q^{-1} : \begin{cases} y^1 = a\cos x^1 \\ y^2 = a\sin x^1 \end{cases} \quad (0 < x^1 < 2\pi)$$

then $(\mathbf{S}^1 - q, \varphi_p)$ and $(\mathbf{S}^1 - p, \varphi_q)$ will constitute an atlas. Since there are no 'singular' points involved, it is clear that φ_p, φ_q and their inverses are C^∞.

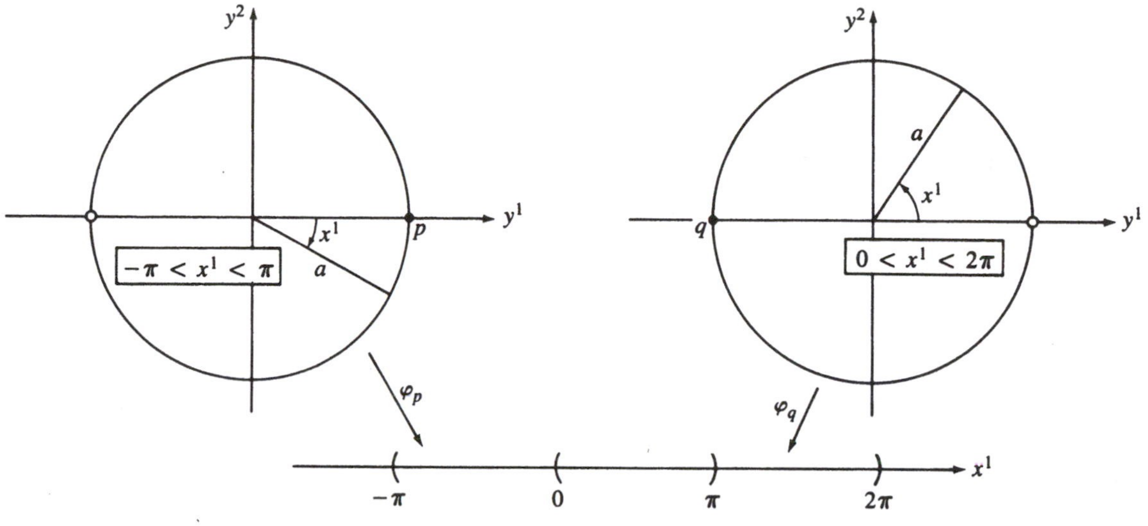

Fig. 13-6

(*b*) Suppose $(\mathbf{U}, \phi)$ covered $\mathbf{S}^1$ $(\mathbf{U}=\mathbf{S}^1)$ and ϕ mapped $\mathbf{S}^1$ to the real line (x^1), with both ϕ and ϕ^{-1} continuous. It is not too difficult to see that ϕ maps the circle to a closed interval **I**: for continuous maps take bounded, closed sets to bounded, closed sets, and connected sets to connected sets; and the only bounded, closed, connected subsets of the real line are closed finite intervals. For any point P on the circle let P' be its diametrically opposite point. The map $g(t) \equiv \phi[(\phi^{-1}(t))']$ takes a real number t in **I**, maps it to a unique point P on $\mathbf{S}^1$, goes to the (unique) diametrically opposite point P', and returns to a unique real number t' in **I**; it is thus a continuous map from **I** to **I**. As such, it must (by a familiar theorem of analysis) have a fixed point:

$$g(t_0) = t_0 \quad \text{for some } t_0 \text{ in } \mathbf{I}$$

But this means that ϕ sends some pair of diametrically opposite points on $\mathbf{S}^1$ to the same real number, denying one-oneness of ϕ.

13.14 A manifold in $\mathbf{R}^4$ is defined by the charts $(k = \ldots, -2, -1, 0, 1, 2, \ldots; x^1 > 0)$

$$\mathbf{r}_{(k)} \ : \begin{cases} y^1 = x^1 \cos x^2 \cos x^3 \\ y^2 = x^1 \cos x^2 \sin x^3 \\ y^3 = x^1 \sin x^2 \\ y^4 = a(x^2 + x^3) \end{cases} \qquad \left((k-1)\,\frac{\pi}{2} < x^3 < (k+1)\,\frac{\pi}{2}\right)$$

(*a*) Show that on each coordinate patch, the mapping $\mathbf{r}_{(k)}$ is one-to-one; hence, $\varphi_{(k)} = \mathbf{r}_{(k)}^{-1} : U_{(k)} \to \mathbf{R}^3$ exists. (*b*) Show that both $\varphi_{(k)}$ and $\varphi_{(k)}^{-1}$ are continuous. (*c*) Show that the manifold is generated by a line in $\mathbf{R}^4$ moving along an axis orthogonal to it, with the axis, in turn, orthogonal to the hyperplane $y^4 = 0$ (use vector geometry in $\mathbf{R}^4$). Verify that the parameter x^1 measures the distance from a given point on the manifold to the axis. (*d*) Show that the parametric section $x^3 = 0$ is a right helicoid (Example 10.4), lying in the hyperplane $y^2 = 0$ ($\mathbf{R}^3$ coordinatized by y^1, y^3, y^4).

(*a*) Assume that $\mathbf{r}_{(k)}(x^i) = \mathbf{r}_{(k)}(u^i)$; we want to show that $(x^1, x^2, x^3) = (u^1, u^2, u^3)$. Now,

$$\left.\begin{aligned} x^1 \cos x^2 \cos x^3 &= u^1 \cos u^2 \cos u^3 \\ x^1 \cos x^2 \sin x^3 &= u^1 \cos u^2 \sin u^3 \end{aligned}\right\} \rightarrow \quad \tan x^3 = \tan u^3$$

But, for $\mathbf{U}_{(k)}$, the argument of the tangent function is restricted to a range of π units; so $x^3 = u^3$. It follows that

$$a(x^2 + x^3) = a(u^2 + u^3) \quad \rightarrow \quad x^2 = u^2$$

Finally, from $x^1 \sin x^2 = u^1 \sin u^2$, we obtain $x^1 = u^1$.

(*b*) From the form of $\varphi_{(k)}^{-1} \equiv \mathbf{r}_{(k)}$, this function is C^∞. To solve for (x^i) in terms of (y^i) (to find $\varphi_{(k)}$), write

$$(y^1)^2 + (y^2)^2 + (y^3)^2 = (x^1)^2(\cos^2 x^2)(\cos^2 x^3 + \sin^2 x^3) + (x^1)^2(\sin^2 x^2) = (x^1)^2$$

or $x^1 = \sqrt{(y^1)^2 + (y^2)^2 + (y^3)^2}$ (since $x^1 > 0$). Then

$$\sin x^2 = \frac{y^3}{x^1} = \frac{y^3}{\sqrt{(y^1)^2 + (y^2)^2 + (y^3)^2}}$$

or, for a suitable branch of the function $\sin^{-1}$,

$$x^2 = \sin^{-1}\left(\frac{y^3}{\sqrt{(y^1)^2 + (y^2)^2 + (y^3)^2}}\right)$$

It is seen that

$$\varphi_{(k)} : \begin{cases} x^1 = \sqrt{(y^1)^2 + (y^2)^2 + (y^3)^2} \\ x^2 = \sin^{-1}\left(\dfrac{y^3}{\sqrt{(y^1)^2 + (y^2)^2 + (y^3)^2}}\right) \\ x^3 = \dfrac{y^4}{a} - \sin^{-1}\left(\dfrac{y^3}{\sqrt{(y^1)^2 + (y^2)^2 + (y^3)^2}}\right) \end{cases}$$

is continuous (in fact, C^∞).

(*c*) The axis orthogonal to $y^4 = 0$ is the vector $\mathbf{e}_4$ in $\mathbf{R}^4$. At the point $y^4 = a(x^2 + x^3) = \text{const.}$ on the manifold, we have (with x^2, x^3 constants and $x^1 = t$)

$$y^1 = t\cos x^2 \cos x^3 \qquad y^2 = t\cos x^2 \sin x^3 \qquad y^3 = t\sin x^2 \qquad y^4 = \text{const.}$$

—a straight line with direction vector orthogonal to $\mathbf{e}_4$. A previous calculation gives the distance from (y^1, y^2, y^3, y^4) on $\mathbf{M}$ to $(0, 0, 0, a(x^2 + x^3))$ as

$$\sqrt{(y^1)^2 + (y^2)^2 + (y^3)^2} = x^1$$

(*d*) Set $x^3 = 0$ and the map reduces to

$$y^1 = x^1 \cos x^2 \qquad y^2 = 0 \qquad y^3 = x^1 \sin x^2 \qquad y^4 = ax^2$$

13.15 Derive the charts of Example 13.11(*a*), using stereographic projection (Fig. 13-7).

As P is a "convex" combination of Q and N,

$$(y^1, y^2, y^3) = \lambda(x^1, x^2, 0) + (1 - \lambda)(0, 0, a) \tag{1}$$

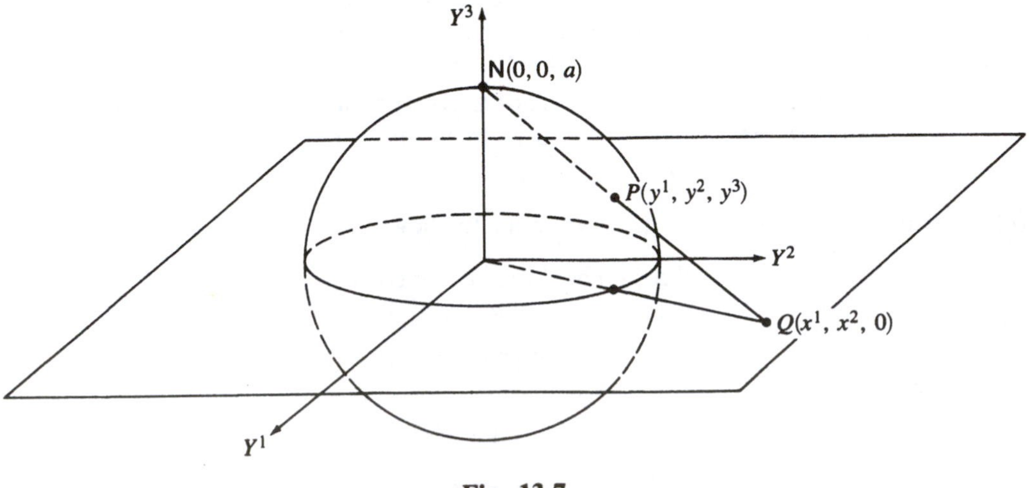

Fig. 13-7

To determine λ ($\lambda > 0$), write

$$a^2 = (y^1)^2 + (y^2)^2 + (y^3)^2 = (\lambda x^1)^2 + (\lambda x^2)^2 + [(1-\lambda)a]^2$$

and solve, obtaining

$$\lambda = \frac{2a^2}{(x^1)^2 + (x^2)^2 + a^2} \tag{2}$$

[Note that λ is less than or greater than 1 according as P lies in the northern or southern hemisphere; $\lambda \neq 0$, so this patch omits the north pole.] Together, (*1*) and (*2*) yield the chart $\varepsilon = +1$; the chart $\varepsilon = -1$ is obtained by changing a to $-a$ in the above (stereographic projection from the south pole).

VECTOR FIELDS ON MANIFOLDS

13.16 The hyperboloid of one sheet $4(y^1)^2 + 4(y^2)^2 - (y^3)^2 = 16$ is a C^∞ 2-manifold **M**, by the coordinatization ($k = 1, 2$)

$$\varphi^{-1}_{(k)} : \begin{cases} y^1 = 2\cos x^1 \cosh x^2 \\ y^2 = 2\sin x^1 \cosh x^2 \\ y^3 = 4\sinh x^2 \end{cases} \qquad ((k-2)\pi < x^1 < k\pi)$$

with $\mathbf{U}_{(1)} = \mathbf{U}_p$ and $p = (2, 0, 0)$, $\mathbf{U}_{(2)} = \mathbf{U}_q$ and $q = (-2, 0, 0)$. Represent the vector field on **M** given by

$$(V^i) = (4\sinh x^2, 4\cosh x^2)$$

in terms of (*a*) a vector basis for the tangent space $T_p(\mathbf{M})$, and (*b*) extrinsically. (c) Describe this field geometrically.

(*a*) By the usual tools of surface theory (Section 10.5):

$$\begin{aligned}
\mathbf{r} &= (2\cos x^1 \cosh x^2, 2\sin x^1 \cosh x^2, 4\sinh x^2) \\
E_1 &= \mathbf{r}_1 = (-2\sin x^1 \cosh x^2, 2\cos x^1 \cosh x^2, 0) \\
E_2 &= \mathbf{r}_2 = (2\cos x^1 \sinh x^2, 2\sin x^1 \sinh x^2, 4\cosh x^2) \\
V &= V^iE_i = (-8\sin x^1 \sinh x^2 \cosh x^2, 8\cos x^1 \sinh x^2 \cosh x^2, 0) \\
&\qquad + (8\cos x^1 \sinh x^2 \cosh x^2, 8\sin x^1 \sinh x^2 \cosh x^2, 16\cosh^2 x^2) \\
&= (4(\cos x^1 - \sin x^1)\sinh 2x^2, 4(\cos x^1 + \sin x^1)\sinh 2x^2, 16\cosh^2 x^2)
\end{aligned}$$

(*b*) From the equations for y^1, y^2, y^3 we may calculate:

$$\cosh x^2 = \frac{1}{2}\sqrt{(y^1)^2 + (y^2)^2} \qquad \sinh x^2 = \frac{1}{4}y^3$$

$$\cos x^1 = \frac{y^1}{\sqrt{(y^1)^2 + (y^2)^2}} \qquad \sin x^1 = \frac{y^2}{\sqrt{(y^1)^2 + (y^2)^2}}$$

so that

$$\begin{aligned}
E_1 &= (-y^2, y^1, 0) \\
E_2 &= \left(\frac{y^1y^3}{2\sqrt{(y^1)^2 + (y^2)^2}}, \frac{y^2y^3}{2\sqrt{(y^1)^2 + (y^2)^2}}, 2\sqrt{(y^1)^2 + (y^2)^2}\right) \\
(V^i) &= (y^3, 2\sqrt{(y^1)^2 + (y^2)^2}) \\
V &= V^iE_i = (-y^2y^3, y^1y^3, 0) + (y^1y^3, y^2y^3, 4(y^1)^2 + 4(y^2)^2) \\
&= (y^3(y^1 - y^2), y^3(y^1 + y^2), (y^3)^2 + 16)
\end{aligned}$$

(using the equation of the hyperboloid). Hence, in terms of the coordinates (y^i),

$$V = \sigma = y^3(y^1 - y^2)\,dy^1 + y^3(y^1 + y^2)\,dy^2 + [(y^3)^2 + 16]\,dy^3$$

(*c*) See Fig. 13-8 and note that the first component is zero in the plane $y^1 = y^2$. Hence, along the curve of intersection, the field is always parallel to the y^2y^3-plane. Similarly, along $y^1 = -y^2$, the field is parallel to the y^1y^3-plane. On the circle $y^3 = 0$ the field is $(0, 0, 16)$, or vertical. Since the third component is $\geqq 16$, there is always a vertical component.

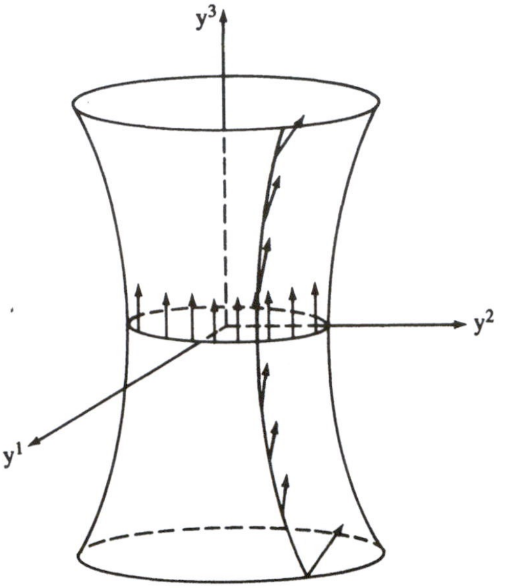

Fig. 13-8

13.17 Show that the restrictions of (*a*) $\sigma_1 = y^1\,dy^2 - y^2\,dy^1$ and (*b*) $\sigma_2 = (y^2 - y^3)\,dy^1 - (y^1 + y^3)\,dy^2 + (y^1 + y^2)\,dy^3$ to the sphere $(y^1)^2 + (y^2)^2 + (y^3)^2 = a^2$ are vector fields. (See Fig. 13-9 for a graph of selected values of σ_1.) By the well-known "Hairy-Ball Theorem" (every head of hair has a cowlick), every continuous vector field on $\mathbf{S}^2$ (and also on $\mathbf{S}^n$, for all even integers n) is zero at some point on the sphere. In fact, the field must vanish at some point of an arbitrarily selected, closed hemisphere. (*c*) Find the zero points explicitly.

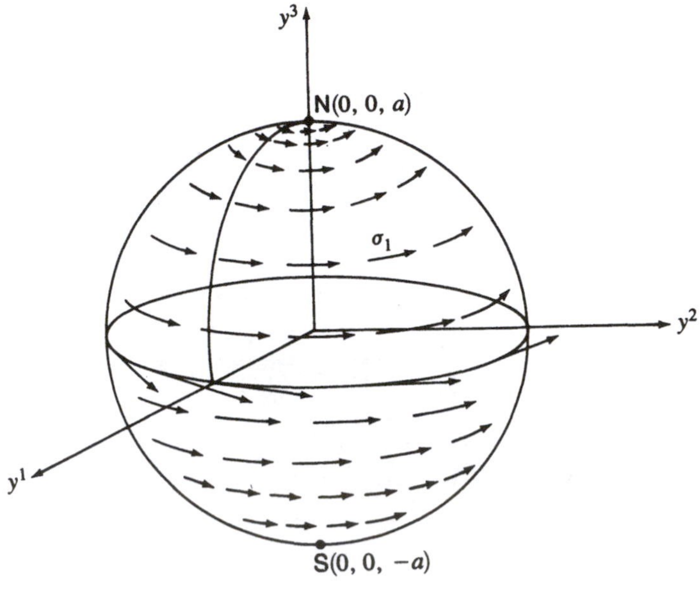

Fig. 13-9

(*a*) The normal vector to $\mathbf{S}^2$ is $\omega = 2y^1\,dy^1 + 2y^2\,dy^2 + 2y^3\,dy^3$ and

$$\sigma_1 \cdot \frac{1}{2}\,\omega = (-y^2)(y^1) + (y^1)(y^2) + (0)(y^3) = 0$$

(*b*)
$$\begin{aligned}\sigma_2 \cdot \frac{1}{2}\,\omega &= (y^2 - y^3)(y^1) - (y^1 + y^3)(y^2) + (y^1 + y^2)(y^3)\\ &= y^1y^2 - y^1y^3 - y^1y^2 - y^2y^3 + y^1y^3 + y^2y^3 = 0\end{aligned}$$

(*c*) If $\sigma_1 = 0$, $-y^2 = y^1 = 0$ and $0^2 + 0^2 + (y^3)^2 = a^2$, or $y^3 = \pm a$. Thus, the zero points are $(0, 0, \pm a)$. For $\sigma_2 = 0$,

$$y^2 - y^3 = y^1 + y^3 = y^1 + y^2 = 0 \quad\rightarrow\quad y^2 = y^3 = -y^1$$

and $(y^1)^2 + (y^2)^2 + (y^3)^2 = a^2 = 3(y^1)^2$, or $y^1 = \pm a/\sqrt{3}$. Hence, the zero points are $\pm(a/\sqrt{3}, -a/\sqrt{3}, -a/\sqrt{3})$.

13.18 Consider a manifold whose coordinatization is not easily determined ($\mathbf{SO}(n)$, of Example 13.2(*e*), is such a manifold in $\mathbf{R}^{n^2}$) for which, therefore, base vectors $\mathbf{r}_i = E_i$ for $T_p(\mathbf{M})$ are unavailable. Develop a reasonable definition of $T_p(\mathbf{M})$ in this situation, which possesses the salient properties of a "tangent space" at point p.

To get an idea of what may be desirable, examine the case when the vectors $\mathbf{r}_1, \mathbf{r}_2, \ldots, \mathbf{r}_n$ *are* available. Each tangent vector has the form $V = V^i\mathbf{r}_i$, and when V is the tangent vector of a curve $\mathscr{C}$ on $\mathbf{M}$—the image of $\mathscr{C}' : x^i = x^i(t)$ in the coordinate space $\mathbf{R}^n$—then

$$V = \frac{dx^i}{dt}\,\mathbf{r}_i \qquad \text{or} \qquad V^i = \frac{dx^i}{dt}$$

Thus (V^i) is a direction vector. Recall that

$$\mathbf{r} = \mathbf{r}(x^1, \ldots, x^n) \equiv \mathbf{r}(y^1(x^1, \ldots, x^n), y^2(x^1, \ldots, x^n), \ldots, y^m(x^1, \ldots, x^n))$$

whence
$$\mathbf{r}_i = \left(\frac{\partial y^1}{\partial x^i}, \frac{\partial y^2}{\partial x^i}, \ldots, \frac{\partial y^m}{\partial x^i}\right)$$

Thus when we write $V^i\mathbf{r}_i$ we are actually indicating the m directional derivatives

$$V^i\,\frac{\partial y^j}{\partial x^i} \equiv \nabla y^j \cdot V \qquad (1 \leqq j \leqq m)$$

with each $y^j(x^1, x^2, \ldots, x^n)$ a C^∞ real-valued function (defined on $\mathbf{M}$ if we identify the points of $\mathbf{M}$ with their coordinates (x^i) in $\mathbf{R}^n$). It is customary to let the directional derivative of a function $f : \mathbf{R}^n \rightarrow \mathbf{R}$ in the direction V be denoted

$$V(f) \equiv \nabla f \cdot V$$

Thus, each vector V *maps a differentiable, real-valued function f to its directional derivative in the direction V*. The properties of this mapping are immediate: If f and g denote any two differentiable functions from $\mathbf{M}$ to $\mathbf{R}$, with fg denoting the ordinary product of two functions, and if a and b are two scalar constants, then

linearity	$V(af + bg) = a\,V(f) + b\,V(g)$
Leibniz' rule	$V(fg) = V(f)\,g + f\,V(g)$

With this in mind, and armed with the knowledge that the directional derivatives of all functions on $\mathbf{M}$ would be enough information to construct the basis $\{\mathbf{r}_i\}$ when $\mathbf{r}$ is known, we frame

Definition 9: By $\mathbf{C}^\infty(p)$ will be understood the real-valued C^∞ functions on $\mathbf{U}_p$, such that any two functions that agree on some neighborhood of p are identified.

Definition 10: The *tangent space* $T_p(\mathbf{M})$ at p is the set of all mappings $V_p : \mathbf{C}^\infty(p) \rightarrow \mathbf{R}$ that satisfy for all a, b in $\mathbf{R}$ and f, g in $\mathbf{C}^\infty(p)$ the two conditions

(i) $V_p(af + bg) = a\,V_p(f) + b\,V_p(g)$

(ii) $V_p(fg) = V_p(f)\,g + f\,V_p(g)$

with the vector-space operations in $T_p(\mathbf{M})$ defined by

$$(U_p + V_p)(f) \equiv U_p(f) + V_p(f)$$
$$(aV_p)(f) \equiv aV_p(f)$$

Any V_p in $T_p(\mathbf{M})$ will be called a *tangent vector to* **M** *at p*. This definition has the advantage not only of dispensing with coordinates, but of enabling one to extend naturally a mapping $F : \mathbf{M} \to \mathbf{N}$ (from one manifold to another) to a mapping $F_* : T_p(\mathbf{M}) \to T_{p'}(\mathbf{N})$ at each point p in **M**, where $p' = F(p)$. Such an extension cannot be accomplished using the more elementary definition.

Remark 2: The vectors of $T_p(\mathbf{M})$ as originally defined, if regarded as mappings on $\mathbf{C}^\infty(p)$, are members of the abstract $T_p(\mathbf{M})$ (Definition 10). In more advanced treatments it is shown that the reverse is true and that $\dim T_p(\mathbf{M}) = \dim \mathbf{M} = n$. Hence, the *two approaches to tangent spaces are equivalent.*

TENSOR FIELDS

13.19 Show that tensor fields always have the property of being bilinear with respect to scalar *functions* (as well as to scalar *constants*), unlike differential operators.

We must show that for any scalar function f on **M** and any tensor T of type $\binom{0}{r}$,

$$T(V_1, \ldots, fV_i, \ldots, V_r) = f\,T(V_1, \ldots, V_i, \ldots, V_r)$$

This is true, since it is true at each point p of **M**:

$$T_p(V_1, \ldots, fV_i, \ldots, V_r) \equiv T(V_{1p}, \ldots, f(p)V_{ip}, \ldots, V_{rp}) = f(p)\,T(V_{1p}, \ldots, V_{ip}, \ldots, V_{rp})$$
$$= f\,T_p(V_1, \ldots, V_i, \ldots, V_r)$$

13.20 Show how to interpret the tangent vector to a curve on a surface S as a (contravariant) tensor of type $\binom{1}{0}$.

Let $\mathbf{c} = \mathbf{c}(t)$ be a given curve on $\mathbf{M} = \mathbf{S}$, with

$$\mathbf{c}_*(t) = \frac{d\mathbf{c}}{dt} = \frac{\partial \mathbf{y}}{\partial x^i}\frac{dx^i}{dt}$$

Define for any one-form $\boldsymbol{\omega} = a_i\,dz^i$ the linear mapping from $T^*(\mathbf{M})$ to $\mathbf{R}$:

$$T(\boldsymbol{\omega}) = a_i\frac{dx^i}{dt} \equiv \boldsymbol{\omega}\left(\frac{d\mathbf{x}}{dt}\right)$$

Under the standard basis $\{dz^1, dz^2, \ldots, dz^n\}$ of $T_p^*(\mathbf{M})$, and with $\boldsymbol{\omega} = dz^i \equiv \delta^i_j\,dz^j$,

$$T^i = T(dz^i) = \delta^i_j\frac{dx^j}{dt} = \frac{dx^i}{dt}$$

(We saw earlier that the dx^i/dt were contravariant components.)

13.21 Show how to interpret the gradient of a function as a tensor of type $\binom{0}{1}$.

Let f have gradient $\nabla f \equiv (\partial f/\partial x^i)$. Define the linear mapping

$$T(V) = V^i\frac{\partial f}{\partial x^i} \qquad \left(\frac{\partial f}{\partial x^i}\text{ fixed}\right)$$

Use the basis $\{E_1, E_2, \ldots, E_n\}$ for $T_p(\mathbf{M})$; then with $V = E_i \equiv \delta^j_i E_j$,

$$T_i = T(E_i) = \delta^j_i\frac{\partial f}{\partial x^j} = \frac{\partial f}{\partial x^i}$$

Supplementary Problems

13.22 The set of all 2×2 matrices of the form

$$\begin{bmatrix} \pm 1 & 0 \\ 0 & \pm 1 \end{bmatrix}$$

where all possible combinations of signs are taken, forms a four-element subset of $\mathbf{GL}(2, \mathbf{R})$. Is it a subgroup?

13.23 Prove that $\mathbf{SU}(n)$, the set of all $n \times n$ matrices over the complex numbers having determinant $+1$, is a subgroup of $\mathbf{GL}(n, \mathbf{C})$. [*Hint*: $\det AB = (\det A)(\det B)$ holds for complex matrices.]

13.24 Show that the operator $L(f) = \int_0^1 f(x)\,dx$ is a linear functional over the set of continuous, real-valued functions on $[0, 1]$.

13.25 In terms of the standard basis $\{dx^i\}$ of $(\mathbf{R}^3)^*$, a new basis is defined by

$$\boldsymbol{\beta}^1 = dx^1 - 2\,dx^3 \qquad \boldsymbol{\beta}^2 = 2\,dx^1 + dx^2 \qquad \boldsymbol{\beta}^3 = dx^1 + dx^3$$

Find the corresponding dual basis $\{\mathbf{b}_i\}$ for $\mathbf{R}^3$ in terms of $(\mathbf{e}_i)$, using (*13.6*). Check your answer by making several calculations of the form $\omega(v) = \bar{\omega}(\bar{v})$ (a change of basis does not affect the value a linear functional assigns to a vector).

13.26 Consider a tensor $T(\boldsymbol{\omega}; \mathbf{v})$ over a vector space of dimension n and its dual, with components $T^i_{\,j\cdot}$. (*a*) Show that the trace $\tau(T) \equiv T^i_i$ is invariant under changes of bases. (*b*) Find $\tau(T)$ for the tensor defined by $T(\boldsymbol{\omega}; \mathbf{v}) = \boldsymbol{\omega}(\mathbf{v})$.

13.27 Show that every metric tensor G induces a one-to-one mapping (which is an isomorphism, since it is linear) $\hat{G} : \mathbf{V} \to \mathbf{V}^*$ from a vector space to its dual, under the definition: For each fixed $\mathbf{u}$ in $\mathbf{V}$, let $\hat{G}(\mathbf{u})$ be the linear functional $\hat{G}(\mathbf{u})(\mathbf{v}) = G(\mathbf{u}, \mathbf{v})$, for all $\mathbf{v}$ in $\mathbf{V}$. This proves for vector spaces *of arbitrary dimension*:

Theorem 13.3: If $\mathbf{V}$ possesses a metric tensor, then $\mathbf{V}$ is isomorphic to its dual $\mathbf{V}^*$.

13.28 Find a convenient atlas showing that the set in $\mathbf{R}^4$ given by the equation

$$(y^1)^2 + (y^2)^2 + (y^3)^2 - (y^4)^2 = a^2$$

can be made into a C^∞ 3-manifold. [*Hint*: Use radicals, as in Example 13.11(*b*); here, 6 charts will suffice.]

13.29 Show that the restriction of $\sigma = y^1\,dy^2 - y^2\,dy^1 + y^3\,dy^4 - y^4\,dy^3$ on $\mathbf{R}^4$ to the sphere $\mathbf{S}^3$ is a nonzero vector field on $\mathbf{S}^3$.

13.30 Extend Problem 13.29 to the sphere $\mathbf{S}^{2k-1}$ $(k \geqq 2)$.

13.31 Show that if there are only two points, p_1 and p_2, on $\mathbf{S}^2$ where a vector field is zero, those points must be antipodal (endpoints of a diameter).

13.32 Show, by geometric reasoning, that there exists a continuous, nonzero vector field on the torus.

13.33 Show that the restrictions of the following one-forms to $\mathbf{S}^4$, $(y^1)^2 + \cdots + (y^5)^2 = 1$, are vector fields on $\mathbf{S}^4$, and find the points where they are zero:

(*a*) $\sigma = y^2\,dy^1 - y^1\,dy^2 + y^4\,dy^3 - y^3\,dy^4$

(*b*) $\sigma = (y^2 - y^3 - y^4)\,dy^1 + (y^3 - y^1)\,dy^2 + (y^1 - y^2 + y^5)\,dy^3 + y^1\,dy^4 - y^3\,dy^5$

13.34 Although no nonvanishing continuous vector field exists on the 2-sphere $\mathbf{S}^2$, there are *three*, mutually orthogonal, *unit* vector fields on $\mathbf{S}^3 \subset \mathbf{R}^4$. These are, in the extrinsic representation of $\mathbf{S}^3$,

$$\begin{aligned}
\sigma_1 &= -y^2\,dy^1 + y^1\,dy^2 + y^4\,dy^3 - y^3\,dy^4 \\
\sigma_2 &= -y^3\,dy^1 - y^4\,dy^2 + y^1\,dy^3 + y^2\,dy^4 \\
\sigma_3 &= -y^4\,dy^1 + y^3\,dy^2 - y^2\,dy^3 + y^1\,dy^4
\end{aligned}$$

Show this. [*Note*: Manifolds with such vector-field bases are called *parallelizable*. The manifolds $\mathbf{S}^1$, $\mathbf{S}^3$, $\mathbf{S}^7$— and no other n-spheres—and the torus are examples.]

13.35 Without resorting to coordinate patches, express extrinsically the collection of tangent spaces $T(\mathbf{M})$, if $\mathbf{M}$ is the hyperboloid of one sheet $(y^1)^2 - 4(y^2)^2 + 4(y^3)^2 = 4$.

13.36 For the manifold $\mathbf{M}$ of Problem 13.35, consider the coordinate patch

$$y^1 = x^1 \qquad y^2 = x^2 \qquad y^3 = \sqrt{1 - (x^1/2)^2 + (x^2)^2}$$

valid for $y^3 > 0$. Find an expression for an arbitrary vector in $T_p(\mathbf{M})$.

13.37 One way to show that two surfaces meet at right angles is to show that along the curve of intersection the normal vector to one lies in the tangent space of the other. Illustrate this idea for the sphere $(y^1)^2 + (y^2)^2 + (y^3)^2 = 16$ and the cone $(y^3)^2 = 9(y^1)^2 + 9(y^2)^2$, the latter coordinatized by

$$y^1 = x^1 \qquad y^2 = x^2 \qquad y^3 = 3\sqrt{(x^1)^2 + (x^2)^2}$$

Answers to Supplementary Problems

CHAPTER 1

1.15 $a_1b_1 + a_2b_2 + a_3b_3 + a_4b_4 + a_5b_5 + a_6b_6$

1.16 $R^1_{jk1} + R^2_{jk2} + R^3_{jk3} + R^4_{jk4}$. The index i is a dummy index, while j and k are free indices; there are 16 summations.

1.17 x_j

1.18 (*a*) n; (*b*) $\delta_{ij}\delta_{ij} = \delta_{ii} = n$; (*c*) $\delta_{ij}c_{ij} = c_{ii} = c_{11} + c_{22} + c_{33} + \cdots + c_{nn}$

1.19 $a_{i3}b_{i3} \quad (n = 3)$

1.20 $a_{ij}x_ix_j \quad (n = 3)$

1.21 $y_i = c_{ij}x_j \quad (n = 2)$

1.22 $a_{1k} \quad (k = 1, 2, 3)$

1.23 $\dfrac{\partial}{\partial x_k}(a_{ij}x_j) = a_{ij}\dfrac{\partial}{\partial x_k}(x_j) = a_{ij}\delta_{jk} = a_{ik}$

1.24 $a_{ik}[(x_i)^2 + 2x_ix_k]$ [not summed on k]

1.25 $(a_{lij} + a_{ilj} + a_{ijl})x_ix_j$

1.26 $a_{kl} + a_{lk}$

1.27 (*a*) $b^i_jT^{rr}_i$; (*b*) $a_{ij}b_{jr}x_r$; (*c*) $a_{ijk}b_{ir}b_{js}b_{kt}x_rx_sx_t$

1.28 (*c*) $a_{ij}(x_i + x_j) = a_{ij}(\varepsilon_jx_i + \varepsilon_ix_j) = a_{ij}\varepsilon_jx_i + a_{ij}\varepsilon_ix_j = a_{ji}\varepsilon_jx_i + a_{ij}\varepsilon_ix_j = 2a_{ij}\varepsilon_ix_j$

CHAPTER 2

2.24 (*a*) and (*b*) $\begin{bmatrix} u^{11} & u^{12} & u^{13} & u^{14} & u^{15} \\ u^{21} & u^{22} & u^{23} & u^{24} & u^{25} \\ u^{31} & u^{32} & u^{33} & u^{34} & u^{35} \end{bmatrix}$ (*c*) $\begin{bmatrix} u^{11} & u^{12} & u^{13} \\ u^{21} & u^{22} & u^{23} \\ u^{31} & u^{32} & u^{33} \\ u^{41} & u^{42} & u^{43} \\ u^{51} & u^{52} & u^{53} \end{bmatrix}$ (*d*) $\begin{bmatrix} 1 & 0 & 0 & 0 & 0 & 0 \\ 0 & 1 & 0 & 0 & 0 & 0 \\ 0 & 0 & 1 & 0 & 0 & 0 \end{bmatrix}$

2.25 (*a*) $\begin{bmatrix} 5 \\ 0 \\ 5 \end{bmatrix}$ (*b*) $\begin{bmatrix} 1 & 2 & -4 \\ 2 & 2 & -2 \end{bmatrix}$

2.29 (*a*) 17; (*b*) 0; (*c*) -1

2.30 (*a*) $-a_{12}a_{21}a_{33}a_{44} + a_{12}a_{21}a_{34}a_{43} + a_{12}a_{23}a_{31}a_{44} - a_{12}a_{23}a_{34}a_{41} - a_{12}a_{24}a_{31}a_{43} + a_{12}a_{24}a_{33}a_{41}$

(*b*) $-a_{12}\begin{vmatrix} a_{21} & a_{23} & a_{24} \\ a_{31} & a_{33} & a_{34} \\ a_{41} & a_{43} & a_{44} \end{vmatrix} \equiv a_{12}A_{12}$

2.32 (*a*) $\begin{bmatrix} 2 & -1 \\ -5 & 3 \end{bmatrix}$ (*b*) $\frac{1}{7}\begin{bmatrix} 1 & 3 & 2 \\ 1 & -4 & 2 \\ 3 & 2 & -1 \end{bmatrix}$

2.33 One need verify only that (i) interchanging a pair of consecutive indices changes the sign of a single factor in the product; (ii)

$$\prod_{p>q} \frac{p-q}{|p-q|} = \prod 1 = 1$$

2.34 $2\pi/3$

2.35 One pair are $(2, 3, 0)$ and $(-3, -2, 5)$.

2.36 $\begin{bmatrix} x \\ y \end{bmatrix} = \begin{bmatrix} -1 \\ 5 \end{bmatrix}$

2.37 $Q = x_1^2 + 2x_2^2 - x_3^2 + 8x_1x_2 + 6x_1x_3$

2.38 $A = \begin{bmatrix} -3 & -\frac{1}{2} & -\frac{1}{2} & 3 \\ -\frac{1}{2} & -1 & 0 & 0 \\ -\frac{1}{2} & 0 & 1 & 0 \\ 3 & 0 & 0 & 0 \end{bmatrix}$

2.39 $\bar{c}_i = c_r b_{ri}$, where $(b_{ij}) = (a_{ij})^{-1}$.

2.40 $g_{11} = 13/49$, $g_{12} = g_{21} = 4/49$, $g_{22} = 5/49$

2.41 $d(\bar{\mathbf{x}}, \bar{\mathbf{y}}) = 3 = d(\mathbf{x}, \mathbf{y})$

CHAPTER 3

3.23 (*a*) $\mathscr{J} = -2\exp(2x^1) < 0$

(*b*) $\mathscr{T}^{-1}: \begin{cases} x^1 = \frac{1}{2}\ln(\bar{x}^1\bar{x}^2) \\ x^2 = \frac{1}{2}\ln(\bar{x}^1/\bar{x}^2) \end{cases} \quad (\bar{x}^1, \bar{x}^2 > 0)$

(*c*) $\bar{J} = \begin{bmatrix} 1/2\bar{x}^1 & 1/2\bar{x}^2 \\ 1/2\bar{x}^1 & -1/2\bar{x}^2 \end{bmatrix} = \begin{bmatrix} \exp(x^1 + x^2) & \exp(x^1 + x^2) \\ \exp(x^1 - x^2) & -\exp(x^1 - x^2) \end{bmatrix}^{-1}$

3.26 $\dfrac{\partial \bar{f}}{\partial \theta} = 0$, so that $f(x, y) = \bar{f}(r) = \bar{f}(\sqrt{x^2 + y^2}) = g(x^2 + y^2)$.

3.29 $\delta^r_s \dfrac{\partial \bar{x}^i}{\partial x^r}\dfrac{\partial x^s}{\partial \bar{x}^j} = \dfrac{\partial \bar{x}^i}{\partial x^r}\dfrac{\partial x^r}{\partial \bar{x}^j} = \delta^i_j = \bar{\delta}^i_j$

3.30 The inverse Jacobian matrix at $(1, 2)$ is

$$\bar{J} = \begin{bmatrix} \bar{x}^2 & \bar{x}^1 \\ 0 & 1 \end{bmatrix} = \begin{bmatrix} 2 & 1 \\ 0 & 1 \end{bmatrix}$$

By Problem 3.14(*a*), covariance of the matrix

$$E \equiv [e_{ij}]_{22} = \begin{bmatrix} 0 & 1 \\ -1 & 0 \end{bmatrix}$$

would imply the matrix equation

$$\begin{bmatrix} 0 & 1 \\ -1 & 0 \end{bmatrix} = \begin{bmatrix} 2 & 0 \\ 1 & 1 \end{bmatrix}\begin{bmatrix} 0 & 1 \\ -1 & 0 \end{bmatrix}\begin{bmatrix} 2 & 1 \\ 0 & 1 \end{bmatrix}$$

or

$$\begin{bmatrix} 0 & 1 \\ -1 & 0 \end{bmatrix} = \begin{bmatrix} 0 & 2 \\ -2 & 0 \end{bmatrix}$$

which is patently false.

3.32 (*a*) $(T^i_j + T^j_i)$ represents a tensor if and only if

$$(T^r_s + T^s_r)\frac{\partial \bar{x}^i}{\partial x^r}\frac{\partial x^s}{\partial \bar{x}^j} = T^r_s \frac{\partial \bar{x}^i}{\partial x^r}\frac{\partial x^s}{\partial \bar{x}^j} + T^s_r \frac{\partial \bar{x}^j}{\partial x^s}\frac{\partial x^r}{\partial \bar{x}^i}$$

which requires that $JT\bar{J} = \bar{J}^T T J^T$. This last relation, in turn, generally requires that $\bar{J} = J^T$; i.e., J must be an orthogonal matrix.

(*b*) $\bar{T} = JT\bar{J}$, so that $\bar{T}^T = \bar{T}$ if $\bar{J} = J^T$.

3.35 As **T** is a tensor (Example 3.4), it is an affine tensor: $\bar{T}^i = a^i_r T^r$. Thus,

$$\frac{d\bar{T}^i}{dt} = a^i_r \frac{dT^r}{dt}$$

showing $d\mathbf{T}/dt$ also to be an affine tensor. Any affine tensor is a fortiori a cartesian tensor.

3.36 (*a*) $$\bar{u}_i\bar{u}_i = (a_{ir}u_r)(a_{is}u_s) = a_{ir}a_{is}u_r u_s = \delta_{rs}u_r u_s = u_r u_r$$

(*b*) No, because distance and angles are not preserved under arbitrary linear transformations. Specifically, consider $\bar{x}^1 = 3x^1$, $\bar{x}^2 = x^1 + x^2$. A scalar product in $(\bar{x}^i)$ is

$$\bar{u}_i\bar{v}_i = (3u_1, u_1 + u_2)\cdot(3v_1, v_1 + v_2) = 10u_1v_1 + u_1v_2 + u_2v_1 + u_2v_2$$

This clearly will not coincide with $u_1v_1 + u_2v_2$.

CHAPTER 4

4.19 Write $[\mathbf{ST}] = (U^{ijk}_{lmn})$. There are $\binom{3}{2}$ ways of choosing locations for the contraction indices u and v among the contravariant indices, and, for each of these, $\binom{3}{2}$ ways of choosing locations among the covariant indices. A given quartet of locations can be filled in 2 *inequivalent* ways. Thus, the desired number is

$$\binom{3}{2}\cdot\binom{3}{2}\cdot 2 = 18$$

4.23 First, use the device of Problem 4.11 to establish that $T^i_{jkl}U^kV^l$ are tensor components for all (U^i) and (V^i); then apply the Quotient Theorem twice.

CHAPTER 5

5.21 $L = a\pi$; semicircle of radius a.

5.22 No: $Q(1, 0, 3) = -1$.

5.23 $L = 2 + e$

5.24 The true distance formula, $\overline{P_1P_2} = \sqrt{(x^1_1 - x^1_2)^2 + (x^2_1 - x^2_2)^2 - 0.2021125(x^1_1 - x^1_2)(x^2_1 - x^2_2)}$, yields 4.751, for an error of $+0.249$.

5.25 $$G = \begin{bmatrix} (x^2)^2 & x^1x^2 & 0 \\ x^1x^2 & 1 + (x^1)^2 & 0 \\ 0 & 0 & 1 \end{bmatrix}$$

5.26 $(U_i) = (0, 1, 0)$, $(V_i) = (x^2, x^1, 0)$

5.27 (*a*) $\|\mathbf{U} + \mathbf{V}\|^2 = (\mathbf{U} + \mathbf{V})^2 = \mathbf{U}^2 + \mathbf{V}^2 + 2\mathbf{UV} = \|\mathbf{U}\|^2 + \|\mathbf{V}^2\| + 2\,\|\mathbf{U}\|\,\|\mathbf{V}\|\cos\theta$

(*b*) Take $\theta = \pi/2$ in (*a*).

5.28 (*a*) $x^2 = C\exp(-2bx^3/a^2)$ (a one-parameter family of spirals on the cylinder $x^1 = a$, $-\pi < x^3 < \pi$)

(*b*) No: the curves of (*a*) have tangent field **V** all along their length; but, for orthogonality, it is necessary only that the tangent *at intersections with the pseudo-helix* be **V**. For example, the curve $x^2 = x^3$ on $x^1 = a$ is also orthogonal to the pseudo-helix at the point $x^2 = -a^2/2b$, $x^3 = a^4/4b$.

5.29 $x^1 = d \exp(-(x^2)^2/2)$ $\quad (d = \text{const.})$

5.30 $f'(\theta_0)g'(\theta_0) = -a^2$ at intersection points.

5.32 (*a*) $g^{i\alpha} = \lambda(\alpha)\delta^i_\alpha$, which is tantamount to $g^{ij} = g_{ij} = 0$ for $i \neq j$.

5.33 $\|\mathbf{V}\| = 1$, $L = \pi/2$

5.34 $x^1 = a$, $x^3 = b \cot x^2 + c$ $(c = \text{const.})$

CHAPTER 6

6.19 $\bar{x}^i = \frac{1}{2} a^i_{rs} x^r x^s + b^i_r x^r + c^i$ $\quad$ (the b^i_j and c^i constants)

6.20 (*a*)

$$G = \begin{bmatrix} 16(x^1)^2 + 1 & 4x^1 - 3 \\ 4x^1 - 3 & 10 \end{bmatrix}$$

(*b*) $\Gamma_{111} = 16x^1$, $\Gamma_{112} = 4$, all others 0

6.22 The values, in (x^i), of the $\partial x^i/\partial \bar{x}^j$ are easiest found by inverting $J \equiv (\partial \bar{x}^i/\partial x^j)$. Final results are:

$$\Gamma^1_{11} = \Gamma^1_{22} = \Gamma^2_{12} = \Gamma^2_{21} = 1$$

6.23 From Problem 6.21(*b*), $\Gamma^i_{jk} = 0$ for $j \neq k$; while, for $j = k = \alpha$ (no summation on α),

$$\Gamma^i_{\alpha\alpha} = \frac{\partial}{\partial x^\alpha}\left(\frac{\partial \bar{x}^r}{\partial x^\alpha}\right)\frac{\partial x^i}{\partial \bar{x}^r} = \left(d_\alpha \frac{\partial \bar{x}^r}{\partial x^\alpha}\right)\frac{\partial x^i}{\partial \bar{x}^r} = d_\alpha \delta^i_\alpha$$

6.26 $-\Gamma_{221} = \Gamma_{212} = \Gamma_{122} = x^1$; $\Gamma^2_{21} = \Gamma^2_{12} = 1/x^1$, $\Gamma^1_{22} = -x^1$

6.27

$$\mathcal{J} = \begin{vmatrix} a^1_1 \exp \bar{x}^1 & 2a^1_2 \exp 2\bar{x}^2 & 3a^1_3 \exp 3\bar{x}^3 \\ a^2_1 \exp \bar{x}^1 & 2a^2_2 \exp 2\bar{x}^2 & 3a^2_3 \exp 3\bar{x}^3 \\ a^3_1 \exp \bar{x}^1 & 2a^3_2 \exp 2\bar{x}^2 & 3a^3_3 \exp 3\bar{x}^3 \end{vmatrix} = [6 \exp(\bar{x}^1 + 2\bar{x}^2 + 3\bar{x}^3)] \det(a^i_j) \neq 0$$

Hence the condition is $\det(a^i_j) \neq 0$.

6.29 $\bar{x}^i = A^i x^1 \sin x^2 + B^i x^1 \cos x^2 + C^i$ $\quad (i = 1, 2)$, with

$$x^1 \begin{vmatrix} A^1 & B^1 \\ A^2 & B^2 \end{vmatrix} \neq 0$$

for a bijection.

6.30 No, because of the presence of Christoffel symbols in (*6.7*).

6.31 $T^i_{jrs,k} = \dfrac{\partial T^i_{jrs}}{\partial x^k} + \Gamma^i_{uk} T^u_{jrs} - \Gamma^u_{jk} T^i_{urs} - \Gamma^u_{rk} T^i_{jus} - \Gamma^u_{sk} T^i_{jru}$

6.36 $\kappa = 1/b$

6.37 (*a*) $\dfrac{d^2u}{ds^2} = \dfrac{d^2v}{ds^2} = 0$

(*b*) $x^2 = p(x^1)^2 + q$ (a two-parameter family of "parabolas")

6.38 (*a*)
$$\frac{d^2x^2}{ds^2} - (\sin x^2 \cos x^2)\left(\frac{dx^3}{ds}\right)^2 = 0$$
$$\frac{d^2x^3}{ds^2} + (2\cot x^2)\frac{dx^2}{ds}\frac{dx^3}{ds} = 0$$

(*b*) $x^2 = \frac{1}{a}s \qquad x^3 = 0$

(*c*) The solution (*b*) represents an arc of a particular great circle ($x^2 + z^2 = a^2$, in the usual cartesian coordinates) on the sphere. By symmetry of the sphere, all great-circular arcs, and only these, will be geodesics.

CHAPTER 7

7.24 $\varepsilon = \begin{cases} +1 & t \leqq \frac{1}{2} \\ -1 & t > \frac{1}{2} \end{cases}$

7.25 $t = 0, 1$

7.26 $L = 8\sqrt{2}/3$

7.27 $t = \pm\sqrt{5}/3$ [$t = 0$, which makes $\gamma \equiv |g_{ij}| = 0$, is disallowed]

7.28 $L = (64 + 11\sqrt{11})/216 \approx 0.465$

7.29 $\theta = i \ln 2$ at $(0, 2, 0)$; $\theta = \cos^{-1}(7/4\sqrt{11})$ at $(5, 2, 3)$

7.30 (*a*) $L = 8(1 + 3\sqrt{3}) \approx 49.57$

(*b*) $x^1 = 3(\sigma s^{2/3} + 4)$, $x^2 = (\sigma s^{2/3} + 4)^{3/2}$, where
$$\sigma = \begin{cases} -1 & -8 \leqq s < 0 \\ +1 & 0 \leqq s \leqq 24\sqrt{3} \end{cases}$$

(*c*) The null points are $t = 0$ ($s = -8$) and $t = 1$ ($s = 0$).

7.31 $L = 8(5\sqrt{5} - 1) \approx 81.44$

7.32 For $s \neq 0$, $\mathbf{T} = (2|s|^{-1/3}, \sqrt{\sigma + 4s^{-2/3}})$ and $\|\mathbf{T}\|^2 = |-\sigma| = +1$

7.33 $N^1 = T^2$, $N^2 = T^1$

7.34 For $s \neq -8, 0$,
$$\kappa = \frac{-2}{3s\sqrt{\sigma s^{2/3} + 4}} \qquad \kappa_0 = |\kappa|$$

7.36
$$\kappa_0 = |\kappa| = \frac{2}{3|s|(s^{2/3} - 4)^{1/2}} \qquad (s \neq 8)$$

At the null point (0, 0), both Euclidean and Riemannian absolute curvatures become infinite; but at the null point (12, 8), only the Riemannian curvature becomes infinite.

7.37 $\mathbf{T} = |1 - 4t|^{-1/2}(1, 2t)$, $\mathbf{N} = |1 - 4t|^{-1/2}(1, 1 - 2t)$, $\kappa = 2\,|1 - 4t|^{-3/2}$

7.38 (*a*) $L = a$. (*b*) $L = 3a/2$. (*c*) Riemannian: $\mathbf{T} = |\cos 2t|^{-1/2}(-\cos t, \sin t)$, $\kappa_0 = (2/3a)(\csc 2t)|\cos 2t|^{-3/2}$; Euclidean: $\mathbf{T} = (-\cos t, \sin t)$, $\kappa_0 = (2/3a)\csc 2t$. (Curve not regular at $t = \pi/4$ for Riemannian metric.)

7.39 (*a*) $\Gamma^1_{11} = 1/2x^1$, $\Gamma^2_{22} = 1/2x^2$, others zero

CHAPTER 8

8.16 By Problem 6.34 and (*8.1*),

$$V^i_{,kl} - V^i_{,lk} = g^{ir}(V_{r,kl} - V_{r,lk}) = g^{ir}R^s_{rkl}V_s$$
$$= g^{ir}(g_{st}R^s_{rkl})V^t = (g^{ir}R_{trkl})V^t = -R^i_{tkl}V^t$$

8.22 $\mathrm{K} = 1/4(x^1)^2$

8.24 (*a*) and (*b*) $\mathrm{K} = \dfrac{x^1 + x^2}{4(x^1)^2 x^2(1 + 2x^2)}$

(*c*) $\dot{\mathbf{U}}_{(2)} = -\mathbf{U}_{(1)} + \mathbf{V}_{(1)},\ \mathbf{V}_{(2)} = \mathbf{U}_{(1)} + \mathbf{V}_{(1)}$

8.25 $\mathrm{K} = 1/a^2$

8.26 Basic sets of nonvanishing terms are:

$$\text{(A)}\quad R_{1212} = -\frac{1}{4}\left(2f'' - \frac{f'^2}{f} - \frac{f'g'}{g}\right),\quad R_{1313} = -\frac{1}{4}\frac{f'h'}{g},\quad R_{2323} = -\frac{1}{4}\left(2h'' - \frac{h'^2}{h} - \frac{h'g'}{g}\right)$$

and

$$\text{(A)}\quad G_{1212} = fg,\quad G_{1313} = fh,\quad G_{2323} = gh$$

so that

(a) $\mathrm{K}(x^2; \mathbf{U}, \mathbf{V}) = \dfrac{R_{1212}W_{1212} + R_{1313}W_{1313} + R_{2323}W_{2323}}{fgW_{1212} + fhW_{1313} + ghW_{2323}}$

(*b*) $R = -\dfrac{2}{fgh}(hR_{1212} + gR_{1313} + fR_{2323})$

8.27 (*a*) $\mathrm{K}(x^2; \mathbf{U}, \mathbf{V}) = \dfrac{-2(\ln|f|)''(W_{1212} + W_{2323}) - (\ln|f|)'^2 W_{1313}}{4f(W_{1212} + W_{1313} + W_{2323})}$

(*b*) $R = \dfrac{4f''f - 3f'^2}{2f^3}$

8.28 Isotropic points compose the surface $x^2 = e^{-3/2}$, over which $\mathrm{K} = 2e^3/27$.

8.29 $\mathrm{K} = -1/4$

$R_{11} = -1, R_{12} = R_{21} = 0, R_{22} = -\sin^2 x^1;\ R^1_1 = -1/a^2 = R^2_2, R^1_2 = 0 = R^2_1;\ R = -2/a^2$

8.33 $R_{11} = R_{22} = R_{33} = 2/(x^1)^2$, others 0; $R^1_1 = R^2_2 = R^3_3 = 2$, others 0; $R = 6$

8.35 $g_{ij} = (x^1)^4\delta_{ij}$ has $R = 0$, $\mathrm{K} \neq 0$ (use Problem 8.27).

8.36 No implication either way.

CHAPTER 9

9.17 (*a*) $u_0 = \pm\sqrt{x^1x^2 + a}$ $\quad$ (a = const.); (*b*) incompatible

9.20 flat, non-Euclidean

9.21 Euclidean

9.22 $(+ + -)$

9.26 With the notation $f_i \equiv \partial f/\partial x^i$, for any function f:

$$G^1_1 = \frac{1}{(x^1)^2} + e^{-\varphi}\left[-\frac{\psi_1}{x^1} - \frac{1}{(x^1)^2}\right]$$

$$G^2_2 = e^{-\varphi}\left(-\frac{\psi_{11}}{2} - \frac{\psi_1^2}{4} + \frac{\varphi_1\psi_1}{4} + \frac{\varphi_1}{2x^1} - \frac{\psi_i}{2x^1}\right) + e^{-\psi}\left(\frac{\varphi_{44}}{2} + \frac{\varphi_4^2}{4} - \frac{\varphi_4\psi_4}{4}\right) = G^3_3$$

$$G^4_4 = \frac{1}{(x^1)^2} + e^{-\varphi}\left[\frac{\varphi_1}{x^1} - \frac{1}{(x^1)^2}\right] \qquad G^1_4 = -\varphi_4 e^{-\varphi}/x^1 \qquad G^4_1 = \varphi_4 e^{-\psi}/x^1$$

CHAPTER 10

10.30 (*a*) The curve lies on a right circular cylinder of unit radius, beginning at the point $(1, 0, 1)$ and rising in helix fashion, approaching ∞ asymptotic to the vertical line $x = \cos 1$, $y = \sin 1$, as $t \to 1$.

(*b*)
$$L = \int_0^{1/2} \frac{\sqrt{(1-t)^4 + 1}}{(1-t)^2}\, dt \approx 1.13209039$$

10.31 16/3

10.32 (*a*) $\mathbf{T} = (-(a/c)\sin(s/c), (a/c)\cos(s/c), b/c)$. Hence the tangent line, $\mathbf{r}(t) \equiv \mathbf{r} + t\mathbf{T}$, has the coordinate equations

$$x = a\cos\frac{s}{c} - \frac{at}{c}\sin\frac{s}{c} \qquad y = a\sin\frac{s}{c} + \frac{at}{c}\cos\frac{s}{c} \qquad z = \frac{bs}{c} + \frac{bt}{c}$$

(*b*) Q corresponds to $t = -s$ and $PQ = \|-s\mathbf{T}\| = s$.

(*c*) The interpretation is that Q can be thought of as the free end of the taut string as it is unwound from the helix. [The locus of Q, $\mathbf{r}^* = \mathbf{r}(s) - s\mathbf{r}'(s)$, is called an *involute* of the helix.]

10.33
$$\frac{\mathbf{T}'}{\|\mathbf{T}'\|} = \frac{t/|t|}{(1 + 25t^8)^{1/2}}(-5t^4, 1, 0)$$

10.34
$$\kappa = \frac{20t^3\sqrt{2}}{(1 + 50t^8)^{3/2}} \qquad \tau = 0$$

10.35 Let the curve $\mathbf{r} = \mathbf{r}(s)$ lie in the plane $\mathbf{br} = \text{const.}$, where $\mathbf{b} = \text{const.}$ and $\|\mathbf{b}\| = 1$. Differentiate twice with respect to s: $\mathbf{bT} = 0$ and $\mathbf{bT}' = 0$; hence, $\mathbf{b}(\kappa\mathbf{N}) = 0$ or $\mathbf{bN} = 0$. It follows that $\mathbf{b} = \mathbf{B}$, the binormal vector, so that $\mathbf{B}' = \mathbf{0}$ and $\tau = -\mathbf{B}'\mathbf{N} = 0$. Conversely, if $\tau = 0$ for a curve $\mathbf{r} = \mathbf{r}(s)$, then $\mathbf{B}' = -\tau\mathbf{N} = \mathbf{0}$ and $\mathbf{B}$ is a constant unit vector. Define the function $Q(s) \equiv \mathbf{B} \cdot (\mathbf{r}(s) - \mathbf{r}(0))$; we have

$$Q' = \mathbf{Br}' = \mathbf{BT} = 0$$

whence $Q = \text{const.} = Q(0) = 0$. Therefore, the curve lies in the plane

$$\mathbf{Br} = \mathbf{Br}(0) = \text{const.}$$

10.38 $E = |x^1|\sqrt{a^2 + 1} = 0$ at $x^1 = 0$.

10.39 $E = a^2\cosh^2 x^1 > 0$, $\mathbf{n} = \dfrac{1}{\cosh x^1}(-\cos x^2, -\sin x^2, \sinh x^1)$

10.40
$$L = \int_1^2 \frac{\sqrt{5t^4 + 1}}{t}\, dt = \frac{1}{2}\left[9 - \sqrt{6} + \ln\frac{2}{5}(\sqrt{6} + 1)\right]$$

10.41 $(v^1, v^2) = (\sqrt{12}, \sqrt{17}$ or $(-\sqrt{12}, \sqrt{17})$

10.43 $\Gamma^2_{12} = \Gamma^2_{21} = \dfrac{x^1}{(x^1)^2 + a^2}$, $\Gamma^1_{22} = -x^1$; all others zero

10.44 $$\text{II} = \frac{f'g'' - f''g'}{\sqrt{f'^2 + g'^2}}(dx^1)^2 + \frac{fg'}{\sqrt{f'^2 + g'^2}}(dx^2)^2$$

10.46 (*a*) (i) $\text{K} = \dfrac{4a^2}{[1 + 4a^2(x^1)^2]^2}$, $\text{H} = \dfrac{4a[1 + 2a^2(x^1)^2]}{[1 + 4a^2(x^1)^2]^{3/2}}$

(ii) $\text{K} = \dfrac{4a^2}{[1 + 4a^2(\bar{x}^1)^2 + 4a^2(\bar{x}^2)^2]^2}$, $\text{H} = \dfrac{4a[1 + 2a^2(\bar{x}^1)^2 + 2a^2(\bar{x}^2)^2]}{[1 + 4a^2(\bar{x}^1)^2 + 4a^2(\bar{x}^2)^2]^{3/2}}$

(*b*) Consistent with the invariance of K and H, the change of parameters $\bar{x}^1 = x^1 \cos x^2$, $\bar{x}^2 = x^1 \sin x^2$—i.e., the transformation from polar to rectangular coordinates in the parameter plane—takes the forms (i) into the forms (ii).

10.49 (*a*) $\mathbf{r}^* = (a \operatorname{sech} x^1, 0, ax^1 - a \tanh x^1)$

10.50 The two FFFs correspond under the mapping $\bar{x}^1 = a \sinh x^1$, $\bar{x}^2 = x^2$.

CHAPTER 11

11.14 (*a*) $v = \sqrt{1 + \csc^4 t} \to \sqrt{2}$; (*b*) $a = \sqrt{1 + 4\csc^4 t \cot^2 t} \to 1$; (*c*) $\max v = \sqrt{5}$, $\max a = \sqrt{17}$ (no minima)

11.15 Rectilinear motion [use (*11.8*) to show that κ must vanish].

11.16 From (*11.7*) and (*10.9*), $\dot{\mathbf{a}} = -\kappa^2 v^3 \mathbf{T} + \dot{\kappa} v^2 \mathbf{N} + \kappa\tau v^3 \mathbf{B}$.

11.17 $a^1 = \dfrac{d^2\rho}{dt^2} - (\rho \sin^2 \varphi)\left(\dfrac{d\theta}{dt}\right)^2 - \rho\left(\dfrac{d\varphi}{dt}\right)^2$, $a^2 = \dfrac{d^2\varphi}{dt^2} + \dfrac{2}{\rho}\dfrac{d\rho}{dt}\dfrac{d\varphi}{dt} - (\sin\varphi \cos\varphi)\left(\dfrac{d\theta}{dt}\right)^2$,

$$a^3 = \frac{d^2\theta}{dt^2} + \frac{2}{\rho}\frac{d\rho}{dt}\frac{d\theta}{dt} + (2\cot\varphi)\frac{d\theta}{dt}\frac{d\varphi}{dt}$$

11.18 Let the center of force be the origin of rectangular coordinates for $\mathbf{E}^3$, with the particle's path given by $\mathbf{r} = \mathbf{r}(t)$. By Newton's second law, $f\mathbf{r} = m\ddot{\mathbf{r}}$, so that

$$\frac{d}{dt}(\mathbf{r} \times \dot{\mathbf{r}}) = \mathbf{r} \times \ddot{\mathbf{r}} = \mathbf{r} \times \left(\frac{f}{m}\mathbf{r}\right) = \mathbf{0}$$

and $\mathbf{r} \times \dot{\mathbf{r}} = \mathbf{p} = \text{const}$. It follows that $\mathbf{p} \cdot \mathbf{r} = 0$.

11.19 $\nabla^2 f = \dfrac{\partial^2 f}{\partial r^2} + \dfrac{1}{r^2}\dfrac{\partial^2 f}{\partial \theta^2} + \dfrac{\partial^2 f}{\partial z^2} + \dfrac{1}{r}\dfrac{\partial f}{\partial r}$

CHAPTER 12

12.34 (*a*) timelike; (*b*) spacelike; (*c*) lightlike

12.35 Yes: travel at 4167 mi/sec $\ll c$. Timelike interval.

12.36 Premultiply $A^T G A = G$ by AG, and postmultiply by $A^{-1}G$.

12.38 (*a*) $t = a_0^0 \bar{t}$, $x^1 = -a_1^0 ct$, $x^2 = -a_2^0 ct$, $x^3 = -a_3^0 ct$. (*b*) $a_0^0 > 0$ if t and $\bar{t}$ have the same sign; that is, if the clocks of the two observers are both turning clockwise or both counterclockwise.

12.39 $$\bar{x}^0 = \frac{5}{3}x^0 - \frac{4}{3}x^1 \qquad \bar{x}^1 = -\frac{4}{3}x^0 + \frac{5}{3}x^1 \qquad \bar{x}^2 = x^2 \qquad \bar{x}^3 = x^3$$

12.40 (*b*) zero

12.41
$$L^* = \begin{bmatrix} 5/4 & -3/4 & 0 & 0 \\ -3/4 & 5/4 & 0 & 0 \\ 0 & 0 & 1 & 0 \\ 0 & 0 & 0 & 1 \end{bmatrix} \qquad R_1 = \begin{bmatrix} 1 & 0 & 0 & 0 \\ 0 & 1 & 0 & 0 \\ 0 & 0 & 4/5 & -3/5 \\ 0 & 0 & 3/5 & 4/5 \end{bmatrix} \qquad R_2 = \begin{bmatrix} 1 & 0 & 0 & 0 \\ 0 & -2/3 & -1/3 & 2/3 \\ 0 & 1/3 & 2/3 & 2/3 \\ 0 & -2/3 & 2/3 & -1/3 \end{bmatrix}$$

$v = (3/5)c$

12.42 $v = \sqrt{\frac{2}{3}}\, c$

12.45 $v = (4/5)c$

12.47 Approximately 25% slow.

12.48 About 45 years old.

12.49 $\approx$17 000 mi/sec

12.50 For constants $\hat{F}$ and $\hat{a} \equiv \hat{F}/m$, and with $\mathbf{F} = (\hat{F}, 0, 0)$ and $\mathbf{v} = (v_x, 0, 0)$, (*12.29*) becomes identical with (*1*) of Problem 12.26.

12.52 Since $\partial \bar{s}^i/\partial \bar{x}^i = 0 = \partial s^i/\partial x^i$ (the equation of continuity), (s^i) may be identified with the vector (S^i) of Problem 12.32.

12.54 (*a*) By analogy with the evaluation of $\frac{1}{2}e_{ijkl}P_{kl}$ in Problem 12.53,

$$[\tfrac{1}{2}\, e_{ijkl}(-\Phi_{kl})]_{44} = \begin{bmatrix} 0 & -\Phi_{23} & \Phi_{13} & -\Phi_{12} \\ * & 0 & -\Phi_{03} & \Phi_{02} \\ * & * & 0 & -\Phi_{01} \\ * & * & * & 0 \end{bmatrix} = \begin{bmatrix} 0 & -H_1 & -H_2 & -H_3 \\ * & 0 & E_3 & -E_2 \\ * & * & 0 & E_1 \\ * & * & * & 0 \end{bmatrix} = [F^{ij}]_{44}$$

(*b*) Let $(a\,b\,c\,d)$ denote a permutation of $(0\,1\,2\,3)$. Then $\Phi_{ab} = -e_{abcd}F^{cd}$ (no summation) and

$$\frac{\partial \Phi_{ab}}{\partial x^c} + \frac{\partial \Phi_{ca}}{\partial x^b} + \frac{\partial \Phi_{bc}}{\partial x^a} = -e_{abcd}\frac{\partial F^{cd}}{\partial x^c} - e_{cabd}\frac{\partial F^{bd}}{\partial x^b} - e_{bcad}\frac{\partial F^{ad}}{\partial x^a}$$

$$= -e_{abcd}\left(\frac{\partial F^{cd}}{\partial x^c} + \frac{\partial F^{bd}}{\partial x^b} + \frac{\partial F^{ad}}{\partial x^a}\right) = \pm\frac{\partial F^{jd}}{\partial x^j} = 0$$

The second set of equations is derivable directly from (*12.45b*), the definition of Φ, and the fact that the g_{ij} are constants.

CHAPTER 13

13.22 Yes; it is isomorphic to the 4-group.

13.26 (*a*) By (*13.6*),

$$\bar{T}^i_j = T(\bar{\boldsymbol{\beta}}^i, \bar{\mathbf{b}}_j) = T(\bar{A}^i_j\boldsymbol{\beta}^j, A^k_i\mathbf{b}_k) = \bar{A}^i_j A^k_i T(\boldsymbol{\beta}^j, \mathbf{b}_k) = \delta^k_j T^j_k = T^j_j$$

(*b*) $\tau(T) = n$

13.27 Suppose that $\hat{G}(\mathbf{u}_1) = \hat{G}(\mathbf{u}_2)$. Then $G(\mathbf{u}_1, \mathbf{v}) = G(\mathbf{u}_2, \mathbf{v})$, or by symmetry, $G(\mathbf{v}, \mathbf{u}_1) = G(\mathbf{v}, \mathbf{u}_2)$ for all $\mathbf{v}$ By nonsingularity, $\mathbf{u}_1 = \mathbf{u}_2$.

13.28

$$\varphi_p^{-1}:\begin{cases} y^1=\sqrt{a^2-(x^1)^2-(x^2)^2+(x^3)^2}\\ y^2=x^1\\ y^3=x^2\\ y^4=x^3\end{cases}\qquad \boxed{\begin{array}{c}\mathbf{U}_p:\ y^1>0\\ p=(a,0,0,0)\end{array}}$$

$$\varphi_{-p}^{-1}:\begin{cases} y^1=-\sqrt{a^2-(x^1)^2-(x^2)^2+(x^3)^2}\\ y^2=x^1\\ y^3=x^2\\ y^4=x^3\end{cases}\qquad \boxed{\begin{array}{c}\mathbf{U}_{-p}:\ y^1<0\\ -p=(-a,0,0,0)\end{array}}$$

$$\varphi_{\pm q}^{-1}:\begin{cases} y^1=x^1\\ y^2=\pm\sqrt{a^2-(x^1)^2-(x^2)^2+(x^3)^2}\\ y^3=x^2\\ y^4=x^3\end{cases}\qquad \boxed{\begin{array}{c}\mathbf{U}_q:\ y^2>0\\ \mathbf{U}_{-q}:\ y^2<0\\ q=(0,a,0,0)\end{array}}$$

$$\varphi_{\pm r}^{-1}:\begin{cases} y^1=x^1\\ y^2=x^2\\ y^3=\pm\sqrt{a^2-(x^1)^2-(x^2)^2+(x^3)^2}\\ y^4=x^3\end{cases}\qquad \boxed{\begin{array}{c}\mathbf{U}_r:\ y^3>0\\ \mathbf{U}_{-r}:\ y^3<0\\ r=(0,0,a,0)\end{array}}$$

13.30 $$\sigma=y^2\,dy^1-y^1\,dy^2+y^4\,dy^3-y^3\,dy^4+y^6\,dy^5-y^5\,dy^6+\cdots+y^{2k}\,dy^{2k-1}-y^{2k-1}\,dy^{2k}$$

13.31 If p_1 and p_2 are not antipodal, there exists a closed hemisphere containing neither one, on which the given (continuous) vector field is nonzero—an impossibility by Problem 13.17(*b*).

13.32 As shown in Fig. 13-10, let a unit tangent vector be constructed to a generating circle; as the circle is revolved to generate the torus, the tangent vector is obviously propagated continuously to all points of the torus.

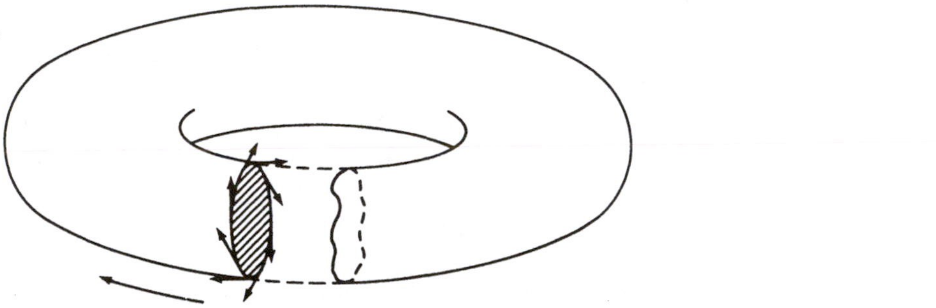

Fig. 13-10

13.33 Zero points are: (*a*) $(0,0,0,0,\pm 1)$; (*b*) $\pm(0,1/\sqrt{3},0,1/\sqrt{3},1/\sqrt{3})$.

13.35 With $\omega\equiv 2(y^1\,dy^1-4y^2\,dy^2+4y^3\,dy^3)$, $\sigma=f\,dy^1+g\,dy^2+h\,dy^3$ must be orthogonal to ω, for C^∞ functions f,g,h. Hence,

$$y^1f-4y^2g+4y^3h=0$$

Replace f, g by $4y^3F$, y^3G, and solve for h. Similarly, replace g by y^1G and h by y_1H and solve for f; etc. All possible tangent vectors are given by one of three distinct types (F,G,H denote arbitrary C^∞ functions of y^1, y^2, y^3):

(1) $\sigma=4y^3F\,dy^1+y^3G\,dy^2+(y^2G-y^1F)\,dy^3$

(2) $\sigma=(4y^2G-4y^3H)\,dy^1+y^1G\,dy^2+y^1H\,dy^3$

(3) $\sigma=4y^2F\,dy^1+(y^1F+y^3H)\,dy^2+y^2H\,dy^3$

13.36 $$U=U^i\mathbf{r}_i\equiv\left(2U^1\sqrt{4-(x^1)^2+4(x^2)^2},\ 2U^2\sqrt{4-(x^1)^2+4(x^2)^2},\ -x^1U^1+4x^2U^2\right)$$

for any two C^∞ functions U^1, U^2 on (x^1,x^2).

13.37 The normal vector to the sphere is represented by $\sigma=y^1\,dy^1+y^2\,dy^2+y^3\,dy^3$; the tangent space of the cone is given by

$$(U_1,U_2,(3x^1U^1+3x^2U^2)((x^1)^2+(x^2)^2)^{-1/2})$$

Set $U^1=y^1$ and $U^2=y^2$.

Index

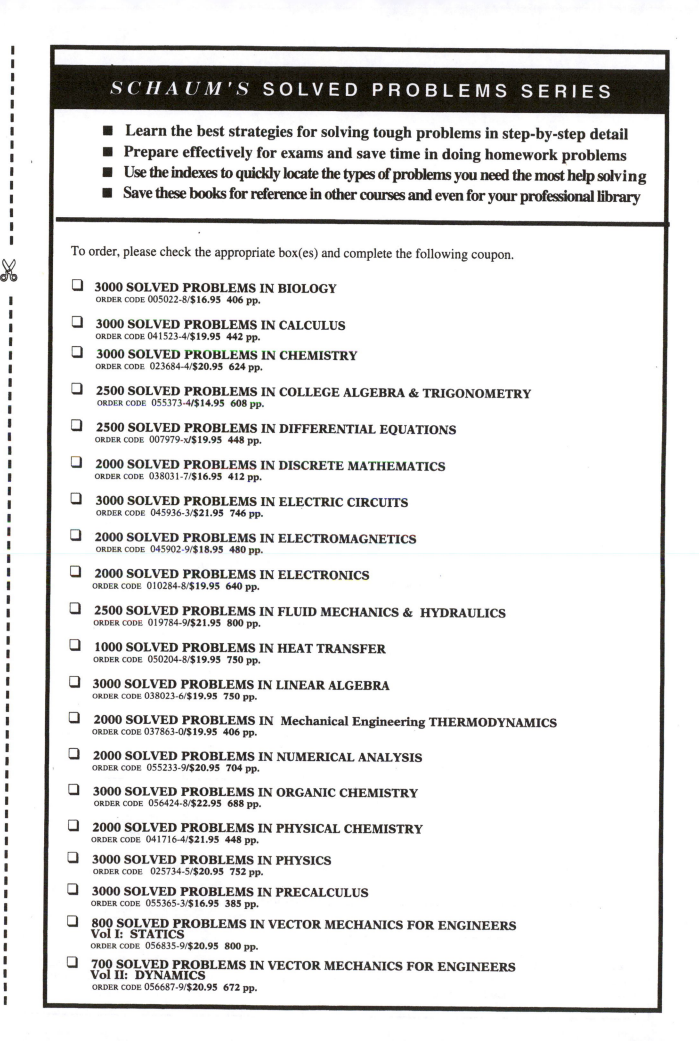